Cultures of Control

Studies in the History of Science, Technology and Medicine
Edited by John Krige, CRHST, Paris, France

Studies in the History of Science, Technology and Medicine aims to stimulate research in the field, concentrating on the twentieth century. It seeks to contribute to our understanding of science, technology and medicine as they are embedded in society, exploring the links between the subjects on the one hand and the cultural, economic, political and institutional contexts of their genesis and development on the other. Within this framework, and while not favouring any particular methodological approach, the series welcomes studies which examine relations between science, technology, medicine and society in new ways e.g. the social construction of technologies, large technical systems.

Other Titles in the Series

Volume 1
Technological Change: Methods and Themes in the History of Technology
Edited by Robert Fox

Volume 2
Technology Transfer out of Germany after 1945
Edited by Matthias Judt and Burghard Ciesla

Volume 3
Entomology, Ecology and Agriculture: The Making of Scientific Careers in North America, 1885–1985
Paolo Palladino

Volume 4
The Historiography of Contemporary Science and Technology
Edited by Thomas Söderqvist

Volume 5
Science and Spectacle: The Work of Jodrell Bank in Post-war British Culture
Jon Agar

Volume 6
Molecularizing Biology and Medicine: New Practices and Alliances, 1910s–1970s
Edited by Soraya de Chadarevian and Harmke Kamminga

Volume 7
Cold War, Hot Science: Applied Research in Britain's Defence Laboratories 1945–1990
Edited by Robert Bud and Philip Gummett

Volume 8
Planning Armageddon: Britain, the United States and the Command of Western Nuclear Forces 1945–1964
Stephen Twigge and Len Scott

Volume 9
Cultures of Control
Edited by Miriam R. Levin

This book is part of a series. The publisher will accept continuation orders which may be cancelled at any time and which provide for automatic billing and shipping of each title in the series upon publication. Please write for details.

Cultures of Control

Edited by
Miriam R. Levin
Case Western Reserve University
Cleveland, USA

LONDON AND NEW YORK

First Published 1996
by Harwood Academic Publishers.
Reprinted 2004,
by Routledge, 11 New Fetter Lane, London EC4P 4EE

Transferred to Digital Printing 2004

Copyright © 2000 OPA (Overseas Publishers Association) N.V. Published by license under the Harwood Academic Publishers imprint, part of The Gordon and Breach Publishing Group.

All rights reserved.

No part of this book may be reproduced or utilized in any form or by any means, electronic or mechanical, including photocopying and recording, or by any information storage or retrieval system, without permission in writing from the publisher.

British Library Cataloguing in Publication Data
A catalogue record of this book is available from the British Library.

ISBN: 90-5823-013-9 (soft cover)

Contents

List of Figures	vii
Preface	ix
List of Contributors	xvii
Acknowledgements	xix
Introduction Thomas P. Hughes	1

Part 1: Conventional Notions of Culture and Original Modes of Control

1. Contexts of Control Miriam R. Levin	13
2. Nature Out of Control: Cultural Origins and Environmental Implications of Large Technological Systems Rosalind Williams	41
3. Measuring Cloth by the Elbow and a Thumb: Resistance to Numbers in France of the 1780s Daryl M. Hafter	69
4. The Meaning of Cleaning: Producing Harmony and Hygiene in the Home Boel Berner	81
5. How the Motor Car Conquered the Road Catherine Bertho Lavenir	113
6. Culture, Technology and Constructed Memory in Disney's New Town: Techno-nostalgia in Historical Perspective Robert H. Kargon and Arthur P. Molella	135

Part 2: Managing Machines

7. How to Make Chance Manageable: Statistical Thinking and Cognitive Devices in Manufacturing Control Denis Bayart	153

8. **Ideology Counts: Controlling the Bodies of Concentration Camp Prisoners** 177
   Michael Thad Allen

9. **Beasts and Systems: Taming and Stability in the History of Control** 205
   David A. Mindell

10. **Liquifying Information: Controlling the Flood in the Cold War and Beyond** 225
    Mark D. Bowles

11. **Striving for 'Optimal Control': Soviet Cybernetics as a 'Science of Government'** 247
    Slava Gerovitch

Index 265

List of Figures

Figure 1	Measures incised on the city wall of Vila Viçosa in Portugal exemplify one manner of setting standard cloth measurements in proto-industrial Europe.	70
Figure 2	'The conscientious housewife and her collaborators polish the sink every week with "Tekå" Zincpolish.' Advertisement in a Swedish cleaning manual from 1923.	82
Figure 3	Here de Saunier attributes responsibility for a dangerous situation to the steeply graded road surface.	120
Figure 4	The car driver is endangered as a result of the railroad employee's actions, not his own.	121
Figure 5	The reader, conceived as a spectator, is confronted with a situation leading directly to an accident.	121
Figure 6	Other users of the road create dangerous situations by failing to take into account the special characteristics of the automobile.	122
Figure 7	Other road users (pedestrians, cyclists, carters, animals and children) are shown here as the source of potential danger to themselves and to the car driver from the point of view of a spectator aware of the entire situation.	122
Figure 8	Westinghouse Singing Cascades, New York World's Fair, 1939–40.	139
Figure 9	Front porches abound on Celebration streets.	140
Figure 10	Celebration's business district, billed as a 'pedestrian friendly downtown' with lake and boat rentals.	141
Figure 11	A view of Celebration's Mulberry Avenue, lined with approved Charleston siderows, modeled after the 'single house' found in Charleston, South Carolina.	142
Figure 12	Celebration bank by Robert Venturi and Denis Scott Brown.	143
Figure 13	Two-theater cinema by Cesar Pelli & Associates.	144
Figure 14	The original idea of a control chart (1924).	160
Figure 15	A control chart for the parameters of a statistical distribution (1926).	162

Figure 16	The standardized form of the control chart (1935).	163
Figure 17	Wiley Post and his Lockheed Vega, *Winnie Mae*, in which he made his around-the-world solo flight with a Sperry Autopilot.	209
Figure 18	The box above the control stick in *Winnie Mae*.	210
Figure 19	Sperry ball turret for the B-17 bomber, 1941.	211
Figure 20	Main battery fire control system from a cruiser including Ford Rangekeeper Mark 8, Arma Stable Vertical Mark 6, and General Electric Mark 34 Director, circa 1940.	212
Figure 21	M-9 Gun director designed by Bell Laboratories during World War II under an NDRC fire control contract.	213
Figure 22	SAGE computerized air defense system. Operators use 'light guns' to designate targets on a computer-driven radar screen.	214
Figure 23	A flood of information.	227
Figure 24	Floating on the information sea.	233

Preface

The Collection's Origins and Rationale

Three threads intertwined to inspire this collection. In part, the idea for the collection grew out of readings, discussions and student research projects in three research seminars I taught at Case Western Reserve University: Cultural Approaches to the History of Technology, History of Communications, and The Idea of Control in the Industrial Era. Although there were connections between all three topics, these seminars revealed that little work had been done on the theme that linked them: communications and control. Nevertheless, there was a body of texts on which research could begin to be built. These texts fell into two categories of scholarship: those that obviously belonged in the history of technology and those that were 'mainstream' history, cultural history and theory. There was thus a historiographic divide to be bridged, and the theme of control and concepts of culture embedded in both categories of texts provided it.

Key among these texts was the seminal essay by anthropologist Clifford Geertz entitled 'Ideology as a Cultural System'. In it he proposes that we not disregard ideologies as 'false consciousness' in the Marxian sense, but analyze these constructs as serious cultural phenomena.[1] In so doing he takes Marx's views on ideology into a new worthwhile byway. Instead of showing up the contradictions of capitalism while offering an alternative explanation for social change rooted in scientific analysis, Geertz argues that we take the explanatory power of ideologies seriously. We should conceive of them as functional belief systems for those who espouse them. Thus, while Marx recognized the contribution made to political culture by the men who made the Enlightenment, he scorned them and their French followers the Ideologues, for their obscurantist and self-serving explanations. Geertz, on the other hand, is interested in how the cluster of values, attitudes and ideas that belong to a group of people and reflect their mode of production form a semiological pattern. In the eyes of those who accept its pattern, the cluster seems rationally integrated with their experience of the external world. It thus takes on the connotation of being 'natural'. For them it is a rational explanation and prescription for dealing with current circumstances, as well as a vision of the future and the means to achieve it. As false consciousness, ideology could be interpreted as interfering with an individual's ability to analyze social and economic circumstances scientifically (i.e. objectively, in accordance with natural law), because one viewed reality through a subjective lens. The beauty of Geertz's proposal is that it recognizes the functional value of ideology in human society. Here in the context of human relationships, some verifiable notion of nature is necessary not solely as a means to 'truth', but as an instrument of stability and self-preservation that empowers adherents and gives them a claim to exercise real power.

Geertz, thus, provided important theoretical insights into an historical situation. Ideology, reified into a liberal formulation by these Enlightenment leaders and ensconced in the Institute de France after the revolution of 1789–95 by their intellectual descendants, was the quintessential culture of control.[2] It was, as I have discussed elsewhere, a system of moral technology that could be fused with forms of material culture.[3] Geertz helped us understand the rationale that made the technological definition of culture so central to the definition of ideology. Ideology dictates and is embodied in patterns of personal, economic and social behavior, values, methods of production, laws, language, material objects, and the institutions which groups of individuals create to give order to and reinforce coherence in their lives. In addition, Geertz's theoretical insights suggested the need to define the meanings of control and its modes, as well as to explore the relationship between ideologies of control, the historical circumstances in which they arose and changed, and the shifting sociology of those who adhered to them. One could see where cultural history, history of *mentalités*, art history and history of technology and science might cross-fertilize.

In the arena of cultural history, Raymond Williams' examination of the relationship between means of production and means of communication, Jürgen Habermas' of communicative action and the creation of a public sphere, and Norbert Wiener's writings on cybernetics suggested that the cybernetics revolution of the twentieth century was perhaps the latest phase in a longer historical process.[4] David Harvey's book, *The Condition of Post-Modernity*, gave a specific face, name and shape to that process. He linked the mind-set and cultural modes of modernism to those of industrial capitalism formed under the aegis of the Enlightenment. He also described their transformation in the late twentieth century. Michel Foucault, Pierre Bourdieu and social geographers forced attention onto the power which control over design of material culture and physical space gave to groups wishing to structure human mentalities and behavior into conforming with ideological forms.[5]

James Beniger's seminal book, *The Control Revolution*, offered a very important introduction to this history. Beniger argues that capitalists introduced electrically-based information control systems to overcome spatial disjunctions in production, transportation and consumption of goods.[6] He restricts himself to contemporary theories of bureaucracy, entropy, and evolution. He draws on them to explain the adoption of these technologies and the contribution they made to the corporate revolution at the end of the nineteenth and beginning of the twentieth century. Especially significant, he argues, is the way the revolution in modes of information communication gave a new class of managers and engineers the power to coordinate the different sectors of the economy into large-scale, socio-technical systems.

In many ways Beniger's book offered a corollary to Marshall Berman's *All That is Solid Melts Into Air*.[7] His proposal that the introduction of new technologies of communication and control *pari passu* engaged a massive restructuring of industrial society is subsumed within Berman's suggestion that capitalists have long been ideologically committed to periodic destruction and rebuilding of the infrastructures they have created in their search to guarantee profits. Beniger's book also made some crucial connections between late nineteenth century scientific theories concerning work and energy exchange and the concerns of industrialists to find remedies for and even a language to describe a society that seemed to them to be falling into chaos. Anson Rabinbach's book, *The Human Motor*, complemented Beniger's.[8] It offered a more complete picture of the anxieties which this disorder created among the bourgeoisie, and the dynamic that galvanized economic and social disorder, class neurosis and theories of energy exchange, motors and fatigue into a new mentalité at the turn of the century.[9]

As for the second thread, I was also led to recognize a need for this collection by my readings and discussions with colleagues in the United States and abroad, some of whom were in the field of the social construction of technology and others who were critical of it. The idea that technology and science were part of a seamless social web warranted thinking about how mainstream history of modern western society and the history of technology in this society were interconnected. It also warranted more research to trace the threads that made up that web. Wiebe Bijker, Michel Callon and Bruno Latour's work on actor-networks forced the issue of identifying the specific social groups which were responsible for inventing and innovating this culture.[10] Thomas Hughes' work on the history of efforts by entrepreneur-inventors and corporate heads to create large technological systems suggested the case study as a way to pin down the link between social-actors, ideological systems and the creation of technological culture, including what is inside the black box. Hughes also marked out very clearly the manner in which large technological systems work, with beautifully lucid explanations of feedback and the messiness and limitations inherent in the actual, as opposed to ideal, systems.[11] That is, what gave meaning and shape to decisions about technological invention and innovation was not the social group *per se*, but the particular way its members thought about large-scale problem solving as closed and open loop systems. Men and machines could be reduced to a common denominator that allowed conventional and social engineers to integrate them into networks that Lewis Mumford would call 'megamachines'.[12]

As for the third thread, control emerged as a largely unexamined theme in the work of many of my colleagues who presented papers at history, French history, history of technology, and history of science conferences and symposia. They recognized the existence of a relationship between power, control and industrial development. There was, however, a need to sort out the socio-technical and political meanings, forms and structures, as well as the actors associated with the history of control culture. Some of these people came together in 1995 at a symposium on Democracy and the Culture of Communications I organized at CWRU. As I listened to them, I began to feel that here was a topic whose elucidation might be worthwhile.[13]

If the history of the culture of control has yet to be written, there is a foundation on which to build. The introductory essay and eleven papers in this collection contribute to this history in a number of ways that the order of presentation is intended to suggest. The function of ideology as a control mechanism is the theme that runs through all these papers. The entire collection of papers is divided into two major sub-themes. Those in the first part establish the historical origins of this culture and the way in which conventional technologies and culture broadly conceived as technology became interconnected and multivalent. The second part contains papers whose subjects are all situated in the twentieth century and focus on the interplay between information systems, control mechanisms and the need to solve complex problems through design.

Thomas Hughes' introduction grounds the book's major themes in an historical context.

The first chapter by the editor identifies control as an historical phenomenon arising from and helping to shape the economic and political, material and psychological conditions of industrialization. I have set out its technological characteristics, intellectual foundations in the Enlightenment and development as a key word in modern and post-modern society, as well as signaled the major figures responsible for giving the idea of control substance and form. The paper by Rosalind Williams contains an important analysis of the culture of control Enlightenment. She places the ideas of two major philosophies into a broad historical setting and suggests that later efforts to build controlled socio-economic environments, based on these men's constructs of nature as a circulatory system, have led to the degradation of the physical environment. Ann Marie Turgot and the marquis de Condorçet espoused what Williams calls an *ideology of circulation* which favored the development of territory to increase the flow of communications and transportation. Daryl Hafter's piece brings our focus onto the specific policies which Turgot, as *Controleur général* of France, aimed at instituting state control over the textile manufacturing. This detailed study presents an object of precise measure as a force inaugurating a new system of control. She reveals how the inauguration of this project generated a host of disturbances in a system that included worker's procedures, merchants' transactions, and the design of technology. These had to be resolved if a new state-governed system was to be put in place. Both these writers concern themselves with the first period in the history of control when the natural world was assumed to be a model of order, harmony and mathematical regularity that human beings should attempt to mimic.

The next three articles, those by Böel Berner, Cathérine Bertho Lavenir, and Robert Kargon and Arthur Molella, present cultures of control as means bourgeois or upper-class social groups have used to dominate nature by making it conform to their own ideals of order. In all three cases it seems that nature is about to get out of their control or requires reigning in through the proper management of space. Berner and Lavenir's subjects, the domestic interior and the automobile respectively, are set at the turn of the twentieth century, and that of Kargon and Molella on Disney's Celebration City at its very end. There are differences, however, that give depth and insight into the interaction between the control cultures associated with capitalist consumer technologies and the historical contingencies that force them to change.

The next two articles, the first in the second part, are concerned with the topic of control in production, each from a very different perspective. Denis Bayart focuses on the important subject of management, centering his attention on the material culture invented by managers to control production. Here he analyzes and assesses the instrumental value of what seems an insignificant object: a simple chart intended to set production goals for workers in factories. In the process he reveals some of the paradoxical results objects tend to inaugurate. Michael Allen's paper is also concerned with management. It considers the role a specific ideology of the twentieth century, Nazism, played in determining the ends and means managers of concentration camps set for these death factories. Allen deals not with paradox, but with the repellent consequences of an ideology that perverted Enlightenment notions of nature and human nature. He argues against those historians who propose that Nazism discredits the Enlightenment because it carried the ideal of instrumental reason to its logical conclusion.

The last three pieces focus on topics that treat the subject of what has conventionally been designated as control technology: information systems, automatic systems, servo-mechanisms and computers. Each author takes the point of view that technologies are material culture whose design is intimately related to special historical contexts. This leads them to redefine the character of these technologies, how they were invented and developed, and what they have come to mean. Both Mindell and Bowles are interested in the paradoxical role language plays in providing conceptual frameworks helping to shape the invention of information control technologies. Gerovitch's piece on cybernetics under Stalin also serves as a closing parenthesis for the collection, reiterating certain themes in the introduction and in Rosalind Williams' essay. At the same time, the essay provides closure for the collection. Its content suggests that we weigh the social dislocations thwarting the creation of effective control in non-democratic command economies against the perversions of human-environment relationships encouraging controlling cultures in democratic capitalist societies.

NOTES

1 Clifford Geertz, 'Ideology as a Cultural System,' in David E. Apter, ed., *Ideology and Discontent* (New York: Free Press, 1964), pp. 47–76.

2   Antoine Guillois, *Le salon de Madame Helvétius; Cabanis et les idéologues* (Paris: Calmann Lévy, 1894); Winfried Busse and Jürgen Trabant, eds., *Les Idéologues: Sémiotique, théories et linguistique pendant la Révolution française: Proceedings of the conference held at Berlin, October 1983* (Amsterdam Philadelphia: J. Benjamins, 1986) Series Foundations of Semiotics, v. 12; Guy de la Prade, *L'illustre société d'Auteuil 1772–1830, ou, La fascination de la liberté* (Paris : F. Lanore/F. Sorlot, 1989); François Joseph Picavet, *Les idéologues; essai sur l'histoire des idées et des théories scientifiques, philosophiques, religieuses, etc. en France depuis 1789* (Paris; F. Alcan, 1891); Antoine Louis Claude Destutt de Tracy, *A Treatise on Political Economy*, trans. Thomas Jefferson. Foreword by John M. Dorsey (Detroit, Pub. by Center for Health Education 1973, reprint of the 1817 ed.); John M. Dorsey, *Psychology of Political Science, With Special Consideration for the Political Acumen of Destutt De Tracy* (Detroit, Center for Health Education, 1973); Rose Goetz, *Destutt de Tracy : Philosophie du langage et science de l'homme* (Genève: Droz, 1993, Series: Histoire des idées et critique littéraire, v. 328); Brian William Head, *Ideology and Social Science: Destutt De Tracy and French Liberalism* (Dordrecht; Boston: Kluwer Academic Publishers, 1985, Series Archives internationales d'histoire des idées, no. 112); Emmet Kennedy, *A Philosophe in the Age of Revolution, Destutt De Tracy and the Origins of 'Ideology'* (Philadelphia: American Philosophical Society, 1978, Series: Memoirs of the American Philosophical Society; v. 129). For discussions of the Ideologues' views on the social-psychology of signs, particularly the language of material culture and environmental design see: Miriam R. Levin, 'The Wedding of Art and Science in Late Eighteenth Century France,' *Eighteenth Century Life*, Special Issue on Art and Politics, May 1982, 54–73; Miriam R. Levin, 'La Définition du caractère républicain dans l'art français après la Révolution,' *Revue de l'Institut Napoléon*, 1981, no. 187, 140–167; Miriam R. Levin, 'The Printed Image in Industrializing France from the Enlightenment Until 1835,' *The Muse and Reason: Science, Technology and the Arts* (Kingston, Ontario: Royal Society of Canada, 1994), 43–61.

3   Term 'moral technology' coined in Miriam R. Levin, 'The Wedding of Art and Science in Late Eighteenth Century France,' *Eighteenth Century Life*, Special Issue on Art and Politics, May 1982, 54–73.

4   Raymond Williams wrote a great deal on this subject. For a cogent presentation of his ideas, see Raymond Williams, *Problems in Materialism and Culture: Selected Essays* (London; New York, NY: Verso, 1997, c. 1980); but also of import are his: *Resources of Hope: Culture, Democracy, Socialism*, ed. Robin Gable (London; New York: Verso, 1989); *Television: Technology and Cultural Form* (New York: Schocken Books, 1975). For Habermas's ideas see, *Jürgen Habermas, Moral Consciousness and Communicative Action*, trans. Christian Lenhardt and Shierry Weber

Nicholsen, introduction by Thomas McCarthy (Cambridge, Mass.: MIT Press, 1990); Jürgen Habermas, *The Theory of Communicative Action*, trans. Thomas McCarthy, 2 vols. (Boston: Beacon Press, 1984); Jürgen Habermas, *La technique et la science comme idéologie*, trans. (Paris: Denoel/Gonthier, 1978); Jürgen Habermas, *The Structural Transformation of The Public Sphere: An Inquiry into a Category of Bourgeois Society*, trans. Thomas Burger (Cambridge, Mass.: MIT Press, c. 1989); *Critique and Power: Recasting the Foucault/Habermas Debate*, ed. Michael Kelly (Cambridge, Mass.: MIT Press, 1994). For Wiener's ideas, see Norbert Wiener, *The Human Use of Human Beings: Cybernetics and Society*, 2nd ed. Rev. (Garden City, New York: Doubleday, 1954); Norbert Wiener, *The Care and Feeding of Ideas*, Introduction by Steve Joshua Helms, (Cambridge, Mass.: MIT Press, 1993).

5 Michel Foucault, 'Space, Knowledge and Power,' reprinted in Paul Rabinow, ed., *The Foucault Reader* (New York: Pantheon, 1984), 239–256; *Discipline and Punish: The Birth of the Prison* (New York: Vintage Books, 1979); *The Order of Things: An Archaeology of the Human Sciences* (New York: Pantheon Books, 1971). Pierre Bourdieu, *In other Words: Essays Towards a Reflexive Sociology*, trans. Matthew Adamson (Stanford: Stanford University Press, 1990).

6 James R. Beniger, *The Control Revolution: Technological and Economic Origins of the Information Society* (Cambridge, Mass.: Harvard University Press, 1986).

7 Marshall Berman, *All That is Solid Melts into Air: The Experience of Modernity* (New York: Simon and Schuster, 1982).

8 Anson Rabinbach, *The Human Motor: Energy, Fatigue, and the Origins of Modernity* (New York: Basic Books, 1990).

9 Other significant works included: Robert L. Heilbroner, *Teachings from the Worldly Philosophy* (New York: W. W. Norton, 1996); Robert L. Heilbroner, *The Worldly Philosophers: The Lives, Times, and Ideas of the Great Economic Thinkers*, 4th ed. (New York: Simon and Schuster, 1972); Eduard Jan Dijksterhuis, *The Mechanization of the World Picture*, trans. C. Dikshoorn, (Oxford: Clarendon Press, 1961); Otto Mayr, *Authority, Liberty, and Automatic Machinery in Early Modern Europe* (Baltimore: Johns Hopkins University Press, 1986); Langdon Winner, *Autonomous Technology : Technics-Out-Of-Control as a Theme in Political Thought* (Cambridge, Mass.: MIT Press, 1977); Simon Schaffer: 'Babbage's Intelligence: Calculating Engines and the Factory System', *Critical Inquiry* (Autumn 1994) v. 21, Number 1, 203–228; M. Norton Wise, 'Mediating Machines,' *Science in Context* (1988) 2: 77–113; and we also took a critical look at the classic history of industrialization of this period by David S. Landes, *The Unbound Prometheus: Technological Change and Industrial Development in Western Europe from 1750 to the Present* (London: Cambridge U.P., 1969).

10 Wiebe E. Bijker, Thomas P. Hughes, and Trevor J. Pinch, eds., *The Social Construction of Technological Systems: New Directions in the Sociology and History of Technology* (Cambridge, Mass.: MIT Press, 1987); Michel Callon, John Law, Arie Rip, eds., *Mapping the Dynamics of Science and Technology: Sociology of Science in the Real World* (Basingstoke: Macmillan, 1986). Bruno Latour, *Science in Action: How to Follow Scientists and Engineers through Society* (Cambridge, Mass.: Harvard University Press, 1987).

11 Thomas P. Hughes: *Networks Of Power: Electrification in Western Society, 1880–1930* (Baltimore: Johns Hopkins University Press, 1983); *Elmer Sperry: Inventor and Engineer* (Baltimore, Johns Hopkins Press, 1971). Renate Mayntz, Thomas P. Hughes, eds., *The Development of Large Technical Systems* (Frankfurt am Main: Campus Verlag; Boulder, Col.: Westview Press, 1988).

12 Lewis Mumford, *Technics and Civilization* (New York: Harcourt, Brace and World, 1963).

13 The Symposium on Democracy and the Culture of Communications was held at Case Western Reserve University, Cleveland, Ohio, in April 1995. The presenters and their paper titles were as follows: Richard R. John, University of Illinois at Chicago, 'The Communications Revolution and the Democratization of American Public Life, 1792–1835.' Cathérine Bertho Lavenir, Conservatoire National des Arts et Métiers, Paris, 'Democracy and the Technical System of Communication in European Space.' Bryan Pfaffenberger, University of Virginia, 'The Poor Man's ARPANET: Usenet and the (Re-) Construction of a Democratic Ethos.' Levent Koker, Princeton University, 'Communicative Democracy and Technocratic Politics. An Inquiry into the Paradoxes and Potentials of Modernity.' Arthur L. Norberg, University of Minnesota, 'Networks: A Force for Democracy or Further Centralization?' Mark Poster, University of California, Irvine, 'CyberDemocracy: Internet and the Public Sphere.' Patricia Aufderheide, The American University, 'Media Wars on the Electronic Frontier.'

List of Contributors

Michael Thad Allen, Georgia Institute of Technology, Atlanta, GA, USA
Denis Bayart, Ecole Polytechnique, Paris, France
Boel Berner, Linköping University, Linköping, Sweden
Mark D. Bowles, History Enterprises, Cleveland, OH, USA
Slava Gerovitch, Massachusetts Institute of Technology, Cambridge, MA, USA
Daryl Hafter, Eastern Michigan University, Ypsilanti, MI, USA
Thomas P. Hughes, University of Pennsylvania, Philadelphia, PA, USA
Robert H. Kargon, Johns Hopkins University, Baltimore, MD, USA
Catherine Bertho Lavenir, University Blaise Pascal, Clermont-Ferrand, France
Miriam R. Levin, Case Western Reserve University, Cleveland, OH, USA
David A. Mindell, Massachusetts Institute of Technology, Cambridge, MA, USA
Arthur P. Molella, Smithsonian Institution, Washington, D.C, USA
Rosalind Williams, Massachusetts Institute of Technology, Cambridge, MA, USA

Acknowledgements

A number of people have shared my interest and offered their support as I prepared this collection. Arthur Molella of the Smithsonian Institution; Arne Kaiser, Royal Institute of Technology, Stockholm; Boel Berner of the Thema for Technology and Social Change, Linköping University, and Jonathan Friedman of the Social Anthropology Department, University of Lund, provided me with the opportunity to present seminars on the topic of control at their respective institutions. At Case Western Reserve University, my home institution, graduate students in my History of Technology seminars and in my seminar for Management Information Decisions Systems provided cogent and stimulating critiques of several papers in the collection, including my own. Michael Grossberg encouraged my embarkation in pursuit of this theme. In addition, Randolph Packard very kindly made it possible for me to use the Emory University library and research facilities during the periods I spent in Atlanta working on the collection. John Krige offered strong encouragement and pithy insights into the making of this collection.

Finally, I wish to thank my father for passing on to me his fascination with Enlightenment ideas and their legacy.

Introduction

Thomas P. Hughes

Miriam Levin and the other authors in *Cultures of Control* show that 'control' and 'technology' have similar connotations and are nearly interchangeable. This has been especially true in the United States and remains so today. The objects to be controlled by technology, however, have changed over the years. Early nineteenth-century Americans used technology to control nature. About the turn of the century machines were the object of control. Today engineers and managers—even the public—try to control large technological, or sociotechnical, systems.

NATURE CONTROLLED

Early nineteenth-century Americans from abroad contended mostly with the natural, not the human-built, world. The majority lived in the countryside. Some of them experienced bitter cold winters, others knew humid and hot summers, and still more knew both. Torrential rains brought flooding, droughts accompanied extreme heat, and fierce winds played havoc with barns and houses. With simple shelter and tools, the settlers were often overwhelmed by nature. Stories of Midwestern droughts and Mississippi floods are heart-wrenching.

No wonder early settlers looked to technology for help in transforming a wilderness into a living space and work place. Persuaded that the new errand in the wilderness was a God-like endeavor and convinced that the wilderness constituted the raw material for an aggressive technological venture, Americans sought to master the wilderness and the natural forces within it. From the pulpit and podium they heard that the '"Architect of the Universe" has instilled within us a divine spark—a portion of His skill and power.'[1] So 'sparked', Americans used machines and instruments to create order out of a nature composed of 'dull, unorganized matter,' 'shapeless masses,' and 'inferior animals.'[2]

By mid-century a profession of civil engineering had spread from England to the United States. Civil engineers concentrated on replacing the natural with the human built. The 1828 charter of the Institution of Civil Engineers (London) defines engineering as the:

> art of directing the great sources of power in nature for the use and convenience of man, as the means of production and of traffic in states, both for external and internal trade, as applied in the construction of roads, bridges, aqueducts, canals, river navigation and docks for internal intercourse and exchange, and in the construction of ports, harbours, moles, breakwaters and lighthouses, and in the art of navigation by artificial power for the purposes of commerce, and in the construction and adaptation of machinery, and in the drainage of cities and towns.[3]

Engineers, formally trained or self-taught, launched a campaign to reorganize nature for society's ends:

> Where she [Nature] denied us rivers, Mechanism has supplied them. Where she left our planet uncomfortably rough, Mechanism has applied the roller. Where her mountains have been found in the way, Mechanism has boldly leveled or cut through them. Even the ocean, by which she thought to have parted her quarrelsome children, Mechanism has encouraged them to step across.[4]

In 1814 a commentator complacently observed that America, which two centuries earlier had been a country of wild beasts and savages contending for unmeasured desert, was now made over into one of commerce, cities, art, letters, and religion.[5] Man's mind had converted a world of deformed nature into one of enlightened culture, of new order and progressive improvement.[6] Another observer described the state of nature before it had been wondrously improved by technology as a world of wilderness, savages, and 'terrific reptiles.'[7]

MACHINE IN THE GARDEN

Did control bring cities of art, letters, and religion; did it transform a wilderness into a garden? Could American's speak of placing machines in a garden? Did Americans create what Leo Marx, in his seminal study aptly titled *The Machine in the Garden,* calls the middle landscape? Marx writes eloquently of Ralph Waldo Emerson's poetic vision of the factory-village and railway set in a marvelous vista where nature and technology are organically integrated. 'The poet,' Emerson felt, 'sees them fall within the great Order not less than the beehive or the spider's geometrical web.'[8]

Marx recalls a minority of Americans, including Emerson, who had faith that technology guided by the moral mind could create a landscape in the New World wherein machine and garden would harmoniously co-exist. This minority of reflective and articulate Americans urged their fellow citizens to negotiate gently with nature and to design a middle landscape, both human-built and natural. As we shall discover, the clamoring of the majority of Americans for other ends drowned out the voices of the minority.

A MACHINE FOR PRODUCTION

To what overarching ends did the majority of nineteenth-century Americans direct nature-controlling technology? They might have made their country into Nature's nation; but instead, Americans enthusiastically embarked on the mechanization of their world. As a result, they created the most productive economy in the history of the world. Their commitment is understandable, if not entirely commendable. They had come from poverty stricken backgrounds abroad and longed for material goods. Yet, in chosing to make of their country a massive machine for production, they laid to waste the nature they controlled and exploited.

THE TWENTIETH CENTURY MACHINE AGE

In the United States the widespread drive to mechanize became so obvious to early twentieth-century public intellectuals, scholars, architects, and artists that they labeled their times 'the machine age' and their society 'the machine civilization.' Mechanization as a technological activity and as a metaphor transfixed and transformed society. Charles A. Beard, the most widely known historian of his era, in his introduction to *Toward Civilization* observed in 1930:

> The battle over the meaning and course of machine civilization grows apace, with resounding blows along the whole front. What appeared to be a few years ago a tempest in a teapot, [has become] a quarrel among hard-headed men of affairs...No theme, not even religion, engages more attention among those who take thought about life as well as living...[9]

Beard maintained that engineers were forcing nature to conform to human will and, in so doing, acting as all-knowing and powerful creators giving shape to clay 'in the mists of creation.'

The age of mechanization is founded, he declared, upon the work ethic long valued by Christians. Beard ventured that Pasteur and Edison would have been canonized had they suddenly appeared with their discoveries and inventions in the age of Innocent III. Beard predicted that without the mechanization of the world—by virtual saints like Edison—grass would grow in the streets of great cities, communications would be cut off, and Manhattan's Great White Way would fall into darkness. Good works in this era were equated with technological transformation.

While acknowledging that machine-age Americans had made shabby and dilapidated cities, spoiled nature with hideous factories, and ruined highways with billboards and 'gas-filling' shacks, Beard's hope for the nation rested with engineers and scientists. They are the one's capable of heroic and highly imaginative enterprises—if set free to create their own set of values, ideas, and symbols. Optimistically, he predicted that they, the real makers of machine civilization, would control the creation of the material world into which they could project positive values.

CONTROL OF MACHINES

While engineers have not by any means fully controlled the material world, they have exercised unprecedented control over machines. Around the turn of the century they exercised more control over machines than ever before, because of the availability of electricity. The rise of the electrical supply industry after 1880 is usually associated with light and power. It should also be thought of as the primary source of control in the machine age.

Inventors and engineers used electricity to control machines in subtle and elegant ways. A prime example of control by electricity is the inventors' use of an elementary device consisting of an electromagnet combined with a mechanical spring. The output of a large electrical generator could be controlled with such a simple device. It was designed so that the tension on the spring represents the desired output from the generator. The actual output passes through the windings of the electromagnet. If the output rises the magnet's force of attraction increases; if the output falls then the force decreases. When the magnet and the spring are played off one against another, then the spring will extend when the electromagnetic force increases and contract if the force decrease. This movement of the spring makes and breaks the contacts of a small electric motor, which changes the position of brushes on the generator's commutator, which in turn raises or lowers the electrical output. Inventors found many variations on this control device and applied it to many other machines.[10]

The use of electrically driven gyroscopes early in this century also added a major dimension to control technology. The gyroscope has the remarkable ability to maintain its orientation despite the movement of a platform or vehicle on which it is mounted. Like the spring under tension in the simple device described above, the fixed orientation gyro provides a reference against which error can be measured. Engineers use gyros to control the flight of airplanes and the heading of ships. Mounted in gimbals, the gyro has freedom of movement independent of the vehicle on which mounted. The gyro's orientation is set to represent the desired heading of the vehicle. The gyro holds its orientation, and the airplane or ship moves relative to the gyro reference. The physical displacement of a point representing and following the vehicle's movement and a point on the gyro opens or closes switches that energize motors that activate control surfaces, such as a rudder, on the vehicle. Once the movement of the control surface realigns the vehicle and the gyro, the controlling motor is switched off.[11] Control has become the measurement of error and the use of feedback signals to indicate when the error had been corrected. Error is the deviation from goal or purpose.

CONTROL OF TECHNOLOGICAL SYSTEMS

Electricity and the gyro greatly enhanced engineers' confidence that they could control the machines proliferating during the machine age. If their predecessors had brought control to the natural world through machines, in the early twentieth century, engineers felt machines, also promised control of the mechanized, human-built world. Engineers seemed to be fulfilling their professional and civic responsibilities as they turned from the control of individual machines to the control of large machine systems.

During the interval between the world wars, load-dispatching centers of regional electric power systems symbolized society's ability to control human built systems. Images of these centers stirred the public's imagination then as images of large main frame computers did shortly after World War II. Society seemed able to control enormous quantities of energy in one case and information in the other.

The typical control center had two or three engineers seated in front of a large board, extending across the wall of a good-sized room. Small lights on a graphic model of the system with its transmission lines, generating stations and outdoor transformer and switch stations indicated the condition, or state, of these subsystems. Telegraph or telephone lines carried information signals from the remote subsystems to the control board. The signals told how much energy, for instance, a generating station was feeding into the transmission lines or the voltage, amperage, and frequency of the power being transmitted along a transmission line. With this information, the load dispatchers at the center altered the physical conditions in the power system so that the energy generated matched the load demand throughout the system. Otherwise the system lost stability and collapsed.

FAILURE OF CONTROL

Events during World War II and in the decade or so following shook the confidence of Americans who had believed that their technology was under control. The horrendous release of energy at Hiroshima and Nagasaki and the growing realization that either intent or chance could unleash a nuclear holocaust signaled that technology had spiraled out of control. The growing awareness of machine pollution and desecration of the natural world by engineered projects heightened anxieties about the impact of technology. Attitudes toward technology altered drastically as the euphoria over the age of mechanization evaporated.

The combination of anxiety about technology being out of control and the remarkable progress in machine and system control that took place during World War II produced an unprecedented interest in control among engineers, scientists, and the informed public. In 1948 Norbert Wiener, a Massachusetts Institute of Technology mathematics professor, raised the level of interest even higher when he published his book *Cybernetics, or Control and Communications in the Animal and the Machine*.[12] His contribution remains fundamental to current discussions of control. He wrote that if the seventeenth and early eighteenth centuries were the age of clocks, and the later eighteenth and the nineteenth centuries the age of steam engines, then the present time is the age of communication and control.

For Wiener, control and communication are inseparable—two faces of the same coin. Wiener and his contemporaries designed electromechanical and electronic systems in which messages from controller devices pass electrically over distances to the controlled machines. Familiar with these systems, Wiener defines communication as the transmission of information that organizes and controls process in a system.

Besides control and communication, feedback figures as a central concept in Wiener's cybernetic theory. Already in 1943, Wiener, Arturo Rosenbleuth of the Harvard Medical School and Julian Bigelow, an MIT engineer who built mechanical models to demonstrate Wiener's concepts, had defined feedback in an impressively succinct and accessible paper entitled 'Behavior, Purpose, and Teleology.' Negative feedback, they explained, is the use of controlling signals to modify the output of a machine or organism so that it will reach its goal. The controlling signals are fed back to the input of the machine or organism; they are generated by a comparison of the machine or organism's interim state with its eventual goal.[13] This definition describes well the functioning of the gyro control devices for airplanes and ships introduced several decades earlier.

In the 1960s notions of communication and control through feedback spread into the organic realm of developmental biology. Developmental biologists advocated a theory of control more complex than that propounded by molecular biologists at the time. Evelyn Fox Keller, a historian of science, writes:

> Molecular biologists, or geneticists...based their hopes not on the harnessing of complex systems but on what has been for natural scientists the more traditional paradigm of control—on the epistemological and technological benefits of *reductio ad simplicitatum*.[14]

By contrast, Fox emphasizes that in their explanations of growth a large number of developmental biologists introduce the circularity of feedback and the distribution of information sources throughout an organic network, or system. They take their metaphors from cybernetic systems, instead of from mechanical machines as did early molecular biologists.[15]

COMMAND AND CONTROL SYSTEMS

The developmental biologists' approach to control in organic systems resembles that of engineers and managers designing and developing large military command and control systems for defense against air attack in the 1950s and 60s.[16] Such projects included systems maintained by Strategic Air Command Control System, the North Atlantic Air Defense Command, the NATO Air Defense Ground Environment, and the World Wide Military Command and Control System.[17]

The availability of large digital mainframe computers deeply shaped the design philosophy of these engineers and managers. The computers they designed and built became known as the 'brains' of centralized command and control systems located in 'Information and Control Centers' that were analogous to load dispatching centers of regional electric power networks. The first of these, the SAGE (Semiautomatic Ground Environment) by 1958 was deploying computers to solve large-scale problems of real-time control.[18]

JAY FORRESTER: CONTROL AND INFORMATION

Designers of large and complex weapons systems such as SAGE, enthusiasts who spread notions about cybernetics, and social scientists who believe that a systems approach enables them to cope with complexity, all came to appreciate the prime importance of communicating and processing information. Increasingly, they all became convinced that the human-built world could not be understood and controlled like a simple mechanical machine. They now needed information about—and means to process information about—the world as systems.

No one in the 1960s conveyed this message to professional engineers and managers and to an interested and informed segment of the public more clearly and persuasively than Professor Jay Forrester of MIT who had conceived the information-processing computer, Whirlwind, that functioned at the heart of the SAGE system. He had the imagination and the technical genius to see that if a computer could process information and make possible the control of a complex air defense system that computers might also function similarly to control far more complex sociotechnical industrial and urban systems.

Over a fruitful decade following his move from the SAGE project to the then recently established Sloan School of Management at MIT, Forrester authored three books, the titles of which suggest his effort to deal increasingly with the control of complex systems: *Industrial Dynamics*, *Urban Dynamics*, and *World*

Dynamics. Like Wiener, Forrester had designed complex gunfire control systems during World War II; unlike Wiener, he had also developed and placed these systems in operation. Both men stressed the vital importance of feedback controls; both understood the essential role that information played in control. They anticipated the information age in which information and control so closely interact.

LAYERED CONTROL

SAGE and other command and control systems displayed our ability to control large technological systems. The mode of control, however, is hierarchical and younger managers and engineers, as well as the public, in recent years have questioned the appropriateness of this kind of centralized control. Control of the Internet today demonstrates another form of control commonly labeled layered or distributed. In this case, control is distributed throughout the Internet among various people and organizations. The control is in fact distributed among the computer nodes that these various people and organizations have under their control. The future of control may be in these messy, complex, loosely coupled, heterogeneous systems.[19]

NOTES

1. I have had the good fortune to have copies of the notes that Perry Miller used in writing *The Life of the Mind in America: From the Revolution to the Civil War* (New York: Harcourt, Brace & World, 1965). I am deeply grateful to Mrs. Miller allowing me to copy these. The short reference in this footnote and in numbers 2, 5, 6, and 7 refer to these notes, which I have in my files. I have placed these references in parentheses. (Fairbanks, Mass. Char. Association Address, p. 3.).
2. (c. 43 Everett, American Manufacturers, pp. 69–70).
3. *The Encyclopaedia Britannia* (New York: Encyclopaedia Britannica Company, 1910) IX, 406.
4. Quoted in Leo Marx, *The Machine in the Garden: Technology and the and the Pastoral Ideal in America*, (New York: Oxford University Press, 1964): pp. 182–3.
5. (33 Appleton's Works, II, pp. 287–88).
6. (45 Darousmont, State of the Public Mind, p. 6).
7. (45 Mitchell, N.Y. Horticultural Society, p. 7).
8. Ralph Waldo Emerson, 'The Poet,' *The Selected Writings of Ralph Waldo Emerson*, ed. Brooks Atkinson, (New York: The Modern Library, 1992): p. 295.

9   Charles A. Beard, ed., *Toward Civilization*, (New York: Longmans, Green and Co., 1930): p. 1.
10 Thomas Parke Hughes, *Elmer Sperry: Inventor and Engineer* (Baltimore, Md.: The Johns Hopkins Press, 1971 and 1993): pp. 45–6.
11 On an automatic gryo ship pilot, see *Sperry: Inventor and Engineer:* pp. 280–83.
12 Norbert Wiener, *Cybernetics, or Control and Communication in the Animal and the Machine* (Cambridge: MIT Press, 1948).
13 Arturo Rosenblueth, Norbert Wiener, and Julian Bigelow, 'Behavior, Purpose and Teleology,' *Philosophy of Science*,10.1 (1943): p. 19.
14 Evelyn Fox Keller, *Refiguring Life: Metaphors of Twentieth Century Biology* (New York: Columbia University Press, 1995): p. 92.
15 Keller, *Refiguring Life:* pp. 79–118.
16 'Command' is defined as decision-making based on a computer-based information system; 'Control' is managing information-gathering facilities (sensors), making decisions, and deploying resources (effectors).
17 Paul N. Edwards, 'The World in a Machine: Origins and Impacts of Early Computerized Global Systems Models,' p. 18. Paper presented at the 'Spread of the Systems Approach' conference held at the Dibner Institute, MIT, Cambridge, Mass. in May 1996. On the management of weapons and warfare, see Manuel De Landa, *War in the Age of Intelligent Machines* (New York: Swerve Editions, 1991).
18 Edwards, 'The World in a Machine: Origins.' p. 16.
19 For more on this type of control, see Thomas Parke Hughes, *Rescuing Prometheus* (New York: Pantheon Books, 1998): pp. 301–5.

Part 1

Conventional Notions of Culture and Original Modes of Control

CHAPTER 1
Contexts of Control

Miriam R. Levin

NATURE, INSTRUMENTS AND COMMUNICATIVE ACTION

Human efforts to exercise control over the physical and social worlds people inhabit are ancient and intertwined. Means of control include a broad range of technologies from tools, machines and artifacts to processes, legal codes and institutions. In fact, artifacts and institutions are parallel with respect to their control functions. If the former project political power into the design of the physical environment, the latter serve the same function in the intangible social environment. For an historical example we can look to the kingdoms of ancient Mesopotamia, whose economies were intertwined with the construction and maintenance of irrigation systems. The belief that it is possible to extend the scale and scope of control to maximize the material and social benefits of human labor for all members of society dates from the Enlightenment, however. In the mid-eighteenth century as industrialization slowly began and capitalism entered a new phase, a significant watershed occurred in the history of efforts at control. The change it engendered was especially marked in France and Great Britain, while in the United States it came a bit later. Most importantly, it generated a culture that is still vital.

Adherents of the Enlightenment's reified control into a belief system that became the distinguishing feature of their world view. This ideology provided a framework of formative principles whose objective was to structure the design of all cultural forms and human behavior into one coordinated system supporting economic prosperity. The framework itself was, thus, technological in character and rested on a unified theory of matter in motion. Although these men drew on the ideas of seventeenth century philosophers Francis Bacon, René Descartes and Blaise Pascal, for example, they extended the mechanistic world view of their predecessors from the heavens to the worldly realm of society and the economy. In contrast to Francis Bacon's vague proposals regarding the social benefits of scientific inquiry, they wedded theirs to specific, practicable programs of action. They and their intellectual heirs for the next two centuries have promoted, implemented and adapted these ideas about the meaning of control, giving it a Janus-faced character. It became both a lens through which they diagnosed the crises of industrial society and a strategy for resolving these crises with variations of cultural forms generously defined as communications media.[1] These two facets of control provide the theme of this collection and the subject matter for the essays included in it.

The culture of control in the modern age has a history, then. It constitutes a master theme in the history of industrial society that embeds the history of technology *cum* tools, machines, and manufacturing processes within it. Its focus is the fortunes of this belief system, as well as the cultural forms, behaviors and attitudes to which it gave birth. If a detailed history of the culture of control has yet to be written, we can begin to outline its features by establishing the Enlightenment as the point of origin, identifying the significant moments of development, and examining changes in the meanings associated with the term 'control' in light of this chronology. We can also begin to give some fullness of feature to this history. For, just as traditional artists worked pencilled images from life into the schema of historical paintings to communicate the texture and meaning of great events, historians' research on a variety of special subjects related to control provide detail, depth and interpretive subtlety to this schema.

HISTORICAL CONTEXT

There are certain significant moments that punctuate the *longue durée* of control, complicating the ideology and the material culture which gave control form and agency. Three major crises of control mark the industrial period in Western society, and they provide the organizational structure for this collection. Historians of technology have generally considered them stages in the process of industrialization; however, I am proposing that the process of industrialization itself is one in which three crises centered on the issue of work in capitalist society arose, and were resolved by our belief in our ability to steer human beings into socially beneficial systems. David Harvey and others going back to Karl Marx have pressed the important point that capitalism is 'creatively destructive' by nature. It also has its phases.[2] The internal contradictions that allow capitalists to accumulate great wealth from the profits of industry also periodically force them to destroy existing infrastructures and introduce new ones as shifting contingencies threaten profits. The logic that informs capitalists' activities thus leads to periodic social psychological and economic disorder. At the same time, adherents have established new forms of order by adapting technologies of control to the new technologies of production. This is the response we find in the following instances:

1. The first crisis of control came with the industrial revolution itself. It ran from the mid-eighteenth to the early nineteenth century. Here the issue was one of incoherence in the marketplace, where local privileges created congestion that prevented the rational distribution of goods, labor, and information. But the crisis also existed in the cities and countryside where growing populations pressed hard against traditional community constraints, such as customs for organizing work, exchanging goods and services and supporting the church. As a result, people were no longer rooted in their towns and villages and under local control. Visions of a Newtonian nature promised harmonious equlibrium if people could be brought under the control of a political economy.

2. The second crisis of control came in the late nineteenth to early twentieth century. It is the one Beniger characterizes as a crisis in the coordination of the three major sectors of industrial economy due to a lack of information.[3] Larger-scale businesses and improved techniques of manufacture and transport increased the capacity of each sector, but ignored the need for information to manage the flow of goods as supply and demand fluctuated. This economic crisis was part of a larger disorder in industrial society in which labor conflicts and middle-class aspirations figured large.

The crisis, however, was not solely economic. Anson Rabinbach has described the feeling of fatigue which struck middle-class Europeans at the nineteenth century's end. People felt their society was running down. They lacked the energy and desire to work, and projected their condition onto the working and upper classes. Considering society a giant heat engine propelled by the forces of its members, these *fin de siècle* cadres feared their own bodies and the body social itself were ceding to entropy. In popular parlance, the language of thermodynamics and that of political economy merged to express their feeling that all of nature tended towards chaos. A belief in human willpower soothed their anxieties about the possibility of imposing rational order and direction on nature. This belief also gave them hope they might find a way to reenergize and coordinate all laboring bodies and minds into a different system.[4] Thermodynamics, the study of heat exchange which theorized that the universe tends towards entropy, suggested that nature was a moving force which human beings had to reign in through the organization of work if they were to have control over at least the portion of the universe that touched them most directly.[5] This period saw the expanded use of automatic machinery and electrical generators, as well as the introduction of electrical technologies for managing systems of production, transportation, and advertising. Powerful entrepreneurs avidly adopted methods based in psychological theory for managing workers, administrators and consumers through modes of communication (the telegraph, telephone, photography, lithography, teletype and the typewriter) integrated into economic activity. National governments took up similar social and economic policies.

3. The third control crisis developed after World War II, and is still with us. Increasing automation of all forms of economic activity, including the

production and distribution of information, created crises in the very heart of industrial society. The product of government-funded research projects during and after World War II, cybernetic systems called the economic value of human labor into question, forcing us to examine the nature of work itself, and even industrial production as necessary for organizing the flow of goods and money. At the same time, the global population explosion, accompanied by mass migration of people and manufacturing, the deterioration of the environment due to earlier methods of control, and the breakdown of cities placed large numbers of people at risk. Randomness and chaos, decentralization and de-industrialization in Western countries seemed connected. In this 'post-industrial society,' solution to the crises proposed to shift the population to an economy dependent upon, rather than supported by, an information industry that would be global in scope. New service jobs would be created, new divisions of labor established, flexible manufacturing introduced, new social hierarchies and interconnections developed, and even work itself redefined through the agency of automatic, computerized systems that had replaced the model of nature.[6] Ironically, we also embarked on the project of restoring the salutary character of the natural environment our control efforts had corrupted.

As this chronology suggests, the eighteenth century was a time when historical circumstances favored the ideological formulation of mechanistic conceptions of human nature, of the physical and biological world, and of society itself, by men whose efforts began to structure what the human environment might be. Fortified by the rise in social prestige of instrumental science, the mathematics of probability, sensationalist psychology, and industrious behavior, proponents of this idea became convinced it was possible to create a truly natural environment through artificial means that would marshal individual energies and orchestrate them into a rational socio-economic system. Recent cultural historians interested in identifying the roots of modernism have attached the name 'Enlightenment Project' to their proposals. This project manifested itself in the form of theories and projects for social and economic reform that constitute the 'High Enlightenment,' as well as in equally important administrative, organizational and economic practices affecting production that constitute the 'Low Enlightenment.'[7]

The conception that all phenomena can be functionally interconnected was embedded in the ideology of a new cadre of social and economic leaders. At its center lay the self-conscious notion that culture itself (artifacts, institutions, modes of behavior, values, the production, distribution and consumption of goods and services, as well as the products of that activity) was a means of control. Culture could be designed, systematized and rationalized to produce pre-determined effects if it were modeled on natural systems. That is culture became coterminous with technology.

We might say that the mid-eighteenth century was a point at which an all-encompassing philosophy of technology that would penetrate every facet of Western society came into its own. Enlightenment ideology, the projects it has spawned in the past two hundred and fifty years, cast the master narrative of industrial society as the story of how our power to control has simultaneously served as the engine of progress and the generator of unforeseen political, economic and environmental complexities from which we cannot extract ourselves. Two major works laid out the foundations of this ideology: the *Encyclopédie ou dictionnaire raisonnée des arts et métiers* (first volume appeared 1751) and *An Inquiry into the Nature and Origins of the Wealth of Nations* (1776). These are long and complex publications: one a multi-volume compilation of articles and printed images solicited by its French editors from over a hundred people; the other a tome written by one man who was personally acquainted with a number of the *philosophes*.[8] Both present systems in which the establishment of a proper relationship between human beings' productive activity and natural laws is crucial to making society function rationally. At the same time both propose the systematic exploitation of natural resources for the public's good, and urge the perfection of existing machines and techniques and the invention of new ones to aid humans in this effort. Adherents and their descendants include men who engaged in academic teaching, writing for publication and art criticism, as well as scientific investigation, government administration, manufacturing and engineering.

A new state of mind nurtured by adherents: is what Jürgen Habermas has called 'instrumental reason.' Reason took on a mechanistic character and became a control mechanism.[9] In fact, the Enlightenment not only privileged the heuristic, but a way of thinking about thinking which turned reasoning into a physical process that mimicked natural phenomena. Capable of working in accordance with natural laws, thinking presented to them an ideal control on individuals' behavior, which they also defined in these same technological terms. This new mind-set perceived the act of thinking itself as a tool that had social utility.

There is no doubt that those most closely identified with the Enlightenment characterized thinking as a technology of control. They based their ideas partly on observations of human responses to external stimuli, informed by a conception of human beings as sensate machines responsive to one another and to their surroundings. But the idea of instrumental reason also derived partly from answers they formulated to their queries into the origins and character of society. What is its purpose? How do human inventions (language, machines, institutions, and the fine and applied arts) figure into these social equations?

The profoundly influential philosopher Immanuel Kant offered useful distinctions between pure and practical reason, proposing that the mind operated like a processing machine. In the case of practical reason, it produced science or knowledge, ethics or rules guiding the will or conduct, and aesthetics or a guiding idea of beauty and harmony.[10] In contrast, the equally influential Jean Jacques Rousseau (1712–1778) reasoned that human beings could be educated to manage their emotional energies into the paths of the public good.[11]

Their answers strongly express their mechanistic view of the physical world as one in which individual parts were in communication with one another, and reflect their observations of both machinery and human behavior. In Scotland,

Adam Smith (1723–1790) proposed in the *Theory of Moral Sentiments* (1759) that '... though reason is undoubtedly the source of the general rules of morality, and of all the moral judgments which we form by means of them...,' it is stimulated by people's feelings, which it then regulates.[12] Sympathy is the glue establishing the interactive and reciprocal relationships that make humans into a society, each person responsive to the situation of others.[13]

> Moral faculties...were given us for the direction of our conduct in this life... [They are] denominated by laws: thus the general rules which bodies observe in the communication of motion, are called the laws of motion. But those general rules which our moral faculties observe in approving or condemning whatever sentiment or action is subjected to their examination, may much more justly be denominated such.[14]

The English philosopher John Locke (1632–1704) proposed that our ideas are formed through our response to external stimuli. Traveling through the landscape, or better yet making drawings of the landscape, engaged the hand and the eye, the body and mind in activities that formed healthy, rational habits for organizing experience, which could be used as standards to guide individuals in political and daily life.[15] Building on the sensation-based psychology of Locke and his French contemporary, Condillac, the editor of the *Encyclopédie*, Denis Diderot, proposed that works of art and architecture which artists designed in accordance with the dictates of the Newtonian order could reify Lockean experience and speed the formation of instrumental reason. Art critic, formulator of educational systems for Catherine the Great, and son of an artisan, Diderot included an entry in the *Encyclopédie* for the term 'machine' which offered a short discourse on the category of history paintings known as 'grandes machines.' Here the author of the entry, the Marquis de Jaucourt (1704–1780), characterized paintings as a sort of moral technology analogous to more conventional mechanisms for doing work.[16] He proposed that because their creators combined veracity of detail with rational composition, these paintings could be coordinated into an ideally designed environment with which the individuals interacted. In this way people could be 'moved' to coordinate their actions through the pleasurable sensations these stimuli aroused in them. Julien Offray de la Mettrie's book *L'Homme Machine* (*Machine Man*, 1747), whose extreme materialism placed the author on the fringes of Enlightenment salon society, provided a clear statement of the technological character implied in Enlightenment discussions of human reason:

> 'To be a machine and to feel, to think and to be able to distinguish right from wrong, like blue from yellow—in a word to be born with intelligence and sure instinct for morality and to be only an animal—are thus things which are no more contradictory than to be an ape....'[17]

As these examples make clear, instrumental reason is inseparable from communication and dependent on it to bring about a functional society through the control of information and ultimately of a wide variety of behaviors. Moreover, communication is a physical process in which information in the form of physical signs and signals is transmitted between things and people and between people. Hence, communications media include everything from human senses and sensibility to spoken and written language to painted or printed words and images to buildings and sculpture, to roads, canals, and the postal service. It would come to encompass the telegraph, telephone, film, radio, television, and internet.

Whether they focused on an external or internalized control mechanism, the men of the Enlightenment wavered over the answer to two questions. They came to no consensus about whether the control imposed a form on human behavior, or rather adjusted or steered it. Nor did they agree about whether its origination and implementation was to be exercised by authority at the head of a social hierarchy or by individuals seeking to regulate their own social relations, or a combination of the two. At the same time, they agreed that there were several arenas of behavior that required control: economic (especially production), public or civic, and domestic. And they all looked to some social cadre, especially scientists and moral philosophers, to take the lead in inaugurating programs that would introduce these fundamental controls into society.[18] In the end, they left us with three definitions of machines as control mechanisms, which were or could be interconnected to create an ideally functional environment: human beings; machines in the traditional sense of tools and engines; and society with its organized institutional structures. In addition, by equating human reason with mechanized decision-making activity, they opened the way for the development of automatic machinery.

Later interpretations of Enlightenment instrumental reason fall into two camps, so far as the character of that control is concerned. In both cases the notion of thinking as a technology of control is deeply imbedded, although not specifically identified as such. The camp of Max Weber interprets it as a means used by bureaucratic authority to exercise power by imprinting a certain character on all human activity within its purview. As R. Bernstein summed up Weber on Enlightenment instrumental reason:

> This form of rationality affects and infects the entire range of social and cultural life, encompassing economic structures, law, bureaucratic administration, and even the arts... [Its growth] does not lead to the concrete realization of universal freedom [but to an] [sic] Iron Age of bureaucratic rationality.[19]

In the other camp, Jürgen Habermas proposes that Enlightenment figures conceived of instrumental reason as a tool individuals could use to verify and adjust the course of their discussions with one another about political and social matters. These discussions took place within what Habermas has termed the public sphere. This was a special space invented during the eighteenth century—as much fixed in the collective bourgeois imagination as in specific geographic locations. The public sphere lay outside the home and purview of the state authorities and included places such as cafes, where the latest ideas circulated via books and journals were discussed. For Habermas, the dynamic of these discussions rested on what he calls communicative action, and reason served as an internal gyroscope helping to balance these reciprocal exchanges among individuals, to ensure that they came to just decisions.[20]

But what of technology in the traditional sense? How does its history fit with the picture historians paint of instrumental reason? There was, after all, a high and a low enlightenment. Neither Habermas nor Weber concern themselves much with physical technology, although Habermas emphasizes how important the printing press and a free press have been to the Enlightenment project. They are interested in the side of the equation that can be called institutional technology. The *philosophes* made communication of ideas a high priority, not simply through the importance they placed on the invention of the printing press and freedom of the press to further the dissemination of ideas. To them, traditional technologies—whether tools, machines, or techniques—were forms of communication media. These media, like those who used them, could all be designed to the specifications of instrumental reason. Writing at a time when industrialization was barely underway, these men talked of perfecting machines, methods and processes, designing them to work in accordance with the patterns of instrumental reason. They were interested in identifying the connections that could make workshops and their products, inventions and inventors, part of a vast communications system—an objective to which their contributions to the *Encyclopédie* bear witness. For Diderot the complicated and trying process of realizing this vast publishing project brought the very abstract plan it promulgated into the realm of practical communications.[21]

Thus, this belief in control became a master theme in the cultural history of the industrial era. The notion of control took on two different political meanings identified with two different models of productive social orders. One was of the culture of control as a set of values, activities and modes imposed by a central authority that leaves little or no room for individual initiative; the other, the idea that control cultures are designed to allow individuals to steer their own energies into productive activity that is mutually beneficial. Underlying the faith in control was the discovery that both the universe and the human body, physical nature and human nature, were energy systems whose flow depended on the principles of control. This discovery promised that systematically organizing human relationships around the activities of production and consumption would increase wealth and with it human happiness. In order to do so, it was necessary to create a properly designed material culture out of the raw physical matter of the earth.

These instrumental proposals for bringing direction and some level of organization were not without moral flaws, political complications and social contradictions themselves. Even in the eighteenth century Adam Smith and the French *philosophes* noted this. In the nineteenth century Saint-Simon, Fourier and Marx criticized and offered alternatives to the liberal bourgeois notion of control. In the twentieth century critiques of control, aimed especially at the political oppressiveness of automatic machinery and corporate operated communication systems became a major theme of Frankfurt School members, post-modernists and post-structuralists.[22] The **obverse**, uncontrollable or out-of-control became a theme as well. Meanwhile, the term 'control' itself acquired ever more technical definitions that testify to the central place this culture continues to hold in our society.

HISTORICAL DEFINITIONS OF CONTROL

Definitions of key words help us to understand the various meanings attributed to control. They offer insight into shared ways of thinking and communicating about the subject by defined social groups at particular historical moments. Definitions also mark the historical development of an idea in light of social, economic and technological changes.[23] Generally speaking definitions of control have shared three characteristics. First, whether as a noun or a verb, control is identified with inventions and techniques that affect economic, social, and psychological existence. It is inherently technological in character. Second, the term is identified with the formation and regulation of systems in which energy is kept circulating efficiently. Control has to do with functions which we now call feedback, reverse salients, throughput, and output—the structuring, co-ordination, and regulation of machine and human energies and the physical environment. Finally, meanings of control subsume the notion that its objects (technological, environmental, social or economic) are capable of operating in a mechanistic way that allows their coordination into large technological systems through the communication of information—is not limited to telegraph and telephone, nor to the written and spoken word.

In the case of control, definitions fall into one of two major socio-economic divisions: the term is associated with financial record keeping and more broadly with the economy; and it is used in reference to human beings' social behavior. They meet at the junction production end-consumption. Thus, it is not surprising to find that the origins of the word lie in the period that prepared the way for modern society—the 15th century when merchant capitalism, and the modern state with its fiscal apparatus began to emerge.

The word "control" is an Anglo-French term, a contraction of the word *contre-rolle* referring to a register kept in duplicate by administrators to serve as a means of verifying accounts by comparing one to the other.[24] It also came to mean the act of verification and became a title given to the person with the authority to make that judgment.[25] For mercantilist states, beginning to rationalize the collection and expenditures of their revenue, the 'contra-roll' provided a way to verify budgets and incomes from production and exchange "control" engaged ideas of human feedback by comparison and readjustment to given standard. In addition, it involved the notion of creating functional interfaces that could serve to align the behavior of those engaged in the circulation of money, credit and taxes. Those who used the term thought of it as a means of accounting to optimize the flow of money. Underlying this function, however, was a growing sense that a dynamic system of social relationships based on exchange could be created and maintained for the benefit of the state and/or the merchants through information control mechanisms. It is also important to note that the associations of control with responsible fiscal behavior carried over into the parlance of moral literature. In the sixteenth century, the great French writer Montaigne mocked those who thought of God as the keeper of ultimate accounts.[26]

While the term kept this meaning of verification, other meanings developed in the midst of a society that saw increasing specialization in belief, knowledge, skills, administration, and the growth of absolutism. By 1580, in the midst of the wars of religion that heightened natural concerns about human inclinations to avoid accountability, the term control took on the expanded meaning of surveillance. By the seventeenth century, it had also become more focused on the nature of the relationship between the verifier and the verified: mastery over something or someone.[27] The French mathematician and philosopher Pascal, for example used it in this way to mean the power and ability of an individual to make a thing or an individual perform in a predetermined way. His own interest in the mathematics of probability, in questions of chance and necessity, and individual ability to chose, reveal a growing awareness at least within his circle that human beings can never make any system work perfectly all of the time, but that any system may be quantified and analyzed with a view to making it as predictable as possible because it is subject to operation in accordance with certain laws. It also shows us that by the seventeenth century, at the moment when members of the state academies of science began to formulate a mechanistic picture of the world, the increasingly refined idea of control began to be extended to human beings and to physical entities as if they were to some extent equivalent phenomena under natural law.

Newton's work produced a testable theory to explain how the parts of the solar system held their relative places and continued to turn as they circulated about the sun. In the next century, gravity became a metaphor for the control mechanism that could be embodied in jurisprudence, administrative finance, architecture, public works, machinery, communications technologies and human social and economic behavior. People may not have known what the 'motive power' was, but they could observe how it worked. The *philosophes* in France, began to think about a social physics. Across the Channel, the Scotsman Adam Smith, who had cast a critical eye on French economic theory after a visit there in the 1760s, inquired into the social ramifications of moral sentiments and the nature and causes of the wealth of nations. In addition, the clock and the atmospheric steam engine provided reformers, social philosophers and economic theorists with highly charged symbols of control in the arena of political economy, as Otto Mayr has shown.[28]

The *Encyclopédie* reveals its editors and contributors' interest in control as an administrative activity associated with the running of a rational state. "Control" was defined first and foremost in juridical terms. It was either the person or office in charge of seeing that laws and regulations that determined what was and was not legal were applied. In this way individuals who held the posts of 'contrôleurs,' as did the influential contributor to the *Encyclopédie*, Anne Marie Turgot, guided a whole range of activities by setting and enforcing standards for production and transport of goods, the collection of taxes and duties and the building and maintenance of roads, canals and communications networks. The entries under "control" testify to the philosophers' desire for effective bureaucratic tools that would encourage the circulation of goods and money. They reveal the monarchy's unending struggle to find the proper design for its fiscal administration.[29]

Yet, in the eighteenth century we also find that control takes on the meaning of unwanted or objectionable, hence unnatural, action when it is imposed by the visible hand of temporal authority. Most notably in the writings of Adam Smith, the concept of society as a self-acting and self-regulating machine composed of industrious self-interested individuals found its true *contre-rôle* in the abstract mechanism of the market. Although Smith does not use the word 'control,' it is embedded in his discussions of market operation in *The Wealth of Nations* and embodied in his notion of the invisible hand:

> The quantity of every commodity brought to market naturally suits itself to the effectual demand. It is the interest of all those who employ their land, labour, or stock in bringing any commodity to market, that the quantity never should exceed the effectual demand; and it is the interest of all other people that it never should fall short of that demand.

Or we find this notion of 'control-less' control underlying Smith's discussion of the role of fixed capita in a situation where the norm of control is to balance by approving productivity and negating increased costs, as in the case of fixed capital:

> In manufactures the same number of hands, assisted with the best machinery, will work up a much greater quantity of goods than with more imperfect instruments of trade... . It is upon this account that all such improvements in mechanicks, as enable the same number of workmen to perform an equal quantity of work, with cheaper and simpler machinery than had been usual before, are always regarded as advantageous to every society.[30]

The plates and legends in the *Encyclopédie* go even farther in spelling out how improved tools act as salutary controls on human production. In the case of print making, the plates depict the makers of prints and their techniques and tools. Plates demonstrate different techniques and accompanying text explain that the marks constitute a language of visual signs communicating natural effects. But in contrast to Smith's approval of tools that steer workers to be more efficient, in the *Encyclopédie* the use of a tool that makes three lines simultaneously is denigrated in favor of a tool that makes one at a time. A tool requiring more human decision making effort is desirable because it constrains its user to translate the artist's effects more accurately into another communications medium.[31]

The texts of Adam Smith and the author of the entries in the *Encyclopédie* on control exhibit a shift in the discussion of control, bringing it into conjunction with more technical subjects and providing more detailed descriptions of how control mechanisms work in specific, practical situations with measurable results. On the eve of the industrial revolution, both also make it clear that human beings operate in a mechanistic fashion by virtue of their physiology, while Smith brings into play intimate observations of the complex relationship which market controls create between capitalists, workers and machines. Definitions of control in the nineteenth century would make these trends even clearer, as industrialization proceeded.

Although the eighteenth century was fascinated with automata, such as Vaucanson's mechanical duck, during the early nineteenth century men interested in industrial development and science began to realize how instrumental reason might provide models for machine design that would eliminate or reduce the need for human intervention. Not surprisingly then, in the nineteenth century the term 'control,' under the influence of English, began to be used in specialized scientific and technological sense, as well as more generally. As the practice of science and engineering matured, so did its language become more technical. Ampère, the French scientist responsible for significant theoretical work in electromagnetism which had important ramifications for the electric telegraph and automatic communication of information, coined the term 'cybernétique' in 1834 in reference to the art of governance in the case of both machines and human beings.[32] We find Charles Darwin in 1875 describing how he used four bladders as a 'control experiment.' By 1865 control is found as the name given to any apparatus (a machine) intended to automatically determine the actions of people or other machines.[33]

The initial stage of industrialization brought a new class of entrepreneurial manufacturers and traders to economic prominence. These men defined machines as a means to their own political and social advancement. The characteristic they most valued in machines was the status they gave to their inventors, as Simon Schaffer has shown in the case of Charles Babbage's differential analyzer. This machine, meant to quickly provide quantitative information that would eliminate the need for human calculators, appealed to Babbage as a means to his own social advancement, for it would allow him to participate in steering the direction of industry. Babbage himself declared the Crystal Palace exposition, where he had the analyzer on view, a marvelous administrative invention whose organization had steered exhibitors and public into a happily ordered socio-economic relationship built upon machine production.[34]

If Babbage and the circle that included Prince Albert and Andrew Ure saw the bourgeois-dominated system of machine control they inaugurated in an optimistic light, after 1870 middle and upper class definitions of control reveal more pessimistic attitudes. Babbage could hold that his machine would help eliminate 'the floating chaos of error' that disjointed all relations in industrial society; but by century's end the bourgeoisie no longer believed nature flowed in an orderly manner below the misguided ship of human miscreance. They had a whole new range of machines and a new power source, electricity, at hand.[35] The growing number of definitions that appear under the entry 'Control' in dictionaries of the period reveal a concern that Nature, including human nature and the machines humans had created, now had to be harnessed and coordinated through a special class of technologies if circulation were to continue. For example, it is at this time that we find the word 'control' used in conjunction with new combinations of men and machines designed to regulate time and space. There were railroad timetables, mechanisms that coordinated signals and switching of tracks, and electrical systems that integrated them, for example. In terms of technologies regulating how the leisure class used machines, in automobile racing by 1900 the term 'control' referred to a system of limits on speed in towns and other places and to the points at which time-keepers were posted. An analogous system existed for airplane racing by 1912.[36]

Concomitant with the rise of bourgeois *fin de siècle* culture, control became associated with specialized machines and specialized forms of human behavior devised to keep the flow of energy in other machines (including the body) in check. By 1914, the term is used to refer to the apparatus by means of which a machine, as an airplane or automobile, and even the parts of a machine such as the carburetor was made to perform. By century's end the term had already come to refer to mastery of the self. It is not surprising to find 'self-control,' 'self-mastery' and 'nervous control' (1915) come into their own in the 20th century, a period that has witnessed the unveiling of those distressingly disordered depths of the human mind which called natural reason into doubt—the unconscious, the ego, the id and super-ego and the will.[37] Alongside this self-imposed control, were the controls on individuals in the work place devised by Frederick Winslow Taylor and Frank and Lillian Gilbreth.

In fact, the term 'control' is socialized and objectified by the 1920s and 30s, as a growing class of professionals imbued with technocratic enthusiasm took charge of industry and government and these institutions grew in size and influence. 'Control,' as a term and a concept, figured in the language of the professionals. They applied it to production, advertising, industrial psychology management and economic theory. By the 1930s the term control overtly expressed their growing class consciousness. Social distinctions were made between a small group of privileged individuals (professionals) who could exercise their will over nature (the subject) and the laboring masses who carried out that will in concert with the machines. The former group included engineers, economists, scientific managers, inventor entrepreneurs, architects and industrial and commercial designers, and home economists.

Engineers, for example, endowed the term with the mystique of a fraternal order's power to make the energies of the physical world circulate within the patterns they had designed. Nowhere is this meaning better exemplified than in the multimedia pageant mounted by The Society of Mechanical Engineers at Stevens Institute of Technology on the occasion of its 50th Anniversary, April 5, 1930. Entitled 'Control', the pageant featured allegorical figures of 'Control,' 'Intelligence' and 'Imagination' who lauded engineers' domination of nature as an expression of and requisite for individual freedom. It is not surprising to find their guest of honor was Republican President Herbert Hoover, a member of the society then leaving his mark on America's river ways.[38]

During this second phase, 'control' appears in the language of industrial psychology and management, economics, production, distribution and advertising, in conjunction with discussions that set domination beside individual liberty. In the interwar period dominated by Fordism, control came to refer to management, as business owners sought ways to systematically integrate all activity within their corporate environments even as they continued to believe in Adam Smith's 'invisible hand' equilibrating the natural market place.[39] More broadly inclusive than Frederick W. Taylor's 'scientific management' or Frank and Lillian Gilbreth's time-motion studies was a widely adopted philosophy known as 'systematic management.'[40] While the former aimed at increasing efficiency among workers on the shop floor through direct constraints on their bodily actions, in the case of systematic management, 'control' referred to 'managerial control—over employees..., processes and flows of materials....' As Joanne Yates has made clear, this system turned all forms of communication into tools whose purpose was to establish and maintain 'flows of information and orders.'[41] The term, with its growing association with automatism, also appeared in conjunction with the development of electrical power and the electrical industry, and the wide range of new electrical products and systems for converting and delivering electricity in which men and machines were reigned into one large 'circuitry.' In the 1930s, at the height of an economic crises, the colloquial phrase, "Everything's under control," meaning everything is in order, was widely used.[42]

One of the most significant adumbrations of the term's association with management came in the field of economics during the 1930s. Huge fluctuations in prosperity that marked the great depression brought attention to bear on the tenets of political economy when capitalist and totalitarian systems confronted a world-wide depression. The figure of greatest economic influence in the years of the depression was John Maynard Keynes, who saw the need for a new type of control mechanism that would regenerate production and with it the flow of money and goods.[43] In his *General Theory of Employment, Interest and Money* (1936), Keynes referred to 'the vital importance of establishing certain central controls [a somewhat comprehensive socialization of investment] in matters which are now left in the main to individual initiative.'[44] Government controls had the function of 'adjusting to one another the propensity to consume and the inducement to invest,' not the destruction of capitalist forms and their dynamic.[45] The two streams of 'control,' the economic and the social, had become a common confluence.

Objectification is reflected in the way control became attached to the technical names of things, and the growing number of things associated with the term 'control.' There were objects identified with a social activity as seemingly insignificant as the card game Bridge. Invented around the turn of the century, this pastime of the upper-middle classes and the wealthy, brought with it not only the use of 'control' to define one of its principle strategies, but to name cards that served a tactical purpose. More overwhelming, were the growing number of objects whose names associated 'control' with electrical, aeronautical and gunnery systems, and with energy conversion, railroad operation, and radio communications. There were control boards, control panels, control cables, control circuits, control-circuit transformers, control columns or 'joy sticks,' control electrodes, control electrodes, control switches, control surfaces, remote-control junction boxes, etc.[46]

The term had become part of compound words that gave it its own space, the 'control room' in the factory or the radio station for example, and its own place, the 'control desk.' Control thus also became associated with physical place, not limited exclusively to the factory. In fact as these terms suggest, the growing number of compound words and phrases that included 'control' in them provides some measure of the extent of socialization. Swiss-born historian Siegfried Giedion, for whom the term 'mechanization' was synonymous with 'control,' wrote in 1948 that it had found its way into the factory, the home, city and regional planning and all forms of material culture that moved to a common rhythm. In the inter-war years of 1918 to 1939, "...what had been germinating from mid-nineteenth century on...at one sweep...penetrates the intimate spheres of life."[47]

The word even began to take on a resonant power of its own. This was the era that marked its deification and vilification in films such as 'Metropolis' and 'Modern Times,' as well as in *Brave New World* and *1984*, books that still exercise a hold on our imaginations. It saw the definition of systems (open, closed) and the first analyses of their characteristics. It was also the period in which physiologists studied human motion with an eye to making labor in factories and offices more cost-effective, casting their subjects as 'motors' and 'heat converters.'[48] If the objects the experts invested with the name 'controls' made automatic some forms and levels of decision making which humans had once performed, these objects also allowed them more precise quantification, measurement and hence regulation of the conversion and flow of energy between people, people and machines and machines and machines. Control had become both a science and a sociological fact.

During the two and a half decades after the Second World War, two changes occurred in the meaning of control. The first constitutes a synthesis of research and ideas that pre-date the conflict; while the second signals the opening of new cultural horizons in the economy of the late twentieth century. After the War the term 'control' became identified with two sets of theories developed by professional scientists anxious to consolidate findings and develop general laws that gave intellectual integrity and social status to the academic study of control and its applications: decision-making machines and operant conditioning of human behavior. The first referred to rationally steering production and regulating circulation in all mechanistic systems (physical, biological, social and economic); the other to rationally steering patterns of consumption and cultivating desires and habits that bore on social and economic systems. These efforts are identified with the names of two individuals: Norbert Wiener (1894–1964), who developed Cybernetics or the science of communication and control; and B. F. Skinner, who laid the foundations for the science of behavioral psychology.[49] Each man took control theory down one of these two related paths.

Norbert Wiener, the MIT physicist, set the term 'control' into conjuction with 'communication,' using the term 'cybernetics' to describe a new science that theorized about, analyzed and tested machines (or servo-mechanisms) designed to steer other machines through what he metaphorically termed an interchange of information. Instrumental reason had become embodied in a machine that made other machines move. As he explained in *The Human Use of Human Beings*:

> When I control the actions of another person, I communicate a message to him.... Furthermore if my control is to be effective, I must take cognizance of any messages from him which may indicate that the order is understood and has been obeyed.
>
> When I give an order to a machine, the situation is not essentially different from that which arises when I give an order to a person.[50]

Moreover, based on an analogy between the 'physical functioning of the living individual and the operation of some of the newer communication machines,' human beings and computers were 'precisely parallel in their analogous attempts to control entropy through feedback.'[51]

While Wiener's theory wove together many threads of scientific research that pre-dated the war, his ideas also opened up new ways of thinking about control.[52] Wiener's definitions revitalized and reconfigured Enlightenment 'instrumental reason,' human-nature-machine analogies, and notions of communication as the exchange of physical experience between any two entities.[53] Like Turgot, Condorcet and Diderot, he generalized communication into a process of exchange and circulation of information, a concept that defines all physical experience as potential information and all modes of transfer and exchange of that experience as communications media. Like them he also conceived of instrumental reason as a rational tool for directing work energy into productive ends. For example, he recognized that the 'automatization of the factory' through the introduction of the computer, which he fully approved, would be one of the major factors in 'conditioning the social and technical life of the age to come.' But he also defined law in terms of a servo-mechanism, an ethical control or 'process of adjusting' the couplings connecting individual's behavior towards one another.[54]

Wiener's definition also took note of post-enlightenment scientific discoveries, most importantly the laws of thermodynamics and quantum mechanics.[55] Moreover, he was born early enough in the century to share the late Victorian view that human beings could only hope to bring order to a small niche within the vast entropic field of nature. And like many of his professional contemporaries, he put his money on instrumental reason to discipline the physical environment over which they might hold sway. Like Diderot and Condorcet he recognized that human emotions were part of the evidence to be studied and mastered; but unlike Diderot and Rousseau he was unable to factor emotional energy into his theory.[56] As the quotation referring to the factory suggests, for purposes of argument Wiener's definition of control may have assumed that a human mind and a computer were analogous decision-making systems, but the same definition implied a social hierarchy in which certain members of society would dictate the uses of servo-mechanisms, others design them, while large numbers of people would be relegated to tasks requiring no decision-making or be entirely factored out of production.[57]

We might say that the Harvard psychologist B. F. Skinner took up Wiener's slack on the issue of human emotions. He developed a theory that dealt with just those components of social and economic systems and human nature which Wiener failed to address. For Skinner, who developed behavioral psychology into a science based on operant conditioning, the active verb 'control' connoted everything from shaping to steering, to dominating a human being. As a noun it referred to the 'power to influence, change, mold...human behavior.'[58] Those who held this power could strengthen their influence by doing scientific studies on now quantifiable human responses to external stimuli. While 'control' retained its Enlightenment connotation of instrumental reason embodied in the minds of the social leaders who monitored feedback, under Skinner the environmental stimulus seemed to become the refined rational instrument of social control yearned for by the philosophes, an instrument in some ways analogous to the servomechanism.[59] The social applications of these controls included the design of teaching machines for classroom education, the introduction of uncertainty in rewards for work to heighten productivity, the use of programmed instruction and questionnaires in industry and the military to increase administrative efficiency, and the measurement of television viewers' looking responses to optimize the effectiveness of advertising designs.[60] In all cases, the controls caused humans to direct their energies into production and consumption channels, keeping the economy moving by cultivating behavioral habits and desires that influenced decision making.

In a 1961 article entitled 'The Design of Cultures,' Skinner extended the definition of control culture to embrace the organizations and institutions that structure and orchestrate society. 'Doing-something-about-human-behavior is a kind of social action,' he wrote. It was highly dependent for its success on the design of institutional systems incorporating more restricted stimuli. Moreover,

Skinner explicitly classified these institutions as technologies because of their instrumental character.

> There is a considerable advantage in considering these [governmental, religious, economic, educational and therapeutic] institutions simply as behavioral technologies. Each one uses an identifiable set of techniques for the control of human behavior, distinguished by the variables manipulated. The discovery and invention of such techniques and their later abandonment or continued use—in short, their evolution—are, or should be, a part of the history of technology. The issues they raise, particularly with respect to the behavior of the discoverer or inventor, are characteristic of technology in general.[61]

While Wiener and Skinner lamented that their theoretical definitions of control had been misapplied, they also warned the public of the perils lying in wait when, as Wiener expressed it, 'human atoms are knit into an organization in which they are used, not in their full right as responsible human beings....'[62] This concern became a theme in popular culture in the 1950s and 60s as 'control' took on connotations of insidious, inhuman or uncaring ultra-rational forces identified with machines. 'Technology out of control' became a critical theme in these last years of modernism.[63] 'The Manchurian Candidate' and Vance Packard's *Hidden Persuaders* both spoke to public paranoia about an insidious presence behind manufactured appearances that made them do things they couldn't control. Books, movies, and particularly television, as well as media advertising helped build credence for these preparative connotations, which carried over to modernist architecture.[64]

In the post-industrial economy that began in the 1970s, new definitions of control increasingly attached themselves to the computer as the embodiment of instrumental reason. It was a servo-mechanism, that is the controller of machines, and even the teacher of machines. It was also an aid to human decision-making. The computer became the *sine qua non* of the postmodernist era.[65] Interestingly, the computer was identified with control in ways that recognized the complex interdependency between individuals' social behavior and the financial activities that kept manufacturing, goods and services flowing. Flexible behavior was the key. David Harvey summarized this adaptation quite well when he pointed out that the computers' prime meaning became identified with a tool for controlling information in two very particular senses. One was associated with '"smart" and innovative entrepreneurialism' that allows for 'control·over information flow and over the vehicles for propagation of popular taste and culture as the vital weapons in a competitive struggle.' The other was far more important. Control via computers meant 'the complete reorganization of the global financial system and the emergence of greatly enhanced powers of financial co-ordination.' These are 'international and instantaneous.' Control carried with it the concern for inaugurating patterns of human and machine behavior that optimized their flexible character, a response that optimized their ability to maintain the flow of goods and money in the face of multiplying contingencies.⁶⁶

In conclusion, since Wiener and Skinner, the definition of control has failed to contain a reference to nature as the ultimate verifier or *contre-rôle*. Control is devoid of that Newtonian vision of dynamic equilibrium embodying an aesthetic of harmony that instrumental reason could mimic and embody in cultural forms. Definitions of 'control' based on computer culture currently refer to a self-generated design for economic and social controls that answer to our self-designated contingencies. Whether they be the internet, telephone, television, post-modern buildings or planned and gated communities, the environment of communications media we have created is an artificial Nature enclosing and subjecting us to its laws, even as we seem increasingly capable of re-designing it to meet our own specifications. Meanwhile, the natural environment we have exploited to create this artificial one is deteriorating, as a result of our effort to embed the patterns of instrumental reason into our surroundings without calculating the consequences.

NOTES

1. For a discussion of changes in the mind-set of seventeenth century philosophers, see E. J. (Eduard Jan) Dijksterhuis, *The Mechanization of the World Picture*, Translated by C. Dikshoorn (Oxford: Clarendon Press, 1961). This study, however, does not take into account the social or economic dimensions of this change in *mentalité*. James R. Beniger, *The Control Revolution: Technological and Economic Origins of the Information Society* (Cambridge, Mass.: Harvard University Press, 1986). The idea of control as a master theme raises the prickly issue of technological determinism, for even if decision-making process rests in the character of human reason, by Enlightenment definitions instrumental reason is inherently technological. See the collection of essays, *Does Technology Drive History?: The Dilemma of Technological Determinism*, eds. Merritt Roe Smith and Leo Marx (Cambridge, Mass.: MIT Press, 1994). Also suggestive in considering the reciprocal mediating relationship between scientific ideas and technological forms is M. Norton Wise, 'Mediating machines,' *Science in Context* (1988) 2: 77–113. I want to thank Prof. Stuart Leslie for suggesting this article to me.
2. I have based this periodization on the following texts which discuss all or a major portion of these 250 years: David Harvey, *The Condition of Post-Modernity: An Enquiry into the Origins of Cultural Change* (Oxford: Basil Blackwell, 1989); James R. Beniger, *The Control Revolution*; David S. Landes, *The Unbound Prometheus: Technological Change and Industrial Development in Western Europe from 1750 to the Present* (London: Cambridge University Press, 1969); Daniel Bell, *The Coming of*

Post-Industrial Society; A Venture in Social Forecasting (New York: Basic Books (1973); Krishan Kumar, *From Post-Industrial to Post-Modern Society: New Theories of the Contemporary World* (Cambridge, Mass.: Blackwell Publishers, 1995); Yoneji Masuda, *The Information Society As Post-Industrial Society* (Washington, D.C.: World Future Society, 1981). On systems see, C. West Churchman, *The Systems Approach* (New York: Laurel of Dell Publishing Col, 1979 rev. ed.); Thomas P. Hughes, 'The Evolution of Large Technological Systems,' in *The Social Construction of Technology*, eds. W. Bijker, *et al.* (Cambridge, Mass: MIT Press, 1989), pp. 51–82.

3. James R. Beniger, *The Control Revolution.*
4. Anson Rabinbach, *The Human Motor: Energy, Fatigue, and the Origins of Modernity* (New York: Basic Books, 1990): 12–18, 126.
5. See Rabinbach, *Human Motor.* An useful source for information about the impact of thermodynamics on religious ideas is Erwin N. Hiebert, 'The Uses and Abuses of Thermodynamics in Religion,' *Daedalus* 95 (1966), pp. 1046–79.
6. See n. 2 above. On flexible manufacturing, see: Michael J. Piore & Charles F. Sabel, *The Second Industrial Divide: Possibilities for Prosperity* (New York: Basic Books, 1984); *Small and Medium-size Enterprises*, eds. Arnaldo Bagnasco and Charles F. Sabel (London: Pinter, 1995). *Markets And Manufacture In Early Industrial Europe*, ed. Maxine Berg (London; New York, NY: Routledge, 1991) and Philip Scranton, *Endless Novelty: Specialty Production and American Industrialization, 1865–1925* (Princeton, N.J.: Princeton University Press, 1997) are also important in establishing that flexible manufacturing was not an invention of late industrial capitalist economies.
7. I believe Robert Darnton was the first to use these terms in his book on the popular press during the Enlightenment: *The Literary Underground of the Old Regime* (Cambridge, Mass.: Harvard University Press, 1982). See also Kenneth Ludwig Alder, 'Forging the New Order: French Mass Production and the Language of the Machine Age, 1763–1815,' thesis presented to the Department of the History of Science, Harvard University, June 1991. Useful introductions to the history of this period, include: Norman Hampson, *A Cultural History of the Enlightenment* (New York: Pantheon Books, 1968); Franck Salaün, *L'ordre des moeurs: essai sur la place du matérialisme dans la société française du XVIIIe siècle (1734–1784)* (Paris: Editions Kimé, 1996); *Wealth And Virtue: The Shaping of Political Economy in the Scottish Enlightenment*, eds. Istvan Hont and Michael Ignatieff (Cambridge [Cambridgeshire]; New York: Cambridge University Press, 1983); Dena Goodman, *The Republic of Letters: A Cultural History of the French Enlightenment* (Ithaca, N.Y.: Cornell University Press, 1994); Peter Jones, 'Hume on the Natural History of Philosophical Consciousness,' in *The 'Science of Man' in the Scottish Enlightenment: Hume,*

Reid, and Their Contemporaries ed., Peter Jones (Edinburgh: Edinburgh University Press, 1989); Jed Z. Buchwald, 'Mediations: Enlightenment Balancing Acts, or The Technologies of Rationalism,' in *World changes: Thomas Kuhn and the Nature of Science*, ed. Paul Horwich (Cambridge, Mass.: MIT Press, 1993). Peter Gay, *The Enlightenment: An Interpretation*, 2 vols. (New York: Alfred A. Knopf, 1969). The classic work in the field is Ernst Cassirer, *The Philosophy of the Enlightenment* (Princeton: Princeton, N.J., 1951). More recent work of importance is Keith M. Baker, *Condorcet: From Natural Philosophy to Social Mathematics* (Chicago: University of Chicago Press, 1975); and *The French Revolution and the Creation of Modern Political Culture*, ed. Keith Michael Baker (Oxford; New York: Pergamon Press, 1987–1989).

8. See n. 15 above as well as the following on French economists and Adam Smith: Robert Heilbroner, *Teachings from the Worldly Philosophy* (New York: W. W. Norton, c. 1996), 47 and 55–126; Robert L. Heilbroner, *The Worldly Philosophers: The Lives, Times, and Ideas of the Great Economic Thinkers* (4th ed. New York: Simon and Schuster, c. 1972), 19–52 and 38–72. Also see 'Introduction,' in Adam Smith, *An Inquiry into the Nature and Causes of the Wealth of Nations*, ed. Kathryn Sutherland (Oxford: Oxford University Press, 1993), ix–xlv.

9. See the discussion of Habermas's ideas on the Enlightenment project of modernity in David Harvey, *The Condition of Post Modernity*, pp. 12–115. While Harvey identifies some of the problems and questions related to control that Enlightenment ideals of reason raised even for those who formulated them, the following works are more helpful in understanding the various interpretations of the way reason functions as a control mechanism: *Critique and Power: Recasting the Foucault/Habermas Debate*, ed. Michael Kelly (Cambridge, Mass.: MIT Press, 1994); Jürgen Habermas, *La technique et la science comme idéologie*, trans. (Paris: Denoel/Gonthier, 1978); Jürgen Habermas, *The Structural Transformation of The Public Sphere: An Inquiry into a Category of Bourgeois Society*, trans. Thomas Burger (Cambridge, Mass.: MIT Press, c. 1989).

10. On Kant, see: Immanuel Kant, *Critique of Practical Reason* (1949); Ernst Cassirer, *Kant's Life and Thought* (1981); Lewis W. Beck, *A Commentary on Kant's Critique of Practical Reason* (1960); and the articles: 'Immanuel Kant and Kantianism: Life and works: The Critique of Practical Reason.' Britannica Online, <http://www.eb.com:180/cgi-bin/g?DocF=macro/5003/49/6.html>; 'Ethics: Western ethics from Socrates to the 20th century: The Continental Tradition: From Spinoza To Nietzsche: Kant.' *Britannica Online*, <'http://www.eb.com:180/cgi-bin/g?DocF=macro/5002/18/32.html'>

11. Jean-Jacques Rousseau, 'Principes de J.-J. Rousseau, sur l'éducation des enfans; ou, instructions sur la conservation des enfans, et sur leur éducation physique et morale, depuis leur naissance, jusqu'à l'époque de leur entrée dans les écoles nationales. Ouvrage indiqué pour le concours, suivant le

décret de la convention nationale, du 9 pluviose dernier' (Paris: Aubry, l'an II de la République Française [i.e. 1794]); *Rousseau on Education*, ed. Leslie F. Claydon (London: Collier-Macmillan [1969]); Jean-Jacques Rousseau, *Emile: or, On education*, Allan Bloom, ed. and trans. (New York: Basic Books, 1979); Jean-Jacques Rousseau, *Politics and the Arts, Letter to M. d'Alembert on the Theatre*, Allan Bloom, ed. and trans. (Glencoe, Ill.: Free Press, 1960). The letter to d'Alember includes d'Alembert's article 'Geneva' from v. 7 (1757) of the *Encyclopédie*, to which Rousseau's letter is a critical response.

12. Adam Smith, *The Theory of Moral Sentiments*, D. D. Raphael and A. L. Macfie, eds. (Oxford: Oxford University Press, 1976) (Liberty Fund Reprint, 1984), p. 320.
13. Ibid, pp. 10–15.
14. Ibid, p. 165.
15. John Locke, *Some Thoughts Concerning Education; And, Of the Conduct of the Understanding*, edited, with introduction by Ruth W. Grant and Nathan Tarcov (Indianapolis: Hackett Pub. Co., 1996); Barbara Maria Stafford, *Artful Science: Enlightenment, Entertainment, And The Eclipse of Visual Education* (Cambridge, Mass.: MIT Press, 1994).
16. See the entry 'Machines,' in the *Encyclopédie*. Also see Miriam R. Levin, 'The Wedding of Art and Science in Late 18th Century France: A Means if Building Social Solidarity,' *Eighteenth Century Life* 8 (1982): 54–73; and Miriam R. Levin, 'The Printed Image in Industrializing France from the Enlightenment to 1835,' in *Muse and Reason: The Relation of Arts and Sciences 1650–1850*, eds. B. Castel, J. A. Leith and A. W. Riley (Kingston, Ont.: Royal Society of Canada, 1994), pp. 45–61.
17. Julien Offray de La Mettrie, *Machine Man and Other Writings*, trans. and ed. Ann Thomson (Cambridge [England] and New York, NY: Cambridge University Press, 1996), p. 35. La Mettrie's dates are 1704–1780.
18. See Harvey, *Condition of Post Modernity*, pp. 13–15, for a summary of some of the problems and contradictions embedded in enlightenment discussions of instrumental reason. The major one he cites was the tension between individual self-guidance and direction by a powerful outside authority. The ideal these men envisioned was a situation where people would effortlessly conduct themselves in accordance with natural laws because they would have had their own natures properly cultivated. Men and Machines would be coterminous with the great machine of Newtonian nature. The term for this circumstance is 'transparency.' This is why the design of the environment was of such concern to them; but the desire to make humans and society be natural, always led them to propose that a special social group, genius administrators, artists, and scientists, had to be in charge of structuring reality. Rousseau, however, preferred to remain equivocal. He proposed both that people freed from the controls of civilization would spontaneously join together in a leaderless democracy and that

people needed educators to train them to make the right decisions. See Levin, 'Wedding of Art and Science,'; Lynn Hunt, *Politics, Culture, and Class in the French Revolution* (Berkeley: University of California Press, c. 1984), and n. 11 above. Another problem has to do with the character of the control instrument itself. There was little agreement, less understanding and no empirical data about how instrumental reason operated in any detailed way.

19. Quoted in Harvey, Ibid., p. 13.
20. Jürgen Habermas, *Moral Consciousness and Communicative Action*, trans. Christian Lenhardt and Shierry Weber Nicholsen, introduction by Thomas McCarthy (Cambridge, Mass.: MIT Press, 1990); Jürgen Habermas, *The Theory of Communicative Action*, trans. Thomas McCarthy (Boston: Beacon Press, 1984).
21. The full title of the Encyclopedie offers some indication of this objective: *Encyclopédie; ou, Dictionnaire raisonné des sciences, des arts et des métiers, par une société de gens de lettres*, eds. Denis Diderot and Jean le Rond d'Alembert, 36 vols. (Lausanne: Société Typographiques, 1780–82 [v.1,1781]). See also D'Alembert's 'Preliminary Discourse,' *Encyclopédie*, v. 1; and Robert Darnton, *The Business of Enlightenment: A Publishing History of the Encyclopédie, 1775–1800* (Cambridge: Harvard University Press, 1979).
22. *French Utopias; An Anthology of Ideal Societies*, ed. and trans. Frank E. Manuel and Fritzie P. Manuel (New York: Schocken Books, 1971); Frank E. Manuel, *The Prophets of Paris* (Cambridge: Harvard University Press, 1962); Frank E. Manuel and Fritzie P. Manuel, *Utopian Thought in the Western World* (Cambridge, Mass.: Belknap Press, 1979); George Friedman, *The Political Philosophy of the Frankfurt School* (Ithaca: Cornell University Press, 1981); Peter Gay, 'The Social History of Ideas: Ernst Cassirer and after,' *The Critical Spirit; Essays in Honor of Herbert Marcuse*, eds. Kurt H. Wolff and Barrington Moore (Boston: Beacon Press [1967]); *Technology, Pessimism, and Postmodernism,* eds. Yaron Ezrahi, et al. (Dordrecht [Netherlands]; Boston: Kluwer Academic Publishers, c. 1994); and the two volumes by Howard Segal: *Future Imperfect: The Mixed Blessings of Technology in America* (Amherst: University of Massachusetts Press, c. 1994); *Technological Utopianism in American culture* (Chicago: University of Chicago Press, 1985).
23. Two good examples of collections of keywords are: Raymond Williams, *Keywords: A Vocabulary of Culture and Society*, rev. ed. (New York: University Press, 1985); and *Keywords in Evolutionary Biology*, eds. Evelyn Fox Keller and Elisabeth A. Lloyd (Cambridge, Mass.: Harvard University Press, 1992).
24. It was a 'rôle, registre opposé' or 'Registre qu'on tenait double dans certaines administrations, pour que l'un servit à verifier l'autre.' From 'Contrôle,' *Dictionnaire Général de la Langue Française du Commence-*

ment du XVIIe siècle jusqu'à nos jours, eds. Adolphe Hatzfeld, *et al.*, 2 vols. (Paris: Delagrave, 1890–93), vol. 1. Also see 'Contro' and 'Contra' in *A Comprehensive Dictionary of the English Language*, ed. Ernest Klein (Amsterdam: Elsever Publishers, 1966).
25. *Dictionnaire Générale de la Langue Française.*
26. 'Contrerôle,' *Dictionnaire de la Langue Française du 16e Siècle*, ed. Edmond Huguet (Paris: Honoré Champion, 1932)
27. *Oxford English Dictionary*, Second Edition (Oxford: Oxford University Press, 1991).
28. *Authority, Liberty, and Automatic Machinery in Early Modern Europe* (Baltimore: Johns Hopkins University Press, c. 1986).
29. For example, the 'contrôl des dépens' or office of expenditures for the courts had originally been the 'controleurs des tiers-reférendaires.' The latter were brought under the aegis of the crown in 1667, suppressed in 1694 and replaced by the 'contrôl des dépens.' In the eighteenth century, this office was joined with the community of court procurers in an agency that combined judicial and budgetary controls. See 'Contrôleur,' in *Encyclopédie*. If the absolute monarchy cherished the ideal of a social system modeled on a clockwork universe driven by a written body of rational laws and regulations, in reality it was unable to design and build the apparatus that would enable the state to realize this dream—to the great frustration of the *philosophes*.
30. *Wealth of Nations*, pp. 172–173.
31. See Levin, 'Defining a Use for Printed Images', pp. 48–49.
32. 'Note: Used in Fr. form cybernétique (= the art of governing) by A.-M. Ampère, *Essai sur la Philosophie des Sciences*, 1834.' Quoted in sv. 'Cybernetics,' *Oxford English Dictionary*, 2nd ed. (Oxford: Oxford University Press, 1991).
33. See 'contrôl' in *Le Robert, Dictionnaire Historique de la Langue Française*, ed. Alain Rey (Paris: Dictionnaires le Robert, 1994).
34. Simon Schaffer, 'Babbage's Intelligence: Calculating Engines and the Factory System' *Critical Inquiry* (Autumn 1994) v. 21, Number 1, pp. 203–228. Also see Charles Babbage, *The Exposition of 1851*, *The works of Charles Babbage*, ed. Martin Campbell-Kelly, 11 vols. (London: Pickering & Chatto, 1989), v. 10, pp. 113–115 especially, but the entire chapter on 'Position' is relevant.
35. Ibid., p. 81.
36. *OED*, 1991.
37. See 'Contrôl' in *Petit Robert*, 1994.
38. *Control: A Pageant of Engineering Progress* (New York: Society for Mechanical Engineers, 1930), See the declamations of 'Intelligence' and 'Imagination' on p. 6 for example.
39. See Alfred Chandler, *The Visible Hand: The Managerial Revolution in American Business* (Cambridge: Harvard University Press, Belknap Press, 1977).

40. See Joanne Yates, *Control through Communication: The Rise of System in American Mangement* (Baltimore: Johns Hopkins, 1989), pp. xv.
41. Ibid, xvi–xvii, 10–11. Typical of studies done during this period related to control on the shop floor are: Jules Amar, *The Human Motor; or, The Scientific Foundations of Labour and Industry* (London: Routledge; 1920. New York Dutton, 1920), Series: Efficiency books and Francis G. Benedict and Edward P. Cathcart, *Muscular Work; A Metabolic Study with Special Reference to the Efficiency of the Human Body as a Machine* (Washington, D.C.: Carnegie Institution of Washington, 1913)
42. See 'Control,' *OED*.
43. For a biographical sketch and good introduction to Keynes's ideas, see Robert Heilbroner, *The Worldly Philosophers* (New York: Simon and Schuster, 1972), pp. 240–278; and Robert Heilbroner, *Teachings from the Worldly Philosophy* (New York: Norton, 1997), pp. 264–296.
44. Quoted in Heilbroner, *Teachings*, p. 292.
45. His theory became a template for New Deal government policies used as tools to engage the entire population in the flow of production and consumption. Ibid, p. 293 and 292.
46. 'Control,' *OED*, n. 5.
47. Sigfried Giedion, *Mechanization Takes Command, A Contribution to Anonymous History* (New York University Press, 1948), p. 41.
48. See n. 41 above.
49. Skinner and his work are enjoying a resurgence of interest. See the biography by Daniel N. Wiener, *B. F. Skinner: Benign Anarchist* (Boston: Allyn and Bacon, c. 1996); *B. F. Skinner and Behaviorism in American Culture*, eds. Laurence D. Smith and William R. Woodward (Bethlehem: Lehigh University Press, 1996).
50. Norbert Wiener, *The Human Use of Human Beings: Cybernetics and Society*, 2nd ed. Rev. (Garden City, New York: Doubleday, 1954), pp. 16–17.
51. Ibid., pp. 26. Between 1948 and 1970 Wiener offered a number of variations on the initial defintion of cybernetics which are listed in the OED (1992 ed.) The following is a good summary of Wiener's accomplishment vis-a-vis computers: 'A major example of Wiener's inventiveness was the statistical theory of communication, an outgrowth of his work during World War II. This was a general mathematical theory, but it would henceforth inform the design of every kind of communication systems.' Steve Joshua Heims, 'Introduction,' in Norbert Wiener, *The Care and Feeding of Ideas* (Cambridge, Mass.: MIT Press, 1993), p. ix.
52. Wiener saw that computers, already being built before the war, could be used to bring organization to the operation of systems and argued for their continued development. William Aspry, 'The Scientific Conceptualization of Information: A Survey,' *Annals of the History of Computing*, v. 7, n. 2, April 1985, pp. 117–140. Servos are automatic control devices. Term applied to a great variety of physical systems, they can be a simple

governor on a steam engine of the mechanical sort, or a thermostat. Regulate—meaning quantifiable. Post-war period developed theory of servo-mechanism expressed in mathematical relationships: A definition of a simple servo-mechanism they are feed back amplifiers or in engineering terms, 'a device for establishing a certain relation between a variable x(t), called the input signal and a variable y(t), called the output signal.' LeRoy A. MacColl, *Fundamental Theory of Servomechanisms* (New York: Dover Publications, 1968), p. 3. Sets the terms of a reciprocal, interactive relationship that is ongoing over time, depends on sequences of calculations and approximations. The input and output variables don't have to signify the same thing, e.g., one can be electrical force and the other an angle at which a gun is aimed. There is energy associated with the output signal (mechanical or electrical) and it is the objective of the servo to see that that energy is properly utilized. A major value of servos is that they calculate and hence regulate much faster and more accurately than human beings. For those in charge, they offer great power; at the same time the reduce the number of people required to work at lower levels of decision making. Among servo-mechanisms, computers figure as the most important post war control devices. In pre-World War II dictionaries, 'computer' (when there is an entry) refers to people, often those who did astronomical calculations, but not to control.

53. *The legacy of Norbert Wiener: a centennial symposium in honor of the 100th anniversary of Norbert Wiener's birth, October 8–14, 1994, Massachusetts Institute of Technology, Cambridge, Massachusetts*, eds. David Jerison, I. M. Singer, Daniel W. Stroock (Providence, R.I.: American Mathematical Society, 1997).
54. *Human Use of Human Beings*, p. 150 and p. 105.
55. See statement by Warren Weaver quoted in Aspry, 'Scientific Conceptualization,' p. 122: 'The word communication will be used here in a very broad sense to include all of the procedures by which one mind may affect another... [it] may include the procedures by means of which one mechanism... affects another...'
56. See *Human Use of Human Beings*, pp. 73–73: 'Emotion...may not be merely a useless epiphenomenon of nervous action, but may control some essential stage in learning.... I definitely do not say that it does, but...those psychologists who draw sharp distinctions between man's emotions and... the responses of the modern type of automatic mechanisms, should be... careful.'
57. See Ibid., p, 112 ff. Wiener also discussed this problem in *Cybernetics, Science and Society*, 2 vols., I, p. 199 and in *Invention: The Care and Feeding of Ideas* (Cambridge, Mass.: MIT Press, 1993), esp. pp. 103–111.
58. Carl F. Rogers and B. F. Skinner, 'Some Issues Concerning the Control of Human Behaviors,' *Science* 1956 (124), 1057–1066, reprinted in *Control of*

Human Behavior, eds. Roger Ulrich, *et. al.*, 2 vols. (Glenview, Ill.: Scott, Foresman and Co., 1966), v. 1, p. 301.

59. Quote from Skinner, p. 301; Alan Turing, a brilliant British mathematician, followed a path parallel to Skinner's in some respects by looking at Pavlovian operant conditioning research from the early twentieth century in order to build a machine for educating a machine to make decisions via a system based on pleasaure-pain responses. Aspray, 'Scientific Conceptualization,' p. 133.

60. *Control of Human Behavior*, eds. Roger Ulrich, *et. al.*, 2 vols. (Glenview, Ill.: Scott, Foresman and Co., 1966), see articles in the section on 'Modification of Behavior in Industry and Advertising,' v. 1, pp. 217–299.

61. B. F. Skinner, 'The Design of Cultures,' *Daedalus* (Summer 1961), pp. 534–546, reprinted in *Control of Human Behavior*, eds. Roger Ulrich, *et. al.*, 2 vols. (Glenview, Ill.: Scott, Foresman and Co., 1966), v. 1, pp. 336–337.

62. *Human Use of Human Beings*, p. 185.

63. Langdon Winner, *Autonomous Technology: Technics-Out-Of-Control as a Theme in Political Thought* (Cambridge, Mass.: MIT Press, 1977).

64. See Harvey, *Postmodernity*, esp. pp. 35–38 and p. 247.

65. The *Thesaurus of Scientific, Technical and Engineering Terms* (Philadelphia: Hemisphere Publishing Co., 1988) and *Websters New World Dictionary of Computer Terms* (comp. Donald Spencer, New York: Prentice Hall MacMillan, 1994) exemplify this shift.

65. See *Control*, p. 301, etc. Responding to criticisms that this kind of control would turn people into slaves of marketers, capitalists and politicians, Skinner and some of his followers argued that under scientists' guidance authorities could design a culture in which 'control' meant stimulating or liberating individual initiative—an approach that eventually led Skinner into utopian social planning. Skinner, Wiener and their followers opened the way for post-war social institutions to adapt the population to a more consumer oriented, flexible economy, in which service jobs, leisure activities, entertainment, and the media would fill the employment gap created by automation.

66. *Condition of Postmodernity*, pp. 159–162.

CHAPTER 2

Nature Out of Control: Cultural Origins and Environmental Implications of Large Technological Systems

Rosalind Williams

The culture of control of modern times is based on the following logic: technology provides more control over nature, and this control permits human beings to have more predictable and fulfilling lives. At the threshold of the twenty-first century, the limits of this logic have become apparent. Technology appears out of control when technological systems are so large and complex that their human creators are unable to predict, evaluate, or influence their effects. Furthermore, because technological systems have become so intertwined with natural ones, nature too may seem out of control. The ozone hole grows; global temperature rises; forests disappear from the land, fish from the sea, birds from the skies. Human efforts to control nature through technology have had the paradoxical result of making us feel we may have lost control over both technology and nature.[1]

This essay seeks to clarify the dialectical relationship in the modern culture of control between the development of technological systems and the destruction of nature. More precisely, it will analyze how the construction of large-scale technological systems to circulate goods, ideas, capital, and people has promoted the degradation of the visible landscape. In order to make this case, it is first necessary to distinguish two types of environmental degradation, and two types of technological systems.

In contemporary discussions of environmental problems, pollution and development are habitually confused. Pollution is chemical or physical degradation that may not be visible, but that damages the ecosystem so it becomes less supportive of human and other forms of life. Pollution may therefore be considered an unintended and correctable consequence of technological change. There may be disagreement about levels of risk to life forms, but in principle the risks can be measured objectively and addressed through scientific and technological means.

Development, on the other hand, while it may not render any physical or chemical damage to the ecosystem, alters the visible landscape in a significant way. Unlike pollution, then, the effects of development are not in principle measurable by scientific standards, but are always mediated by culture. A highway that for one person is an emblem of modernity appears to another person as a scar on the landscape. In matters of pollution, nature is considered as an environmental system; in matters of development, as a visible landscape. While pollution is in principle correctable through technological means, development is not. Technology cannot correct a problem if it is perceived as the problem.[2]

Contemporary public discourse about environmental problems focuses upon pollution, and consequently its rhetoric focuses on global models, scientific evidence, and invisible but insidious threats. On the other hand, private discourse often laments the loss of beloved places—the meadow turned into a mall, the city neighborhood cut through by an expressway, the beachfront property sliced by condominiums. Being based on an assumption of property rights, the American legal system recognizes the public discourse of pollution, but not the private discourse that may value beauty and community over property rights. As a result, the quasi-scientific language of pollution has to be used in legal efforts to resist development that may be unwanted primarily for cultural reasons. The construction of a shopping mall may be fought in court on the basis of the groundwater pollution caused by parking lot run-off, when the deeper objection to it has to do with the desecration of the meadow. Few of us well understand, much less empathize with, environmental models. But most of us care deeply about places that are intertwined with human lives, especially our own. We often find the most troubling evidence of environmental damage not in scientific data but in visible personal experience. We apprehend nature not as a system, but as places.

Just as there are two ways of apprehending environmental destruction, there are two ways of using technology to control nature. In the past decade or so, historians of technology have increasingly relied on the concept of the *system*, rather than that of the *machine* or the *invention* (or *inventor*) as the central category of analysis. Because much of the theorizing about technological systems has been done by social scientists, these systems are often defined as sociological models, as systems 'socially constructed' to transform energy and materials into goods and services for human consumption. In this context, human labor, with the aid of machinery, 'consumes ecosystemic energy flows in the process of performing physiological and mechanical work.'[3] A classic example is the 'American system' or, more narrowly, the 'Fordist system' of automobile production. The definition of such manufacturing systems includes much of what used to be called 'social context'—management techniques, modes of labor control, even techniques of mass consumption.[4] Even natural 'actors' may be considered part of the system in three ways: as resource, as sociologized actor ('cooperating in' or 'resisting' its role), and as dumpsite (see Langdon Winner's critique of the 'social construction of technology').[5]

Besides manufacturing systems, however, there are also connective systems organized not for the direct production of goods but for the circulation of goods, ideas, or people. To be sure, in the economic process there are no firm boundaries separating production, circulation, and consumption. Obviously connective systems are closely connected with manufacturing ones, since manufacture depends on connections (the railways that carried steel to Ford plants, and the phone lines that carried orders there), while communication may fruitfully be interpreted as a form of production. Still, the distinction points to the existence of two broadly different types of engineering practice—mechanical and civil engineering—and two broad types of engineering products—machines and structures.[6]

Furthermore, the distinction points to two rather different ways of thinking about engineering. The concept of manufacturing systems, as we have seen, is fundamentally a sociological construct in which a number of radically different entities—technological, human, and biological—are related as abstract social actors. From this perspective, nature is 'denatured' by being assimilated to what is fundamentally a social system. The language of biology may be used to describe 'inputs,' 'outputs,' and 'flowthroughs,' but these are used as metaphors in reference to a fictitious superorganism, the system.[7]

The concept of connective systems, on the other hand, is primarily phenomenological rather than sociological. These constructions are thought of as tangible structures existing in geographical space, and their components are related primarily in physical rather than in social terms. When engineering involves the creation of such structures, it looks more like a 'mirror twin' of landscape architecture or of urban planning than of science. Technological systems of connection therefore bear a markedly different relationship to nature than do manufacturing systems. In the creation of structures, nature is understood primarily as *space*, and the system as a *means of organizing space*. Nature is not a means to the creation of a product, for the 'product' in this case is the creation of a second nature, of a cultural landscape from the given physical one. 'In the cultural landscape man "builds" the earth, and makes its potential structure manifest as a meaningful totality.'[8] In analyzing technological systems of connection, then, the language of cultural geography is more appropriate and useful than the vocabulary of sociology.

According to cultural geographers, in both the rural and the urban environment, the main elements of the cultural landscape are *centers*, or settlements of differing scale (houses, villages, cities), and *pathways* connecting the settlements.[9] Heidegger put it even more succinctly when he said that the two basic constructions are dwellings and bridges.[10] The category of technological structures thus includes two distinct subsets: architectural structures (skyscrapers, for example) that organize space by creating settlements or places, and connective structures that organize space by creating pathways, which are actual or metaphorical roadways.

The outstanding feature of the modern cultural landscape is the dominance of pathways over settlements. In the city, the central element in urban organization is now the street or highway, as opposed to a square, market, forum, or particular buildings. Outside the city, the dominant element of organization is the roadway—whether actual roads like the 'strips' or interstate highways, or other transportation and communication networks such as railway lines and electric power lines.[11] Engineering the landscape—like any act of engineering—is a process that both reflects and defines human values and relationships. What human values and relationships are represented in the cultural landscape of the late twentieth century, especially in the dominance of pathways over settlements?

Many students of American history and culture would argue that the high value placed on mobility—freedom of movement for people, goods, and ideas—is particularly and prototypically American. In his opening chapter to *Brooklyn Bridge*, Alan Trachtenberg reminds us that the oldest of all the ideas associated with America was that it was a road to elsewhere—a passage to India. 'To many, roads fulfilled fervent dreams of the West as a new Garden of Eden, as the long-sought passage to the Orient.' From the early days of the republic, the theme of 'manifest destiny' was linked to roadways. While Hamilton and Jefferson may have had different visions of their purpose—Jefferson hoping they would protect the agrarian republic, Hamilton expecting they would promote manufacturing—the two statesmen shared the ideal of 'opening up' the West and binding it to the East. Trachtenberg concludes that 'Not the land, not the garden, but the road, from Jefferson's own national turnpike to the latest superhighway, has expressed the essential way of American life.'[12]

The idea of a 'passage to India' may be uniquely associated with America, but the search for that passage, after all, emerged from European culture—as did the Enlightenment values expressed by both Jefferson and Hamilton. The main element of American exceptionalism is not the high value put on mobility, but the particular pattern in which mobility was conceptualized. The main lines of the mental map of Americans ran from east to west, the Eastern seaboard being identified with the built environment and the West with unfilled open space.[13] Europeans, on the other hand, used a mental map more like a grid, with criss-crossing lines linking capital cities with hinterlands and Europe with the rest of the world. The difference in spatial patterns, while significant, is still less important than the shared, powerful cultural assumption that it is the destiny of the West to promote circulation through technological systems-building.

This emphasis on circulation is a key contribution of the Enlightenment in shaping modern patterns of technological development. 'The Enlightenment is known to us primarily as an intellectual and cultural revolution, a breaking of the fetters of religious superstition and ancient dogma,' Albert Borgmann has written. 'It is generally accepted that [the Enlightenment] had reverberations beyond the realm of culture and the intellect, but these are almost exclusively seen in the political area...' Yet it is in the Enlightenment that 'the promise of technology' was first formulated not 'at the center of attention' but as 'the obvious practical corollary of intellectual and cultural liberation.'[14]

I would go further than Borgmann to argue that Enlightenment *philosophes* were technological determinists. They assumed that the progress of civilization depended upon the construction of technological systems of connection. Most interpretations of the Enlightenment stress the temporal dimension of the idea of progress—that is, progress as a sequence of historical stages culminating in a utopian, or semi-utopian, future. In the eighteenth century, it is said over and over, utopia ceased being defined as another place (the distant island, the lost valley) and instead become another time—the future.[15] 'In time we shall have utopia...'[16] But 'the Enlightenment project' (Jurgen Habermas's useful term) also has a crucial spatial dimension. The *philosophes* identified the spread of Enlightenment with diffusion of ideas in space through the intensification and acceleration of global systems of communication. Social progress was assumed to depend upon the construction of connective systems: communication and transportation grids, layer upon layer of roadways for the circulation of people, goods, and ideas.

The Enlightenment concept of progress is based not so much on an ideology of production as on an *ideology of circulation*. The idea of progress has undergone significant evolution since then. Beginning in the nineteenth century, progress came to be identified with economic growth, usually defined as rising productivity, or as a rising standard of living based on rising productivity. When defined in this way, social progress was assumed to depend upon the growth of manufacturing systems. In the eighteenth century, however, the prevailing assumption was that social progress depended most upon the growth of connective systems. The crucial index of progress was not output, but intensity of circulation—whether of capital, of ideas, or of goods, because these three forms of circulation were assumed to reinforce each other.

Let us examine more closely this ideology of circulation underlying the proliferation of technological systems that have shaped the landscape of modern life. The origins of those systems will be traced in the interplay between the developing capitalist world-economy and the historical theories of the Enlightenment. Possible connections between those theories and nineteenth century systems-building will be sketched. The ambiguous character of those technological systems—how the circulation patterns they establish both liberate and constrain—will be analyzed. Finally, this essay will explore the challenge of recovering control over places we care for. As we have optimized control over the circulation of goods, ideas, money, and people, we have greatly diminished control over the character of particular places. This loss may involve physical destruction, or it may entail a more subtle withdrawal of economic, political, and cultural meaning and power from localities in favor of space-leaping systems. In either case, the loss of control may be painful for human beings whose lives take place in particular earthly dwellings.

CIRCULATION: FROM ECONOMICS TO IDEOLOGY

In a capitalist economy, profit ultimately derives not from the production of goods, but from the mobility of money.[17] As Karl Marx understood so clearly, the single unitary principle underlying all the changes and insecurities of capital is that of '"value in motion," or, more simply, the circulation of capital restlessly and perpetually seeking new ways to garner profits.'[18] The constant imperative of capitalism, its 'elementary determinism,' is to increase the rate of circulation.[19] The rate can be increased in time (by reducing the period required to complete the economic circuit) or in space (by extending the net of circulation and therefore increasing the flow-through). In the early modern period, as a capitalist economy was taking shape, movement in space (rather than location in place) became the key to social reproduction.

The importance of these principles was well understood by early capitalists. The concept of an economic circuit was firmly established in the seventeenth century, when merchants habitually thought of trade routes as completed circuits.[20] The simplest circuit was the round trip: the journey out, the journey back. Other circuits were more complicated, such as the famous 'triangle trade' comprising a voyage from some European port to the coast of Africa, to the Caribbean, and back to Europe. Immanuel Wallerstein has estimated that the circuits of the sixteenth century European world economy could be completed in about sixty days, using the best available means of transport.[21] Any technological improvement that could reduce that time, or that would allow a trader to cover more space in sixty days, was highly prized and would yield significant competitive advantage. The 'elementary determinism' of capitalism therefore strongly favored technological development in these directions.

Wallerstein's *Modern World System* and subsequent publications are of particular importance in understanding how the spatial arrangements of capitalism differ from those of other, earlier economic systems. The world empires of antiquity were primarily political units. As a means of economic domination, they were inefficient because their centralized political structure required a large bureaucracy, which tended to absorb profits. In modern capitalism, however, economic space is distinct from political space. Rather than being organized around one political center, it is organized around a multiplicity of economic centers. With no damage to the overall system, geographical connections can be strengthened or left to decay as the changing flow of capital warrants. Economic surpluses are efficiently directed to the geographic core from peripheral areas without the wastefulness of a cumbersome political superstructure.

Wallerstein's analysis suggests how many modern technological systems differ in their purpose from ancient systems that they may resemble externally—for example, Roman roads or aqueducts, built primarily to enforce political authority, not to facilitate the flow of capital. Such systems were impressive but rigid; in an empire where 'all roads lead to Rome,' the political core cannot change without the fall of the empire. In modern systems, on the other hand, technological construction is less dependent on political boundaries, purposes, and fortunes. To be sure, in the spatial arrangements of the modern world, there is constant tension between political and economic organizations of space—between tentacular lines of economic power connecting cores and peripheries, and blocks representing political power.[22] The 'distinctive historical geography'[23] of capitalism involves the coexistence of these two organizations, with the blocks ultimately being subordinated to the lines.

Philosophes such as Turgot and Condorcet emphasized the circulation of ideas rather than of capital—but with that crucial difference, the cultural geography of Enlightenment looks much the same as the 'distinctive historical geography' of capitalism: networks of tentacular lines superimposed on political blocks. As we shall see, Enlightenment *philosophes* considered political boundaries relatively unimportant compared to the lines connecting cultural centers with peripheries. They also stressed the importance of quickening the pace of circulation and of extending its reach through construction of technological systems. Above all, their ideology of circulation depended upon key assumptions about knowledge and communication. With some exceptions (most notably Rousseau), the *philosophes* assumed that scientific knowledge is capable of being circulated in the same way as capital: it consists of ideas that are universal counters, the same in all places, independent of personality. As a consequence, they thought of communication as a form of transportation. Knowledge, like merchandise, could be accumulated and bundled in one place, transported to another, and consumed there without any alteration in character.

Fernand Braudel reminds us that 'Cultures...are ways of ordering space just as economies are,' and adds the warning, 'The cultural map and the economic map cannot simply be superimposed without anomaly.'[24] What is striking about the era of Enlightenment is the way the dominant economic map—'the octopus grip of European trade,' in Braudel's words[25]—also became the dominant cultural map. The grid became not only a description of economic reality, but also a cultural ideal. In the eighteenth century, the extended grid replaced the enclosed garden as the dominant image of utopia.

Observing a structural similarity does not explain it: there is no automatic (or semi-automatic) process by which economic practices become represented on a cultural level. An economic system can organize and produce space, but it cannot by itself organize and produce meaning. The production of meaning is instead the role of culture, and more specifically of ideology. One crucial function of the ideological process is to articulate the meaning of economic practices in terms that give those practices broad and persuasive social significance. Enlightenment *philosophes* extrapolated the concept of circulation from the economic sphere to express it as a powerful general theory of historical progress. But in proposing a cultural geography that mirrored the geography of capitalism, Turgot and Condorcet were not only articulating a defense of capitalist practices; they were also, simultaneously, expressing a critique of them, because the cultural map, for all its similarities, was not identical with the economic one.

These *philosophes* highlighted the transmission of ideas rather than of goods, and emphasized communicative relationships that were not necessarily instrumental in character. While confirming the historical geography of capitalism, they also set forth a vision of social and intellectual liberation that might not coincide with capitalist practices. In this ambivalent and complicated way, Turgot and Condorcet provide the ideological link between the trade routes of early capitalism and the large technological systems of late capitalism.

CIRCULATION AND PROGRESS: TURGOT

Anne Robert Jacques Turgot (1727–1781) has three claims to historical significance: as an economic thinker identified with the Physiocrats; as an administrator, first as an *intendant* in Limoges and later, in his brief but ambitious tenure as minister of finance under Louis XVI (1774–1776); and, finally, as author of what is 'by general agreement the first important statement in modern times of the ideology of progress'[26]—the *Discours sur les progrès successifs de l'esprit humain* (A Philosophical Review of the Successive Advances of the Human Mind)—a speech delivered by Turgot at the Sorbonne in 1750, when he was only twenty-three.[27]

Turgot's emphasis on circulation as the engine of progress derives from his economic principles as a Physiocrat. While Turgot always demonstrated a good deal of independence in interpreting those ideas, he did accept the Physiocrats' fundamental emphasis on the unique productivity of land. More specifically, he accepted the belief that land is the only form of production capable of yielding a disposable surplus over necessary costs, and therefore of yielding taxable income.[28] These principles led Turgot and other Physiocrats to place great emphasis on economic circulation. Precisely because all economic value comes from one source—agriculture—commercial and industrial classes were essentially 'paid' by agricultural ones, and as a result the wealth created by agriculture had to be cycled through the entire society as efficiently as possible.[29]

With the Physiocrats, then, we can already discern links between an economic theory of circulation, a more general theory of social progress, and a technological program of system-building. The Physiocrats thought of economic circuits not as abstract patterns but as physical connections in geographical space. The city-country circuit, being the shortest, was the most important to nurture. The Physiocrats particularly urged the improvement of rural roads that would spread wealth in areas around cities, rather than the construction of roads linking large cities. Leaders of the school such as Francois Quesnay (1694–1774) and Pierre-S. Du Pont de Nemours (who eventually edited Turgot's papers) also urged the state needed to take a more active role in creating a dense network of canals.[30]

In his 1750 *Discours*, Turgot extends the principle of circulation from economics to history in general, and from regions and states to the entire globe. He begins by presenting a startlingly novel definition of history. First, history is global: it is the story of 'the human race,' which to the 'eye of a philosopher' appears 'as one vast whole.' Second, history is intellectual. What really matters are not transient political events (motivated by 'self-interest, ambition, and vainglory'), but the gradual and enduring enlightenment of the human mind. Turgot immediately connects the spatial and temporal dimensions of history in stating that the slow process of enlightenment proceeds to the extent that 'separate nations are brought closer together.'[31]

The rest of the *Discours* elaborates on the civilizing process by which groups of human beings become more enlightened in proportion to their contacts with other groups. Turgot's historical starting point is the original diaspora, the great dispersal of human societies after the failure to construct the Tower of Babel, when people scattered to dwell in largely isolated groups. During the long centuries that followed, human progress depended upon spatial contact as more civilized groups came into contact with less civilized groups, whether through military, economic, or cultural exchanges. Turgot repeatedly uses the language of fluid dynamics—of ebb and flow, of overflowing rivers, of underground streams, of deluge and irrigation—to describe the way these groups interacted with each other.

As Peter Gay points out, however, for a so-called optimist, Turgot's discourse is notably uncheerful, with its 'startling' emphasis on the costs and difficulties of progress.[32] Over and over again, humanity gets stuck in routine and repetition, pulled back to the infinitely repeating cycles of nature. Genius appears in one group, but it disappears beneath the waves of other groups. The Greeks enjoy a splendid age, but then succumb to internal vices and generals from the Middle East. Rome extends the boundaries of the civilized world, but then 'Roman liberty was extinguished in waves of blood.' The pattern keeps repeating: genius emerges in one place, thanks to nature's random distribution of talents, and then 'Circumstances either develop these talents or allow them to become buried in obscurity; and it is from the infinite variety of these circumstances that there springs the inequality in the progress of nations.'

The crucial turning points in human history are ones that have decisively changed human 'circumstances,' so that humankind could not only innovate but also accumulate innovations. These turning points are technological ones—great inventions. The first and greatest, which Turgot mentions in the very beginning of his speech, is language: 'The arbitrary signs of speech and writing, by providing men with the means of securing the possession of their ideas and communicating them to others, have made of all the individual stores of knowledge a common treasure-house, which one generation transmits to another, an inheritance which is always being enlarged by the discoveries of each age.'

Next, the discovery of writing, which 'unites places and times...' meant that genius, until then at the mercy of local oblivion, could reach a global audience and therefore immortality. 'Priceless invention!—which seemed to give wings to those peoples who first possessed it, enabling them to outdistance other nations.'

The climactic invention, of course, is that of printing—and again Turgot uses the image of flight to describe its historical significance. 'What new art is suddenly born, as if to wing to every corner of the earth the writings and glory of the great men who are to come?...At once the treasures of antiquity, rescued from the dust, pass into all hands, penetrate to every part of the world, bear light to the talents which were being wasted in ignorance, and summon genius from the depths of its retreats.' With this ode to the printing press, Turgot concludes his discourse. The scientific revolution would never be safe in one country alone; thanks to printing, it is so widely diffused that its survival is ensured. '...if each day adds to the vast extent of the sciences, each day also makes them easier, because methods are multiplied with discoveries, because the scaffolding rises with the building.' Progress depends upon circulation. Enlightenment in time depends upon its extension in space. Turgot is a technological determinist. Great inventions—above all the printing press—have alone made it possible for civilizations to keep from sliding back into cycles of repetition. If progress is now self-sustaining, if further discoveries are now ensured, it is not because genius appears more often than ever; it is because key inventions have reorganized space so that henceforth discoveries will accumulate, never again to be buried in unmarked, isolated graves.

CIRCULATION AND PROGRESS: CONDORCET

If Turgot's 1750 discourse marks the opening of the age of Enlightenment, the death in 1794 of Marie Jean Antoine Nicolas Caritat, marquis de Condorcet (b. 1743) marks its end. Condorcet had been a pilgrim to Ferney to visit the aged Voltaire (and wrote a biography of Voltaire in 1787); a friend and protége of Jean d'Alembert, who convinced him to contribute to the *Encyclopédie*; but his most important friend, mentor, and model was Turgot. For the brief but decisive period when Turgot was Controller-General, Condorcet threw himself into support for his reforms.[33] In despair when Turgot fell from power, Condorcet wrote a biography of his master (published in 1786) and even developed some of Turgot's unpublished sketches.[34] After the deaths of Turgot, Voltaire, and d'Alembert, the younger Condorcet saw himself as carrying on their intellectual and political mission. This was the mission he expressed in the most famous and influential statement of the Enlightenment idea of progress, the *Esquisse d'un tableau historique des progrès de l'esprit humain* (Sketch for a Historical Picture of the Progress of the Human Mind), composed in 1793 while Condorcet was in hiding from the Jacobins.

Like Turgot's *Discours*, Condorcet's *Esquisse* defines history as the global extension of scientific knowledge among humanity as a whole. The 1793 essay, however, has a sense of urgency, a breathlessness, not found in Turgot's speech delivered almost a half-century earlier.[35] 'Everything,' he writes in the introduction, 'tells us that we are now close upon one of the great revolutions of the human race.'³⁶ Much of that urgency, of course, comes from knowledge that time was running out for him. (Suspecting that his hiding place was being watched, he fled to search for another, was captured, and was found dead the next day, apparently a suicide.)

Another reason for Condorcet's impatience, however, is more intrinsic to his historical vision. For Turgot, the main danger to progress had been inertia, routine, passivity. Condorcet sees a far more active and therefore more dangerous opponent: the oppressive, mystifying class of priests. From the very first stage of human development, this 'institution...has had contrary effects upon human progress...I refer to the separation of the human race into two parts; the one destined to teach, the other made to believe.' The main reason error has proved so tenacious through history is the 'crude cunning' of imposters on the one hand, and the 'credulity' of dupes on the other.

In achieving social control, the priests' most powerful tool has been a linguistic one. Condorcet proposes that early in the age of agriculture, priestly castes developed a double language: a primitive, allegorical, image-laden language that they reserved for their dealings with the common people, and an abstract, non-allegorical language that they used among themselves. Accordingly, Condorcet explains, when the priests 'used some expression and meant by it a quite simple truth, the people understood by it heaven knows what absurdity.' Condorcet's conspiracy theory is an important reminder of the political agenda behind the Enlightenment ideology of circulation. The diffusion of scientific knowledge is a heroic struggle to overcome priestly error: the conquest of space is necessary to conquer the place-based power of the clergy. Condorcet's theory of two languages foreshadows twentieth century critical theories of false consciousness, which also propose that a calculating elite maintains social power by hypnotizing the credulous masses with powerful imagery.³⁷

Condorcet therefore concludes—like Turgot, but with even more force—that the key to progress in time is diffusion of scientific information in space. Condorcet explicitly distinguishes intensive progress—that is, progress within a particular science—from the extensive kind, 'which is to be measured partly in terms of the number of people who are familiar with the more obvious and important truths and partly in terms of the number and nature of these truths.' Intellectual revolution will never be safe in one country alone. Because progress is assured only when it is extensively disseminated, it depends upon technological invention.³⁸

The titles of the ten chapters of the *Esquisse* correspond to what Condorcet sees as the major steps in technological progress. The first crucial invention was the alphabet. Once spoken language could be reproduced as enduring and transportable signs, the 'progress of the human race [was assured] forever.' Still, there were many obstacles along the way; for example, Condorcet argues that the progress of the Greeks was 'lost to later nations' because they lacked means of communication and thus succumbed to the tyrannical domination of Rome. The eighth chapter of the *Esquisse* describes the momentous effects of the invention of printing, when, for the first time, the human mind was truly freed from spatial limitations: 'Men found themselves possessed of the means of communicating with people all over the world...The public opinion that was formed in this way was powerful by virtue of its size, and effective because the forces that created it operated with equal strength on all men at the same time, no matter what distances separated them.'

In the chapter on the tenth epoch, Condorcet reaffirms Turgot's conviction that progress will now continue indefinitely because, for the first time in history, technological innovations prevent historical regression. 'The strength and the limits of man's intelligence may remain unaltered; and yet the instruments that he uses will increase and improve, the language that fixes and determines his ideas will acquire greater breadth and precision and, unlike mechanics where an increase of force means a decrease of speed, the methods that lead genius to the discovery of truth increase at once the force and the speed of its operations.'

In its main lines, then, the cultural geography of Turgot and Condorcet repeats the distinctive historical geography of capitalism. According to these *philosophes*, the historical record traces the evolution of a intellectual world-system, based on rational thought, that transcends local and national lines. This system has a hierarchical, though dynamic, arrangement of cores and peripheries, linked together in tentacular fashion by lines of transportation and communication. While the cores of civilization change over time (for example, Rome succeeds Greece as the cultural core), the constant tendency is for the world-system to incorporate more and more of the world's surface into these networks. A second tendency is for the rate of circulation to become increasingly rapid and easy.

In both senses, then, technological innovation determines cultural progress. The basis of Enlightenment is language, defined as the clear, non-metaphorical articulation of universally valid information. The most important innovations are ones that facilitate the transmission of this information: writing, the alphabet, the printing press. Techniques that disseminate this language are the necessary tools of universal Enlightenment.

TECHNOLOGICAL REIFICATIONS OF THE IDEOLOGY OF CIRCULATION

France, home of Enlightenment philosophy, is also home of systems engineering, and both emerged during the eighteenth century. Economic and technological historians have finally ceased to puzzle over the non-problem of 'French backwardness,' which had been fostered by a far too exclusive focus on British industrial practice as a model. If the focus is not on manufacturing systems but on systems of connection—on structures rather than machines—it is clear that French military and civil engineers have long excelled at the latter. In a comprehensive overview, Cecil O. Smith has traced the continuity of the civil engineering tradition in France 'from the highway system built by the Corps [des Ponts et Chausées] in the eighteenth century through the waterways and railroads of the nineteenth century to electric power and economic planning in the twentieth.' As Smith shows, ever since the birth of the Corps in the eighteenth century, 'French state engineers have promoted the complementary notions of rational public administration in the general administration and planning on a national scale.'[39]

In his study of arms and Enlightenment in France, Kenneth Alder has forcefully argued that even in the realm of mechanical engineering, French practitioners led the world in adopting a systems approach. Alder focuses on military engineering—specifically, the manufacture of artillery carriages and muskets—to demonstrate the extent to which military engineers of the late eighteenth century were thinking in terms of developing technological systems, including the development of rational systems of language. In elucidating the origins of the systems approach, Alder argues the need to connect what he so happily calls the 'high Enlightenment' of intellectual and political historians and the 'Low Enlightenment' of engineering practice. The coincidence between Enlightenment philosophy and the systems approach in engineering practice is striking; the problem is to explain it.[40]

Alder's work suggests that French systems engineers saw their work as part of a radical republican program; Smith, on the other hand, emphasizes the origins of the systems approach in civil engineering in the 'enlightened despotism' of the eighteenth century French monarchy. Certainly future work in this area should try to reconcile this apparent contradiction by illuminating the political context of French engineering in the tumultuous half-century framing the Revolution. Both Alder's interpretation and Smith's, however, present a puzzle: how is it that a systems-oriented engineering tradition, whether 'the creature of enlightened despotism' (Smith's words) or of radical republicanism, not only survived but flourished after the fall of both the despotism and the First Republic? Smith describes the continuity in the French national style of engineering rather than explaining it. Alder says that the systems approach did go out of favor as the republic turned into the empire, but he does not explain how it would later be revived.

The problem of historical continuity is less perplexing if we interpret the construction of technological systems as primarily a means of spreading Enlightenment, a goal which could be identified either with the monarchy or with republicanism but which was understood to transcend any national political system. Such a connection between the 'high' and 'low' Enlightenments is evident when we consider that Turgot and Condorcet themselves were systems-builders as well as historical and economic theorists. As Controller General of Finances for Louis XVI, Turgot introduced in 1776 the famous Six Edicts, which, among other things, abolished the *corvée* (the traditional, but inefficient method of getting roads repaired by commandeering peasant labor) in favor of a general tax to support better road repair service.[41] Turgot also asked Condorcet—then his unofficial science advisor—to undertake hydrodynamic experiments to arrive at engineering principles for canal construction, and also undertook studies to evaluate the feasibility of building canals in various areas.[42] After his fall from power, when he no longer was able to command the resources for expensive civil engineering projects, Turgot became more involved with conceptual rather than physical systems of communication. In particular, he 'toyed with inventions of cheap processes for the reproduction of writing, in order to multiply communications and extend progress among those elements in society which were still beyond its pale.'[43]

Condorcet followed a similar pattern in his career. Besides helping Turgot plan canal construction, he aided Turgot's efforts to reform and standardize weights and measures on a scientific basis.[44] After his mentor fell from power, Condorcet became more and more deeply involved with projects to establish a common language for moral and physical sciences based on the calculus of probabilities. In 1791 he was a member of a committee of academicians involved in proposing a decimal-based system of scientific measurements; the work of this committee played a key role in the establishment of the metric system. Even after he had fallen out of favor with the Jacobins, Condorcet continued to seek means of establishing a standardized scientific language. At the very end of the *Esquisse*, he proposes in some detail the creation of a universal language that could be used to express a scientific theory or the invention of a procedure. The language would be learned along with science itself—'the sign...at the same time as the object'—so that all people, not only a few, could understand it. In Alder's words, Condorcet was deeply involved in devising 'A Language of the Machine Age.'[45]

The crucial link, however, between eighteenth century ideas of systems and subsequent French engineering is the Comte Claude Henri de Rouvroy de Saint-Simon, a disciple of Condorcet's. There is a clear line of intellectual genealogy linking Turgot, Condorcet, Saint-Simon, and *his* disciples, collectively known as the Saint-Simonians.[46] The Saint-Simonians were the first coherent group of engineers to propose, and in some cases actually to build, the technological systems Turgot and Condorcet had advocated to disperse and thereby to maintain universal Enlightenment.

Saint-Simon's view of history depended upon the same geographical model as did Turgot's and Condorcet's, but Saint-Simon himself was to give a more romantic, less intellectual cast to the model of global circulation. Whereas Turgot and Condorcet had talked primarily of the circulation of ideas, Saint-Simon emphasized the global circulation of feelings—the gradual spread of assocationism, the gradual retreat of antagonism. In short, with Saint-Simon the diffusion of love replaced that of scientific reasoning, and universal enlightenment was recast as universal association.[47]

Saint-Simon, however—like any engineer of genius—combined a fertile imagination with intense practicality. He was the first to address financing as an integral part of the creation of a large technological systems. A bank at the center of the whole industrial mechanism would permit overall accounting of all incomes and debits, and a central planning agency would consider the credit needs of all branches of industry and development. The Saint-Simonians thus reconnected the cultural ideal of circulation with its economic origins in the circulation of capital. They extended the vision of the *philosophes* in two seemingly opposite directions: towards passion, and towards finance.[48]

Saint-Simon is conventionally described as a utopian, but with relatively minor changes (namely dropping the more romantic, especially sexual theories of the cult) his disciples readily made the transition from utopian fervor to the business realities of mid-century France. Consider, for example, Michel Chevalier, who began his chequered career by controlling the funds and administering the affairs of the Saint-Simonians. After serving some months in prison under the July Monarchy, he began a speedy rehabilitation by conducting a mission of inquiry into the administration of public works in the United States and Mexico, and ended up as a prominent senator of the Second Empire and main organizer of the Crédit Mobilier, the most important development bank of the time.[49]

Backed by these financial resources and those of the Pereire brothers, the Saint-Simonians of the mid-nineteenth century pushed two main lines of technological development. First, they were the major promoters of railroad-building in France (they were responsible for merging smaller companies into the dominant Paris-Lyons-Marseilles organization) and more generally on the Continent. Second, they were the prime organizers behind the construction of the Suez Canal in Egypt, a project they pursued with enormous tenacity and vigor.[50]

The Saint-Simonians thus incarnate what Kenneth Clark has called the 'heroic materialism' of the nineteenth century, when cubic tons of earth were moved to dig canals, railroads, tunnels, roadways, electrical power networks, and the like. These systems of connection—the pathways of modern life—transformed the natural landscape in ways that were immediately visible and often dramatic. But they also reorganized space in ways that were less obvious but equally significant. Let us now look at some of those ways, and consider the cultural and political implications of this vast reorganization.[51]

CONQUEST OF SPACE, DEVALUATION OF PLACE

No modern ideal is more revolutionary than the Enlightenment ideology of circulation, which proclaims the liberation of humanity from place. Until the eighteenth century, the fixed position of most human beings in a geographic locality—so closely related to their fixed position in the social order—was accepted as an inevitable part of human destiny. Enlightenment *philosophes* understood that physical mobility was closely related to social and intellectual mobility, and few of us in the late twentieth century would reject that mobility. As Langdon Winner reminds us, however, 'Each technically embodied affirmation may also count as a betrayal, perhaps even self-betrayal.'[52] From the vantage point of the late twentieth century, we can see how the 'conquest of space' and liberation of place has created different but also powerful restraints on human liberty.

Obsessed, by his own admission, with spatial metaphors, Michel Foucault is the most famous critic of modern relationships between power, knowledge, and space.[53] In analyzing the micropolitics of power in modernity, Foucault devoted special attention to physical layouts that imprison the body—the ultimate irreducible fact of human existence—in spaces of social control. In these spaces, bodies either submit to authority (by no means necessarily state authority) or manage to carve out spaces of resistance and freedom, however limited, in this repressive world.[54]

Foucault paid particular attention to the period of the Enlightenment. According to his analysis, the power of the *ancien régime* was overturned 'only to be replaced by a new organization of space dedicated to the techniques of social control, surveillance, and repression of the self and the world of desire.'[55] Prisons, asylums, and factories were all examples of totalitarian institutions—spaces designed to function as 'containers' of social power—established during the Enlightenment and representing its pervasive drive for ever more efficient social control. They are all variations on the paradigmatic model of spatial control, the Panopticon, the Benthamite 'model prison' which used the efficient technology of a central cockpit to permit an unrelieved and universal one-way gaze from the seat of power. Despite the rhetorical emphasis on individual freedom, the Enlightenment therefore subjected individuals to forms of social control more systematic and thorough than the 'crueler' but limited and arbitrary pre-modern techniques.

The systems of connection described in this essay are quite different in quality from a Panopticon. They are primarily routes, not buildings—systems, if you will, not devices. However, these spatial arrangements also permit a high degree of social control. This is not the control of surveillance, and people are not forced onto the systems. Instead, people 'choose' to use the systems because they are the fastest, most efficient, 'rational' way of circulating ideas and goods. To borrow the terminology of Anthony Giddens, these structures are both enabling and constraining. They overcome distanciation (which Giddens defines as 'simply a measure of the degree to which the friction of space has been overcome to accommodate social interaction') and simultaneously commit the user to the discipline of the grid.[56] Once the 'choice' has been made, one is constrained to use the system as it has been organized. In the case of communications, for example, this means framing the message in a way that conforms to the technical features of the system.

Thus systems of connection may be even more efficient modes of social control than prisons, because they maintain discipline without cumbersome arrangements of surveillance and forced incarceration. Furthermore, these systems are also highly effective means of discipline because they reduce (if not eliminate) the possibility of place-based resistance. You can imagine taking over a prison, but how do you seize control of an entire highway system? Local rebellion becomes impossible, and general revolution necessary, when protest against systems requires the construction of other systems.

In focusing on discrete 'containers of power' modeled on the Panopticon, then, Foucault diverts attention from another, larger form of spatial discipline: technological systems of connection based on the Enlightenment ideology of circulation. As the Age of Enlightenment faded into the Age of Improvement, the spatial basis of Western society began to be reorganized along ever-extending networks of transportation and communication, which are also networks of economic, political, and intellectual power. As the physical networks continued to be laid down in the nineteenth and twentieth centuries, in layer upon expensive layer, they have legitimated themselves in a self-reinforcing cycle of hegemony. The pathways of modern life are also corridors of power, with power being understood in both its technological and political senses. By channeling the circulation of people, goods, and messages, they have transformed spatial relations by establishing lines of force that are privileged over the places and people left outside those lines.

The project of conquering space inherently and inevitably entails the devaluation of place. That devaluation has three dimensions: economic, political, and intellectual. In economic terms, a place without connection is a place without value, because it remains outside the market system. Leo Marx has noted that this is the point of John Locke's famous statement in the *Second Treatise on Civil Government*: 'In the beginning, all the world was America.' Here, Locke was referring not to the lack of settlement in the new world, but to the fact that land there was economically worthless until it acquired the status of a commodity in the marketplace.[57]

In political terms, power relations became based on access to connective systems rather than local sources of power. The Enlightenment ideology of circulation was specifically framed to challenge the locality-based power of priests and aristocrats and to legitimate instead the political power of the bourgeoisie that 'knew no place.' For the *philosophes*, the ideal, utopian community was not based on geographical contiguity, which they considered inherently reactionary. It was to be a community without boundaries based on 'technologies without boundaries.'[58] There were two models for such non-contiguous communities, both of them dating from the seventeenth century. One was the 'universal class' (Wallerstein's term) of capitalist merchants controlling the circulation of capital; the other, the international fraternity of natural philosophers, or 'scientific community' as it would now be called, controlling the circulation of scientific ideas. In both cases the fraternity is united by relationships that are 'disembedded' (Giddens's term) from the natural landscape. They are united instead by common values, language, and information—specifically, since the Enlightenment, by the shared discourse of universal and instrumental reason, which is assumed to be the same everywhere just as the laws of nature are the same everywhere.

In theory, everyone has access to that discourse. In practice, the degree of access differs radically depending on race, sex, income, education, and other factors. (For example, both the merchant and the scientific communities were overwhelmingly male.) While *philosophes* like Turgot and Condorcet were committed to revolutionary republicanism, they were incorrect in forecasting that the development of systems of connection would necessarily have a democratizing effect. Those systems can readily (if not inevitably) create new hierarchies. The dynamic organization of space is closely linked to the investment of power in a meritocratic elite.[59] Conversely, those who remain settled are disempowered, for traditional communities—social organizations gathered in space and developed over time—are devalued. People without access to the infrastructure may be deprived of a place-based neighborhood when social and economic resources are directed away from settlements to networks.

At the same time, local knowledge is devalued. Significant knowledge comes to be defined as information which can be circulated on technological systems, as opposed to that which can only be communicated face-to-face. Types of discourse that do not fit the information model became devalued as 'emotional' and 'feminine.' Truth becomes identified with information that is mobile, universal, contextless. In sum, technological systems of connection both incarnate and reinforce an ideology that accords economic, political, and intellectual power to the global market, to a meritocratic elite, and to information.

CULTURAL RESPONSES TO DESTRUCTION OF PLACE

Destruction of place is therefore not a regrettable side effect but a central goal of modernity. This includes destruction in the obvious visible sense, and also in the less obvious, but highly significant sense of a general withdrawal of economic, political, and intellectual meaning. By the late nineteenth century, destruction of place in Europe in both senses was evident and troubling. The proliferation of technical systems—a proliferation that partially defines 'the second industrial revolution' of that epoch—was rapidly transforming both the urban and the rural landscape. This transformation was far more complicated than a shift in balance between city and country, for the proliferation of technological systems renders that distinction obsolete. On the one hand, those systems permitted the country to 'invade' the city, as village-dwellers migrated to the metropolis in huge numbers. On the other hand, systems originating in the metropolis extended further and more densely into formerly rural areas. Thus rural pathways extended into the city as well as the reverse.[60]

In the arts of late nineteenth century Europe, response to the destruction of place was expressed in two reciprocal and self-reinforcing ways: technological systems were endowed with vitality, while places disconnected from those systems were perceived as moribund. Just as the construction of technological systems renders obsolete any sharp distinction between city and country, so do the systems blur the distinction between organic and mechanical. This blurring is inevitable when circulation, rather than production, becomes a technological ideal, for the concept of circulation comes from biology. (It was introduced in the first half of the seventeenth century, as a result of William Harvey's discovery of the circulation of the blood, published in 1628.) In France, furthermore, the term 'network' (*réseau*) was first used to refer to organisms, then to river patterns, and only then to canals and later to railways.[61]

In late nineteenth century European culture, these organic images became highly conscious and elaborated. The most common metaphor was to describe networks of any kind—roads, railroads, power lines—as 'tentacles' that enmeshed and sucked the lifeblood of the countryside. This image of systems suggests a sort of technological animism that both expresses and protests the fetish of circulation.[62] It is therefore misleading to think of 'systems' and 'life world' as opposites, as Habermas (borrowing the vocabulary of phenomenonology) typically does. Systems too have a life, and at the center of mechanism beats the heart of the superorganism. Much of what we think of as 'technological determinism' derives precisely from this attribution of vitality to technology: its growth cannot be stopped.

On the other hand, the vitality of these systems could also be celebrated. The early works of Jules Verne—the poet laureate of Saint-Simonianism—awakened generations to the thrill of circling the world in eighty days, or at twenty thousand leagues below the sea. Somewhat later, Jules Romains announced the birth of a new social organism, the *unanime*, from the ebb and flow of urban traffic, and lauded the extension of *unanimiste* consciousness into the countryside through highways and telegraph lines. As for the Futurists, they could be collectively described as Enlightenment *philosophes* on speed. They developed new aesthetic practices based on their fascination with systems and with the vehicles on them, and based on new experiences where environment and object merge in fields of force, where the mind becomes a control panel and the body rides new waves of power.[63]

It is one of the self-contradictions of modernism that other European artists at the same time were lamenting the death of place. These laments go deeper than reactionary laments over the uprooting of the peasantry. Indeed, some of the most powerful expressions of grief are found in the poetry of Belgian poet Émile Verhaeren, a fervent socialist. The very titles of his trilogy of the 1890s express his anguish: *Les Campagnes hallucinées* [hallucinated fields] (1893), *Les Villages illusoires* [illusory villages] (1894), and *Les Villes tentaculaires* [tentacular cities] (1895). Verhaeren's poetry forcefully illustrates Anthony Giddens's observation that when 'the truth of experience no longer coincides with place,' places are perceived as dead, unfertile, deserted, void, phantasmagoric.[64] These are the metaphors that obsessively recur in Verhaeren's work, and, though often less vividly, in the work of countless other writers of that time and place (for example, Romains, who experienced the countryside as 'dead nature' except for the roadways and telegraph lines running through it). The land had become unreal because people who lacked a productive relationship with it could no longer understand or interpret it. Those who visited the countryside only on Sundays and holidays had no 'local knowledge' to draw upon in 'reading' the landscape.

As the city too was being spatially transformed, there too the visible environment seemed mysterious and opaque because 'the truth of experience no longer coincides with place.' Over and over, the city is described as a puzzle, a labyrinth, a maze. Whether in urban or rural surroundings, then, a major cultural project of the late nineteenth century was to comprehend one's new relationship to space, and by extension to economic, political, and cultural power. Even when they did not transform the landscape in a direct way, the construction of technological systems created a pervasive sense of homelessness, of disconnection from one's earthly surroundings. The cultural challenge was to overcome this spatial alienation.[65]

One method was the deceptively simple activity of taking a walk. Thanks to the work of two generations of cultural geographers (notably Kevin Lynch and his disciples), walking has come to be appreciated as a significant act of cognition, in which knowledge is created from the active involvement of the entire body.[66] While the human mind can apprehend technological systems as geometric or formal patterns, the mind (and some more than others) also learns in a figurative or sequential way. What looks like a grid to the *philosophe* loftily surveying humanity as a whole is experienced by the individual on the ground as a pathway. Tracing these 'felt paths' is a crucial act in understanding the spatial organization of the grid.[67] The Romantics's 'sublime' walks in the mountains, the *flânerie* of late nineteenth century dandies upon urban boulevards, and turn-of-the-century Sunday strolls in the park may all be interpreted as acts of learning, as efforts to construct a new cognitive map of western Europe.

In this case, there is a deep and necessary relationship between the creation of spatial knowledge and literary creation. Walking in space and constructing a narrative in time are both efforts to organize knowledge through sequential organization. As modes of learning, they both resemble ritualistic or musical acts. Countless nineteenth century narrations, whether in poetry or in prose, trace 'felt paths' at once spatial and linguistic. By interpreting literary narratives as cognitive acts, we can understand more clearly their role in overcoming the spatial dissonance of the late nineteenth and early twentieth centuries. Writing and walking were both active means of creating new forms of local knowledge.

In recent years, postmodernists such as Pierre Bourdieu and Michel de Certeau have interpreted walking not only as a cognitive act but also as a political one. If, as Foucault asserts, the body is the irreducible element in social organization, then carving out an unpredictable, individual trajectory through the creation of an alternative pathway is an implicit act of defiance. De Certeau emphasizes the defiance of nomadism, when one deliberately jumps off the beaten path. Instead of submitting to 'the technological system of a coherent and totalizing space,' walkers erect a 'pedestrian rhetoric' of paths with 'a mythical structure,' narrating 'a story jerry-built out of elements taken from common sayings, an allusive and fragmentary story whose gaps mesh with the social practices it symbolizes.' Thus de Certeau refutes Foucault's pessimistic assessment of the disciplinary imperatives of modernity. Walking, says de Certeau, is an example of a popular practice with liberating potential existing 'at the heart of the contemporary economy.'[68] In the same spirit, in his seminal article 'Postmodernism, or the Cultural Logic of Late Capitalism,' Fredric Jameson draws upon Kevin Lynch's studies in urban topography to argue for the possibility of a 'practical reconquest of a sense of place.' Jameson proposes that this reconquest might result from the activity of 'cognitive mapping,' which he defines as the 'construction or reconstruction of an articulated ensemble which can be retained in memory and which the individual subject can map and remap.'[69]

The cognitive mapping praised by de Certeau and Jameson (as well as Bourdieu) is part of the larger postmodernist emphasis on localism, particularity, and difference. Yet a certain despair lies in this postmodern celebration. The burden of creating freedom is placed on individual human beings—specifically, on their imaginative capacities—because the possibility of reshaping human surroundings in a physical way has been abandoned. Human beings may retain their flexibility and creativity, but the landscape is set in concrete, as they say. The endless circulation of linguistic codes, the endless game of establishing connections between cultural artifacts, is seductive because other patterns of circulation seem beyond intervention. The freedom to deconstruct in linguistic terms is highly valued when freedom to deconstruct in any physical sense seems so remote. The freedom to reject the grand narrative systems of the past seems more important when the freedom to reject the technological systems seems so limited. The central form of protest is no longer political, but aesthetic—the capacity to apprehend differently, to create a different cognitive map.

The seeds of this technological pessimism of postmodernism exuberance lie in the technological optimism of the Enlightenment. Its central *humanistic* goal, the diffusion of Enlightenment, needs to be detached from the twin assumptions that knowledge is information, and that information is best disseminated by large technical systems. Jurgen Habermas and Raymond Williams have argued that instead of abandoning the Enlightenment project, we should seek to understand why its goals have not been fulfilled and continue the struggle to fulfill them.[70] Both Habermas's 'theory of communicative action' and Raymond Williams's 'long revolution' take up the Enlightenment call for the proliferation of increasingly complex and efficient communicative relationships. When Williams defines communication, however, he does so not in terms of exchanging information but of 'describing, learning, persuading and exchanging experiences.'[71] While both Habermas and Williams are keenly aware of the technological possibilities for enhancing such relationships, they keep their focus on the whole range of communication, and do not equate it with the construction of particular technological systems.

62 *Cultures of Control*

Immense damage has been done to human beings and to non-human nature alike by envisioning our globe as empty space across which to string systems for the circulation of capital, goods, and information. Yet we must also recall what motivated the Enlightenment ideology of circulation: an awareness of the immense suffering caused when human minds and bodies are trapped in one place, and when fierce ethnic conflicts arise because human identity is so dependent upon relations to particular places. Is there a way to reaffirm the values of locality and place against destruction by abstract circulation patterns, without getting caught in a reactionary trap of blood and soil? Answering this question is a central challenge of the postmodern age.

NOTES

1. Langdon Winner, *Autonomous Technology: Technics-Out-of-Control as a Theme in Political Thought*, (Cambridge, MA: The MIT Press, 1977).
2. See the vigorous discussion of the concept of development in Marshall Berman, *All That is Solid Melts Into Air: The Experience of Modernity.* (Harmondsworth, England, 1988 [1982]), pp. 63–79.
3. William Cronon, 'Modes of Prophecy and Production: Placing Nature in History,' *Journal of American History* 76:4 (March 1990):1122–31.
4. David Harvey, *The Condition of Postmodernity: An Enquiry into the Origins of Cultural Change* (London: Blackwell, 1989).
5. Langdon Winner, 'Upon Opening the Black Box and Finding It Empty: Social Constructivism and the Philosophy of Technology,' Presidential address, Society for Philosophy and Technology, Mayaguez, Puerto Rico, March 1991.
6. David P. Billington has complained that historians of technology have been preoccupied with machines—devices that are short-lived, relatively mobile and ubiquitous, and largely independent of their environmental context—at the expense of constructions, which are highly visible, massive, permanent, slowly evolving, and site-specific. As examples of machines, Billington mentions pumps, turbines, shops, television sets, and motors; structures are roads, bridges, terminals, dams, harbors, waterworks, power plants, office towers, and public housing blocks. David P. Billington, 'Structures and Machines: The Two Sides of Technology,' *Soundings* 57 (1974): 275, cited in Albert Borgmann, *Technology and the Character of Contemporary Life: A Philosophical Inquiry* (Chicago and London: University of Chicago Press, 1984), p. 66. See also Billington, *Structures and the Urban Environment* (Princeton, NJ, 1978), esp. pp. 149–53.
7. According to Carolyn Merchant, 'The dominant paradigm in ecology draws heavily on economic metaphors such as producers, consumers, productivity, yields, and efficiency. Nature is cast as a computerized network of energy inputs and information bits that can be extracted from the envir-

onmental context and manipulated according to a set of thermodynamic equations.' The value of instrumental reason, the need for unimpeded circulation, the need for expertise—all these are now-familiar themes being applied to natural systems as well as to technological ones. Merchant, *Ecological Revolutions: Nature, Gender, and Science in New England* (Chapel Hill, NC and London, England: University of North Carolina Press, 1989), p. 9.

8. Christian Norberg-Schulz, *Genius Loci: Towards a Phenomenology of Architecture* (New York: Rizzoli, 1980 [1979]). The term 'second nature' is used by William Cronon in *Nature's Metropolis: Chicago and the Great West* (New York: W.W. Norton, 1991).

9. Norberg-Schulz, pp. 10, 52. Norberg-Schulz credits Paolo Portoghesi for the term 'field,' and further explains that a center generates a field when, for example, 'a circular *piazza* is surrounded by a concentric system of streets. The properties of a "field" are hence determined by the centre, or by a regular repetition of structural properties' (p. 61).

10. Martin Heidegger, 'Building Dwelling Thinking,' in *Poetry, Language, Thought*, trans. Albert Hofstadter (New York, 1971), pp. 152–53, cited in Borgmann, p. 200.

11. J. B. Jackson, *The Necessity for Ruins, and Other Topics* (Amherst, MA: University of Massachusetts Press, 1980).

12. Alan Trachtenberg, *Brooklyn Bridge: Fact and Symbol* (New York: Oxford University Press, 1965), pp. 9, 21 (quotes); also pp. 10–12. As Trachtenberg notes on p. 15, it was Henry Nash Smith who referred to the passage to India as 'the oldest of all ideas associated with America.'

13. Leo Marx, 'The American Ideology of Space,' Working Paper Number 8, Program in Science, Technology, and Society, MIT (1990), esp. p. 5.

14. Albert Borgmann, *Technology and the Character of Contemporary Life: A Philosophical Inquiry* (Chicago and London: The University of Chicago Press, 1984), p. 35.

15. 'With Turgot and Condorcet, *eutopia* becomes *euchronia*.' Krishan Kumar, *Utopia and Anti-Utopia in Modern Times* (London: Basil Blackwell, 1987), p. 45.

16. Frank E. Manuel and Fritzie P. Manuel, *Utopian Thought in the Western World* (Cambridge: Belknap [Harvard University Press], 1979), p. 120.

17. Fernand Braudel, *The Wheels of Commerce*, vol. 2, 'Civilization and Capitalism, 15th-18th Century,' trans. Sian Reynolds (New York: Harper & Row, 1982 [1979]), pp. 138–148.

18. Harvey, p. 107.

19. Fernand Braudel, *Civilization and Capitalism, 15th-18th Centuries*. Vol. 3, *The Perspective of the World*. Trans. Sian Reynolds (New York: Harper & Row, 1984 [1979]), p. 582.

20. Pierre Dockes, *L'espace dans la pensée économique du XVIe au XVIIe siècle* (Paris: Flammarion, 1969), pp. 424–425.

21. Immanuel Wallerstein, *The Modern World System: Capitalist Agriculture and the Origins of the European World-Economy in the Sixteenth Century*, text edition (New York: Academic Press, 1976), p. 16.
22. See the discussion of this tension in Harvey, p. 109.
23. Harvey, p. 343.
24. Braudel, p. 65.
25. *Ibid.*, vol. 3, caption to map on p. 29. See Braudel's general comments on the reception of Wallerstein, pp. 68–69.
26. Kumar, p. 43.
27. Manuel and Manuel, p. 468. See also pp. 467–81 for a helpful discussion of 'Turgot on the future of mind.'
28. See the helpful discussion by Ronald L. Meek in Meek (ed.), *Turgot on Progress, Sociology and Economics* (Cambridge: Cambridge University Press, 1973), pp. 14–27, especially pp. 26–27. See also Dockès, p. 295.
29. In his *Tableau économique* (1758), for example, Francois Quesnay (1694–1774) carefully analyzed various ways of shortening the circuits of revenue from the productive class to proprietors, then to the salaried classes, and back to the cultivators.

    In the view of Quesney and other Physiocrats, two things were necessary above all to facilitate the circulation of grain: the suppression of taxes on trade (both internal and external taxes: the Physiocrats were ardent free traders), and the improvement of roadways and canals. Quesnay advised proprietors and local officials as well as the king to use tax revenues on projects that would reduce the price of transport and open new areas to production. He argued that these projects, while initially expensive, would pay off handsomely by providing new markets for rural areas and thus raising farm prices. Dockès, pp. 277–78, 286.
30. *Ibid.*
31. Meek (ed.), pp. 40–41. Subsequent quotations from this edition of Turgot's *Discours* are found on pp. 41–59, *passim*.
32. Peter Gay, *The Enlightenment: An Interpretation*. Vol. II: The Science of Freedom (New York: Alfred A. Knopf, 1969), p. 110.
33. Keith Michael Baker (ed.), *Condorcet: Selected Writings* (Indianapolis: Bobbs-Merrill, 1976), p. xi.
34. Manuel and Manuel, p. 459.
35. *Ibid.*, pp. 504–505.
36. Antoine-Nicolas de Condorcet, *Sketch for a Historical Picture of the Progress of the Human Mind*, trans. June Barraclough, intro. Stuart Hampshire (New York: Noonday Press, 1955 [1795]), p. 12. Subsequent quotations from this translation of the *Esquisse* are found on pp. 7, 17–18, 37, 100, 120, 178.
37. See Gay's rather jaundiced view of Condorcet's anti-clerical conspiracy theory (pp. 112–122).

38. See Manuel and Manuel, p. 501.
39. Cecil O. Smith Jr., 'The Longest Run: Public Engineers and Planning in France,' *American Historical Review* 95:3 (June 1990): 658–59. Smith presents a picture of a struggle in France between enduring, if not eternal, engineering spirits: that of rational, centralized statism, and the liberal proponents of anti-statism and decentralization. The statist centralizers have largely prevailed in directing policy, however, except for the period between the 1880s and World War II.

    In Harvey's terms, the two groups represent primarily the tension between political and economic organizations of space. It would be a mistake, however, to present this tension as one between those who want to build systems and those who don't. As Smith demonstrates, the arguments were rather about which type of systems should be built, and who should pay for them. It is not a tension between system and none, but between two sets of values incarnated as two types of systems.
40. Kenneth Ludwig Alder, 'Forging the New Order: French Mass Production and the Language of the Machine Age, 1763–1815.' Thesis presented to the Department of the History of Science, Harvard University, June 1991. Later published as *Engineering the Revolution: Arms and Enlightenment in France, 1763–1815* (Princeton, 1997).
41. Dockès, p. 277. For a summary on the state of the road system in eighteenth century France, see Dockès, pp. 206–13. It is also true that the intendants who directed *corvée* labor tended to use it to repair major roads between cities instead of the smaller, less noticeable, but to the Physiocrats more important local routes connecting rural areas with neighboring towns [Baker (ed.), p. xi].
42. Baker (ed.), p. xi.
43. Manuel and Manuel, p. 479.
44. Baker (ed.), p. xi.
45. See Alder, 'Forging the New Order,' Chap. 5, 'The Rise and Decline of the Metric System: The Language of the Machine Age,' pp. 403–97. See also pp. 16, 30–3.
46. Although the Saint-Simonians are famed for their role in promoting some of the most ambitious engineering projects of the nineteenth century, Smith scarcely mentions them, and then only to minimize their influence in awakening French administrators to the importance of railway-building. The reason, I would guess, is that the Saint-Simonians do not fit into the Smith's argument that there is a sort of eternal tension in French engineering between economic liberalism and state-sponsored centralization. The Saint-Simonians did not easily fit into either category. Smith, p. 667.
47. As with Condorcet and Turgot, Saint-Simon assumed that the progress of Enlightenment would take place in geographical space; the difference was that he defined enlightenment in affective rather than intellectual terms, emphasizing 'progressions of love' rather than 'successive advances of

66 *Cultures of Control*

mind.' 'World history thus became the study of the general diffusion of love and the contraction of antagonism...' The key historical process was the gravitation of humanity toward universal association. Manuel and Manuel, pp. 626–27.

48. Paul Rabinow, *French Modern: Norms and Forms of the Social Environment* (Cambridge, MA and London, England: The MIT Press, 1989), pp. 29–30. It was he who proclaimed—using a metaphor drawn no doubt from his military service, 'It is we, artists, who will serve you as avant-garde. What a most beautiful destiny for the arts, that of exercising over society a positive power, a true priestly function, and of marching forcefully in the van of all the intellectual faculties in the epoch of their greatest development!' Quoted in Daniel Bell, *The Cultural Contradictions of Capitalism* (New York, 1978), p. 35. Saint-Simon proposed a three-chambered parliament: a Chamber of Inventions that would be the central planning agency dominated by engineers and artists, and that would 'arrive at a master plan for public works, emphasizing circulation'; a Chamber of Review, comprised of pure scientists, would examine these projects; and a Chamber of Deputies, an executive body of industrialists, would coordinate and finance them. Manuel and Manuel, 630–31.

49. Manuel and Manuel, p. 638.
50. Consult Zeldin, France, vol. 1, p. 431.
51. If we want to take a grand overview of systems of connection as they have been built since the beginning of the nineteenth century, we can see a gradual process of dematerialization. Beginning with telegraph lines, however, the construction of connective systems began to become decoupled from massive civil engineering projects. Although one should not minimize the physical effect of telegraph wires strung across the landscape, it is still considerably less than the effect of laying a railway line. A similar point could be made of power lines and telephone lines. The process of decoupling became much more pronounced with the advent of radio and other uses of electromagnetic radiation, which travel across space with minimal disruption of the earth's surface. In even more recent times, large-scale computer networks have been built that are nearly invisible to most observers, because they have so often been piggy-backed onto existing telephone links. As with so many other major shifts in capitalism, the radical dematerialization of connective systems can be dated around 1973—the time when computer links began to be established, and also the time when the great era of interstate highway building drew to a close in the United States.
52. Winner, 'Upon Opening the Black Box,' p. 19.
53. Harvey, p. 205; see also Rabinow, p. 14.
54. Harvey, p. 213.
55. *Ibid.*
56. *Ibid.*, p. 222. A prototypical study of 'the discipline of the grid' might be Wolfgang Schivelbusch, *The Railway Journey: The Industrialization of*

Time and Space in the 19th Century (Berkeley, CA: University of California Press, 1986 [1977]).
57. Marx, 'The American Ideology of Space,' p. 6.
58. Ithiel de Sola Pool, *Technologies without Boundaries*, ed. Eli M. Noam (Cambridge, MA: Harvard University Press, 1990). See especially Chapter 11, 'The Ecological Impact of Telecommunications.'
59. Rabinow, p. 2.
60. In relating material and cultural change, I have tried to avoid the model of a technological base and cultural superstructure, especially by emphasizing the cultural origins of technological systems in Enlightenment ideology. Instead, I have suggested substituting a layered model of coexistent grids by stressing the similarity in geographic pattern between the capitalist world-picture and the historical world-picture of the Enlightenment *philosophes*. This pattern of tentacular networks was reified in the construction of large technological systems. But we can again layer the technological grid with a cultural one by using the same pattern to select and interpret cultural examples from late-nineteenth-century Europe. The technological systems of that epoch provide a 'grid of intelligibility' (Rabinow's expression), or theoretical framework, on which to site cultural examples by relating them to the reorganization of space. David Harvey has demonstrated most persuasively how changes in the spatial and temporal qualities of human experience provide a common ground where technological changes may be related to cultural ones (Harvey, p. 89).
61. André Guillerme, 'Le Rayonnement des écoles françaises d'ingénieurs dans l'aménagement des territoires au début du XIXe siecle,' paper given at the International Conference for the History of Technology, Conservatoire des arts et métiers, Paris, July 1990. Guillerme mentions that the term was used in medicine before being applied to rivers; it then was used to refer to the water supply system for Paris before being applied to canals—at which point the transition from organic to technological frames of reference was complete.
62. In the terminology of Horkheimer and Adorno, nature and technology trade places (see, for example, *Dialectic of Enlightenment*, trans. John Cumming (New York: Herder and Herder, 1972 [1944]), pp. 12, 149.
63. Paul Virilio, *L'esthétique de la disparition* (Paris, 1985); *Speed and Politics* (New York, 1986).
64. Giddens continues, 'The primacy of place in pre-modern settings has been largely destroyed by disembedding and time-space distanciation. Place has become phantasmagoric because the structures by means of which it is constituted are no longer locally organized.' Anthony Giddens, *The Consequences of Modernity* (Stanford, CA: Stanford University Press, 1990), p. 108.
65. One of the clearest descriptions of spatial alienation is found in E. M. Forster's seminal short story 'The Machine Stops' (1909). The ima-

ginary society of that story dwells largely underground—an environment that permits no sense of direction or distance. The hero Kuno explains to his mother how he prepared for a journey to the surface of the earth: 'You know that we have lost the sense of space. We say 'space is annihilated' but we have annihilated not space, but the sense thereof. We have lost a part of ourselves. I determined to recover it, and I began by walking up and down the platform of the railway outside my room. Up and down, until I was tired, and so did recapture the meaning of 'near' and 'far.' 'Near' is a place to which I can get quickly *on my feet*, not a place to which the train or the air-ship will take me quickly. 'Far' is a place to which I cannot get quickly on my feet; the vomitory is 'far,' although I could be there in thirty-eight seconds by summoning the train. Man is the measure. That was my first lesson. Man's feet are the measure for distance, his hands are the measure for ownership, his body is the measure for all that is lovable and desirable and strong.' Forster, 'The Machine Stops,' in *The Eternal Moment and other Stories* (New York: Harcourt Brace Jovanovich, 1928), p. 167.

66. See the eloquent argument of Bruce Chatwin in *The Songlines* (New York: Penguin, 1987), esp. 'From the Notebooks,' pp. 163–206. For summaries of the literature in cultural geography, see the bibliographies in Peter Gould and Rodney White, *Mental Maps*, 2nd ed. (Boston: Allen & Unwin, 1986 [1974]), pp. 165–68, and in Roger M. Downs and David Stea, *Maps in Minds: Reflections on Cognitive Mapping* (New York: Harper & Row, 1977), pp. 264–72.
67. Jeanne Bamberger, *The Mind Behind the Musical Ear: How Children Develop Musical Intelligence* (Cambridge, MA and London, England: Harvard University Press, 1991).
68. Harvey, p. 214, quoting de Michel Certeau, *The Practice of Everyday Life*, trans. Steven Rendall (Berkeley, CA: University of California Press, 1984).
69. Jameson, 'Postmodernism, or the Cultural Logic of Late Capitalism,' *New Left Review* 146 (July/August 1984): 89. See also Jameson, 'Cognitive mapping,' in C. Nelson and L. Grossberg (eds.), *Marxism and the Interpretation of Culture* (Urbana, IL: University of Illinois Press, 1988).
70. See the helpful summary of the debate in Patrick Brantlinger, *Crusoe's Footprints: Cultural Studies in Britain and America* (New York and London: Routledge, 1990), pp. 185–188.
71. Williams, *Communications* (Harmondsworth: Penguin, 1962), p. 18; quoted in Brantlinger, p. 185 (cf. p. 59).

CHAPTER 3

Measuring Cloth by the Elbow and a Thumb: Resistance to Numbers in France of the 1780s

Daryl M. Hafter

Putting mathematical controls on work processes and goods had much appeal to eighteenth-century French administrators. Numerical information helped to extend the influence of the central government over the engine responsible for the state's health. To know how the separate elements of agriculture and manufacture worked satisfied the century's hunger for understanding of the natural world. But more than that, it collected essential data for an administration aimed at reforming the institutions responsible for the economy. How else could social activities be brought into conformity with the laws of nature? One had to learn the steps that led to manufacturing problems and distribution bottlenecks in order to set things right. By the same token, normative standards could only come from gathering information about work patterns all over the kingdom. For this reason, inspectors of manufacture paid close attention in their circuits around their regions to the average daily output, raw materials used, and wages earned by craft workers; the Paris Bureau of Commerce was collecting these statistics to create a mammoth table of information for all of France.

The administration was not reliant only on the reports from inspectors of manufacture—for investigating new inventions, for testing the quality of products, and for validating all sorts of industrial processes—the government looked to the Royal Academy for help. The founding of the Academy owed part of its rationale to the potential benefits that organized science could offer, as Roger Hahn has written, 'for the improvement of man's well-being and in the service of the crown.'[1] The Academy increased its cooperation with government officials throughout the eighteenth-century as it incorporated the utilitarian ideals of its time. Reforming administrators like Turgot saw the group of scientists who could mediate regulations as a potential substitute for the guilds which he suppressed in 1776. Condorcet envisioned connecting the provincial academies with

70 *Cultures of Control*

Paris as a separate department of state, giving the central government much stronger control to bring about change.[2] And one of the first tasks would likely have been to formulate a national set of standards based on the metric system.

In order to create a nationwide system of weights and measures, Turgot and his associates would have to discredit or suppress the provincial powers that were linked to the various measurements. But by using the authority of science, the government could impose new measurements on top of the old institutions, without abolishing them. Scientific explanations of the new standards helped to validate these administrative changes. In Keith Baker's words: 'By instituting a uniform system on the basis of a natural measure, scientifically established, Turgot was aiming to transform a political problem into a cognitive one, thereby invoking scientific knowledge in the exercise of political will.'[3]

Of course the idea of bringing economic production up to national standards had an earlier history in France. The seventeenth-century version of uniformity

Figure 1 Measures incised on the city wall of Vila Viçosa in Portugal exemplify one manner of setting standard cloth measurements in proto-industrial Europe. Photograph: author.

saw Colbert generating codes of manufacture from textile guilds, and then insisting that the whole industry apply them. In this way, goods would be channeled into a set number of cloth varieties that were comparable in any French region and their production would be of a sufficiently high caliber to succeed in the export market. Gail Bossenga interpreted eighteenth-century regulations as equally rational, representing 'the desire to standardize products and predict outcomes. Far from being anachronistic, the attempt to control the productive environment through regulation might be seen as a "scientific"bent of mind, a form of rationalization.'[4]

As the eighteenth-century's ideals came to the fore, belief in a proactive administration was grafted onto other reforming goals. Now 'enlightened' administrators wanted to use the government to suppress the system of regulation, believing that freedom would generate industrial creativity and increased production. Defining their impositions as scientific standards would justify these officials' work, removing the popular criticism of bureaucrats as covert and ruthless actors.[5] The popularity of national standards, and of using mathematics or science to articulate them, was another way to insist on the political integrity of the sovereign and on his right to use an active administration. To be effective, however, these standards had to be translated into the market-day behavior of ordinary craft workers and merchants. In the thicket of dealing with concrete issues, the mathematical and scientific labels introduced practical problems that had unintended results.

Once in the public domain, both the seventeenth-century regulations and the later rules provided means by which groups of workers could try to defend themselves against competitors. The struggle of Lyon's silk workers against the merchants who were in the process of pushing them into a proletarianized condition is a classic case. These silk weavers clung to the industry's complex regulations which only they were skilled enough to follow as a shield against the merchants' use of non-guild workers. A different example comes from the merchants of Lille who supported cloth regulation by urban guilds to choke off the competition from rural weavers.[6] Even the metric system, which might be considered an inevitable reform, foundered on the rock of ordinary usage. '[H]eralded as the universal language of the new national market and the patient tutor of the new rational consumer,' as Ken Alder wrote, 'it was abandoned by Napoleon, who complained that it had thrown local markets into disarray.'[7]

Although reforms of measurement might seem to be the most innocuous of changes those seeking to spread the advantages of simpler, more flexible, or more accurate systems of calculating have often been faced with unexpected opposition. The reformers are sure that their innovation will increase prosperity; the ones for whom the reform is designed are even more convinced that it would undermine their gains. Thus it was with the two systems of linear measure which competed for influence in eighteenth-century France. With the old system called *auner par pouces*, the cloth merchant measured an ell from nose to outstretched fingers and threw in an extra thumbsworth or inch for good measure. With the new standard, known as pejoratively as *ras le bois*, the worker used a pre-measured wooden rod in such a grudging manner that it 'shaved the wood' with exactness.[8] These two kinds of measurement symbolized the contrast between old fashioned traditional business practices and the new methods of discipline and standardization that the French administration was trying to introduce into textile trade. Officials thought that cloth traders would be glad to adopt a system in which they knew just what they were getting and where they no longer had to 'tip' the buyer, but merchants and manufacturers alike opposed it because it would cost them money. The merchants counted on compensating their expenses by the extra inches given in wholesale trade. The manufacturers worried that a stricter measure would conflict with the way they mounted their looms.

Here was a classic confrontation between a reform that was designed to simplify commerce by standardization and the practitioners who had learned to make an advantage out of the traditional complications.[9] As officials ordered the merchants to use the exact yardstick measure, they were astonished by the negative reaction:

> At first [the merchants] raised their voices to an indecent level against the directives of a law as wise as it is just. They claimed that ever since the regulations of 1738 they were accustomed to buy cloth only in lengths of sixty ells. Not only was this more convenient for their trade and simplified their invoices considerably, but they would rather stop buying altogether than to require the manufacturer to have lengths of more than sixty ells cut. The quality of the cloth is less important than its length, [they said]. If the length of a piece of fabric is reduced by this strict measure [*ras le bois*], they would be deprived of the extra inch per ell, which they have always profited from.[10]

In the face of this opposition, why did the government lean so heavily on getting the yardstick measure accepted? For an answer, we must turn to the relationship between the government's belief in reform from above and the workers' understanding of pre-industrial craft. We need to consider how the administration was introducing machines into cloth manufacture and what steps it was taking in regulating the production of textiles.

Throughout the eighteenth century French administrators took part in stimulating the economy, whether by manipulating tariffs or actively encouraging manufacture. This was one field in which the absolute monarchy effectively marshalled its forces of financial aid and technical assistance. Following the new framework of state activism begun in earlier centuries, kingship in the Age of Reason included the active pursuit of prosperity and social well-being. No longer creating simply a public capacity to pay the taxes to underwrite its king's military activities, the goal of government had become the welfare of its people. And this strategy encompassed a new, infinitely increasing horizon. As Keith Baker put it: 'Public welfare and social utility were no longer conceived as inherently limited. On the contrary, they were regarded as indefinitely improveable through the activities of a state apparatus now conceived as an active instrument for the achievement of social progress.'[11] With an increasingly competitive world market, the parameters of social welfare included competitive manufacture–both to export and to serve the domestic market. Skill in making machines to fit modern industry was a benefit that administrators hoped to find in some workers and to teach others. Large new technologies imported from Holland, England, and Germany were worth the price that the state was willing to pay foreign workers and industrial spies. The inspectors of manufacture were well placed to 'translate' the new skills into a French workers' idiom. They offered hands-on assistance with the crucial task of adjusting machines which tended to be clumsy and irregular into smooth, efficient tools. Ken Alder's claim that the Old Regime state policed the market to make it equally accessible to all, but 'the Thermidorian state sought to define and police a public marketplace of information about technology,' is not born out by the documentation concerning the inspectors of manufacture.[12] They were already bringing applied science and practical innovation into rural manufacture. This process of invention advanced machines by comparison and standardization, and showed laborers how make their products competitive.[13] Gathering information did help the state to govern more effectively, but the eventual aim of officials was to inculcate the habits of precision and induction through the length and breadth of France. This was the homely manifestation of the impulse to map weather systems, make an accurate census and promote the metric system. Translated into the world of industry, one reform took the guise of exact measurements for cloth length. At the same time, the economic bureaucracy was debating the advantages of reimposing industrial regulations and inspections, or continuing the widespread freedom of cloth manufacture, which had started in the 1770s.

Two forms of control, standardization of dimensions and standardization of manufactured process, were under scrutiny. They grew out of assumptions linking economic practice with mathematical precision. The first, that 'perfect' textiles would appeal to world trade more than goods with manufacturing defects suggested a return to the traditional, seventeenth-century regulations. The second, that industry-wide standards should be accepted to make technique and products identical throughout the kingdom, also could have been advanced through rigorous application of industrial rules. But the third, standardization of dimensions, might lead to different conclusions.

Most royal administrators felt uncomfortable with unregulated production. They took satisfaction in imposing articulated measures on the number of threads per inch, the length and other qualities of the goods; they deplored the haphazard applications of standards, occurring in the 1770s and 1780s, which allowed regional customs to supersede national rules. As bureaucrats they were biased in favor of industry-wide consistency in the manufacture of the varieties of cloth. Goods conforming to exact specifications were more successful in export trade, they believed. Even merchants within the kingdom could trade with more confidence if standards for the numbers of threads per inch and the sorts of thread used for each kind of textile were the same from Picardy to Languedoc. Underlying these apparently sensible reasons was an aesthetic predilection that considered symmetry on a national scale, symmetry which had scientific exactness, to be the key to efficiency and progress.

But the weavers and merchants lived in another sort of world, where they knew that the manufactured product did not necessarily conform to industry standards. Rural weavers in particular took the irregularities of handicraft in stride since it was the normal way that goods emerged from the workshop. Even the experienced inspector-general Nicolas Desmarest gave his imprimatur to papermakers whose sheets were of good quality, but not the exact dimensions laid down in the industrial regulations. In the cloth trade, merchants and weavers had long been accustomed to adjust the irregularities by adding an inch or two when they were cutting cloth for a buyer. That extra measure constituted equity in textile exchange, since the loom and the sizing process seldom gave an assured length or width. Identical standards entered into the lawbooks were a spurious balance. They looked convincing on paper, but they did not make up for the disparities caused by the various widths of spun thread and different kinds of cotton, flax, and wool.

To achieve standard measures in clothmaking, the manufacturers would have to adopt far more effective machines. The inspectors of manufacture lent their efforts to accomplishing both goals. With bounties for proven new machines, the inspectors tried to encourage inventiveness throughout the ranks of industrial workers. The officials focused attention on solving major technical problems that hindered cloth production: spinning thread more quickly and evenly, and weaving brocade without auxiliary workers to advance the pattern. Technical giants like the inspector general of manufacture, John Holker, circulated throughout spinning and weaving areas, promoting new machines, suggesting improvements for those already in place, and generally encouraging the country workers to move toward more efficient production.

The inspectors of manufacture, however, were up against the cloth makers' hostility to change, as we read in one analysis of obstacles to increasing the output of woolen cloth in cottage industry. It happened that entrepreneurs began to require pieces of cloth in lengths of twenty ells, but the peasant weavers made them only eighteen to eighteen and a half ells. As the inspector of manufactures, Jubié, wrote:

> This abuse appeared easy to overcome since [the adjustment] would not get the worker's back up (*ne contrairieoit point l'ouvrier*). He would receive for his longer stretch of cloth, a price that was higher relative to its length. However, [the worker] resists all enticements, even the prospect of seeing consumption of the product increase. So much does routine reign over manufacture that it will be defeated only by force, that is, by an ordinance or an *arrêt du conseil* which will compel the change. Such action would bring a real advantage to commerce, which finds a reliable outlet for the merchandise.[14]

What the inspectors did not realize was that altering the *length* of a fabric would require significant changes to the loom, costly in time and effort. The warp threads then needed to be longer when the weaver first wound them on the warping mill. The size of the warping board was also a consideration, and of course, the weaver had to advance the warp proportionately more often, to compensate for the changes in the angle of the beat and produce an even surface to the textile.[15] Jubié's comment suggests an important barrier in understanding between the officials and the craft workers. The misunderstandings bore on several steps in the administration's program. The inspectors wanted to impose mathematical criteria on the weavers' product. They conflated the process of making cloth with that of making it to mathematical measure, thinking that once the weavers were accustomed to weaving to a set measure, it would be easy for them simply to extend their weaving for a few additional centimeters. The weavers' reluctance to extend their 'coupons' of cloth was nothing more than obstinacy, to the official mind. Therefore, they explained rural workers' industrial discipline, not as a hard-headed practice of analyzing the balance between effort and profit, but as a psychological reluctance to change.

As the inspectors circulated through the countryside, their reports constituted a veritable anthropology of technical readiness which they linked to geographical setting. The 'taste for work' considered to be inherent to manual laborers did not predispose workers to switch from one activity to another.[16] Habits formed by generations of farming were not necessarily appropriate for craft work. Farming allowed looser attention to the task than industry and a more flexible work week for peasants. 'Their native laziness and the abundance of foodstuffs disincline people from adopting an agreeable work process (*un travail doux*), but one that was too constricted to conform to their taste for dissipation.'[17] In this candid acknowledgment that ingrained habit could derail a project like industrial development and task-oriented labor, the influential intendant of commerce, Tolozan, sketched an important obstacle to introducing machines.

Industrial workers could also be classed as indolent and dull when they clung to routine and shrank from undertaking new techniques. Diderot's condemnation of guildsmen's lack of initiative found an echo in the inspectors' reports. According to the inspector general, John Holker, the French 'always work by hand without searching for inventions which could substitute for them.'[18] As a tract from the Amiens Chamber of Commerce put it, a lot of cloth made locally was not worth much. This product 'provided the occupation for common people (*gens grossiers*), slaves of routine, whom poverty always condemns to ignorance ...'.[19]

The primitive state of new machines already in the hands of workers tended to discredit the adoption of any new devices. While the inspectors were keen on introducing English spinning machines throughout the countryside, they could not help hearing workers' complaints that English spinning wheels were clumsy and inaccurate. In 1784, inspector Bruyard reported on the cotton thread used in great quantities by the stocking knitters of Troyes. This thread was provided by 'fourteen or fifteen thousand' female spinners who worked with machines located in the outskirts and countryside around the city. The thread was annoyingly uneven, but the knitters claimed that without the machines, they would never obtain enough spun cotton. It was the common belief that 'the manufacture of goods spun *à la mécanique* would never be as perfect as the hand product.'[20] The spinners in Soissons echoed this evaluation. Of the forty English spinning machines that had been distributed through the centers of industry here, only six remained in 1789. 'They were considered inconvenient to use; most thread is made now by the big [hand-turned] wheel that gives a more even result.'[21]

While machines needed improvement, the more general task of inculcating work skills and dissipating workers' antagonism to new devices were equally significant problems. Publicizing its competition for six prizes to spinners of linen thread who used the wheels installed in the St. Maclou workshop, the Rouen Bureau of Encouragement declared, 'if the contest succeeds, we will be impressed less by its expense than by the utility of fixing the ideas of the people on the possibility of using the machines and thus destroying the first germ of prejudice.'[22]

The workers had to be brought along gradually to the idea of using mechanical devices and then introduced step by step to more advanced machines. Not every woman knew how to spin well, nor did everyone abandon rudimentary machines voluntarily. One inspector suggested that cloth manufactured at Tarascon would be better if they could do without the local thread which was thick and uneven. Unfortunately, 'it is not possible to persuade the Provençals to make use of the wheel because they simply won't be confined to staying in one place. For this reason, they prefer to spin with the distaff, and they spin very badly.'[23] In order to improve it, inspectors studied the spinning technique and noted the advantages of different wheels and appendages.[24]

Holker and others set up schools for spinning, led by skilled female workers from abroad or from France. The instruction aimed at introducing more efficient machines in place of the primitive rural devices. As one inspector noted of the technical education in his area: 'The goal of these schools is to succeed in the course of time to have the spinners adopt spinning wheels with pedals, which are preferable to those with *manivelles* that are in use now in the land ...'.[25] As the enlightened inspector Isaac de Bacalan wrote of the Beauvais spinners who abandoned the large wheel as too tiring, 'they should [instead] have taken children as workers and accustomed them to it.'[26]

The concerns associated with imposing strict measuring were intertwined in the debate whether or not to reimpose the industrial regulations which specified the various kinds of cloth. These rules had been honored in the breech for some twenty years when director general Necker surveyed the cloth makers in 1780 to find out if they preferred industrial regulations or a system of free manufacture. Necker might have opted to suppress the regulations, but the conflicting opinions he received convinced him to offer both possibilities—weaving cloth under the regulations and weaving it without them—as legal alternatives.

Although the dual system of regulation and freedom provided a space for conflicting practices to exist side by side, many uncertainties remained. Without regulations as a universal guarantee of standards, how could consumers be protected from fraud? If the purpose of the new law was to encourage mechanical innovation among those manufacturing cloth without the regulations, what common practices could be expected from the various ateliers? Relaxing the old rules also implied that there would be less supervision of craft workers, either in guilds or at inspection stations. The inspector of manufactures Lazowsky expressed widely-held views in 1783 when he challenged the viability of the dual approach without first agreeing upon product specification:

> If unlimited liberty and rigorous regulations are not two irreconcilable extremes, one might perhaps prepare tables of manufacture which prescribe the number of warp threads proportionate to their weight, to their degree of fineness, and to the length that [a particular cloth] must have after its sizing and bleaching. But then you would need to adjust (*régler*) the hands of the spinners and to supplement a bit the salary of the winders, which is impractical. A law should be enough. Under whatever penalties seem appropriate, it would request that the cloth called 'demi hollande' would be [one width] on the looms, and that cloth called 'taffetas' would be another.[27]

In other words, Lazowsky speculated that the strict measure might require awkward adjustments in work and pay that rural and urban workers would resist. He understood the objections weavers and merchants made against *ras le bois*, a method now made even more problematic by the need to measure the cloth once on the loom and a second time after it had been sized and bleached. Despite his knowledge, he took it on faith that a detailed chart of cloth specifications (and 'appropriate penalties') would bring weavers to heel.

It was no wonder that businessmen continued to lament the strict measure when it was made law:

> The honorable judges of Boulouère, Thorigne and St. Calais have notified the manufacturers and merchants in their jurisdiction of the new code, and told them to conform to it. The merchants cried out against them and have unanimously declared the impossibility of following the new laws if measuring by the yardstick was the way they had to conduct their commerce ... The poor manufacturers whose existence depends on the sale that they make to the merchants beseeched them with clasped hands to allow the continuance of measuring as usual, and when the judges made it clear that was out of the question, there was a very lively unrest (*une fermentation très vive*.)[28]

But the complaints of merchants and weavers would not be effective in convincing the administration to retain the old way of measuring. *Ras le bois* solved too many problems to be dislodged.

The exact measure could only be achieved with exacting machines. By making one law establishing regulatory norms that could be satisfied only by reliable tools, the government believed it stimulated workers to search out the machines they needed and learn to use them. Self-interest would surely promote mechanical technique as the door to prosperity, without any further administrative cajoling. And the machines themselves would, in the long run, provide workers with discipline superseding guild control and government rules.[29]

More immediately, uniform measures provided the key to dispense with industrial regulations in safety. The issue of controls for workers who were not technically adroit had been a weak point in the rhetoric against regulations. Now administrators could argue that universal measure, along with the discipline of their machines, would keep workers up to the mark. Among skilled workers, craft distinction might be expressed in technical and design originality, while the system of standard measurement provided assurance to the consumer. Thus, in one coup, the system of strict measure could help to bring national standards to bear and provide the possibility for laissez-faire economics to serve the modern state. Once having been a sign of good faith and generous trading, by the nineteenth century, measuring by the ell and a thumb would become George Eliot's emblem of self-serving, dishonest exchange.[30]

NOTES

1. Roger Hahn, *The Anatomy of a Scientific Institution: The Paris Academy of Sciences, 1666–1803* (Berkeley, 1971), p. 117.
2. Ibid., p. 122.
3. Keith Michael Baker, *Inventing the French Revolution: Essays on French Political Culture in the Eighteenth Century* (Cambridge, 1990), p. 157.
4. Gail Bossenga, *The Politics of Privilege: Old Regime and Revolution in Lille* (Cambridge, 1991), p. 136.
5. Baker, *Inventing*, pp. 158–163.
6. Bossenga, *Politics*, p. 136.
7. Ken Alder, *Engineering the Revolution: Arms and Enlightenment in France, 1763–1815* (Princeton, 1997), p. 338.
8. Witold Kula showed that people using anthropometric measures like the ell, the cubit, or the stride, were especially resistant to pressure by officials to change to exact standards like the yardstick. *Measures and Men*, trans. R. Szreter (Princeton, 1986), pp. 24–28, and *passim*. I am indebted to Dr. Sophie Desrosiers for information about cloth measurements.
9. Kula, p. 105, points to the French grain merchants called *blatiers* whose trade was made possible because they acted as agents handling the varieties of grain measures among the small markets.
10. Archives Nationales [hereafter AN], F 12 677A. Brunot to Abeille, 8 August 1781. 'The manufacturers especially want to keep the old form of measurement when the new letters patentes, requiring that the cloth be measured before and after sizing, went into effect. They said: 'The new practice would deprive us of our tradition of compensating ourselves for the price of the linen that we have always been given in the diverse

operations of the Blanchissage, especially before folding the cloths to carry them into the shop.'
11 Baker, *Inventing*, p. 159.
12 Alder, *Engineering*, p. 315.
13 See Liliane Hilaire-Perez, 'Invention and the State in 18th-Century France,' *Technology and Culture* 32 (1991): 930.
14 AN F 12 677C. Jubié, 'Généralité,' 1785.
15 See the helpful explanations in Deborah Chandler, *Learning to Weave* (Loveland, CO, 1995). I am also indebted to Sarah Lowengard for information about loom warping.
16 See Cynthia A. Koepp's discussion of workers' natural affinity for their profession in 'Before Liberty: The Ideology of Work, Taste, and the Social Order,' in *Naissance des libertés économiques*. ed. Allain Plessis (Paris, 1992), pp. 38–40.
17 AN F 12 677C, 'Etat de situation ... Généralité de Trois Evaches,' 1786.
18 Cited by Hillaire-Perez, 'Invention,' p. 912.
19 Tract dated 1778, Archives départementales de la Somme, C 350, cited by E. Levasseur, *Histoire des classes ouvrières et de l'industrie en France avant 1789* (Paris, 1901) II: 661, note 6.
20 AN F 12 650, ' Mémoire,' 1784.
21 AN F 12 678, 'Etat général ... Soissons,' [1789].
22 AN F 12 658, 12 July 1788, Bureau d'Encouragement de Rouen to M. Boyetet.
23 AN F 12 677, 'Mémoire ... Provence,' 1781.
24 For instance, after actually trying out the process himself, inspector of manufacturers Latapie commented, 'The hard thing with spinning is to proportion the thread that you allow to come through the fingers to the degree of speed of the roving and to hold the cotton lightly enough so that it is not too loose in one place and tight in another.' AN F 12 560, 'Voyage de Rouen,'
25 In this effort at technical education, the government distributed Flemish spinners through the villages and opened seven schools, one of which had the distinction of training eighty spinners to produce thread fine enough for delicate manufacturers in only two months. AN F 12 677B, inspector of manufactures Lazowsky, "Précis ... Soissons,"1783. For further information about technical schooling in the eighteenth century see Antoine Léon, *La Révolution française* et l'éducation technique (Paris, 1968).
26 AN F 12 650, 'Observations faites par M. De Bacalan ... en 1768.' Koep cited the advice of Abbé Noël-Antoine that children of craftsmen should 'see large machines in operation, for such sights could "inspire in them a taste for mechanics,"' (Spectacle de la nature) (1750–52), 6:300, cited in 'Before Liberty,' pp. 40–41.
27 AN F 12 677B, Lazowsky, 'Précis sur le commerce de la généralité de Soissons,' 1783.

28 AN F 12 677A, 24 June 1781, from de Tournai. He added, 'I tremble sir, lest the novelty [in measuring] should overthrow this kind of manufacture and occasion some violent revolution, which will result in no advantage for the good of commerce. Tolerance in regard to measuring as in the past cannot bring the least injury to executing the regulation for manufacture, while measuring by the ell occasions no hindrance for the manufacturer.'
29 Gail Bossenga suggests that this was an ideal of some liberal reformers, but that 'this route, pioneered by the English, did not make inroads on French production before the Revolution.' 'Capitalism and Corporations in Eighteeenth-Century France,' in ed. Alain Plessis, *Naissances des libertés économiques* (Paris, 1933), p. 18.
30 See George Eliot's use of the rapscallion Bob Jakin, in *The Mill on the Floss*, Book 5, Chapter 2.

CHAPTER 4
The Meaning of Cleaning: Producing Harmony and Hygiene in the Home

Boel Berner

Towards the end of the nineteenth century cleaning and maintaining orderliness in the home took on increasing moral significance. Dust and dirt became signs of previously unknown risks to health and property. Disorder in the home was seen as a threat to the foundation of society. Just as the engineer should shape the parameters of urban society in public life, the housewife was to keep the household machinery in order, and create a dust-free environment at home. To master the numerous newly discovered tasks required knowledge, vigilance and organizational skills. In this essay I will discuss what these requirements meant for middle-class women in turn-of-the-century Sweden.[1]

The essay takes issue with a certain conception of Victorian women as weak, emotional and insecure.[2] They may have been expected to show these traits—but more important was to show *domestic competence and control.* The meaning of cleaning was to combat nature's threats to health and possessions. The necessity for such control was part of middle class anxiety about social standing: a clean and well-ordered home expressed breeding, distinction and self-control. The harmonious surroundings created were also to serve as a refuge for the man of the house from the disorderly public world outside. The women's control of everyday life was, however, a fragile one. The wife may have been the organizer of the home, but she was not its master. Laws and traditions gave the power to the husband. He had the income, she was dependent on his contribution and goodwill. The discourses on cleaning in turn-of-the-century accounts express the contradictions of women's position; they reflect both the anxiety of women's dependence and the counter-power involved in their exclusive mastery of household activities.

In what follows I will focus on the threats to women's control of the harmony and hygiene of the home; on the enemies to domestic order but also on the

82 *Cultures of Control*

means to restore it. Paradoxical demands appeared in discourses on cleaning: for harmony and hygiene, authority and submission and against too much or too little cleaning zeal. At stake was the creation of a new *domestic culture* of habits, preferences, ideals and organized competences. This culture helped pave the way for the later construction of the home as a site of domestic technology, rational organization and public planning, and thus to integrate it into the socio-technical networks of modern times.

The Discourses of Cleaning

My analysis of the emerging domestic culture is based on Swedish pamphlets, handbooks and articles on the art of homemaking. Such texts began appearing in large numbers from the 1870s onwards. Authors were mostly middle class and urban, and so were the readers. Most household texts were by women, who often presented themselves as 'experienced housewives': a few were by male professionals spreading the gospels of science and engineering to the home.

Figure 2 'The conscientious housewife and her collaborators polish the sink every week with "Tekå" Zincpolish.' Advertisement in a Swedish cleaning manual from 1923.

Together, the texts identified a problem and a task not previously dignified with a separate discourse, and whose social significance was now created and elaborated upon.

The time period studied goes from the 1870s when the texts began to appear, to the early 1920s when women's social status changed and the rationalization of homework was put on the public agenda. We can discern a shifting focus. What women were expected to control in the 1870s was, above all, the 'Order' of the home. Towards the end of the century medico-scientific arguments for improved hygiene led to somewhat of a 'cleaning craze.' In the early twentieth century, a more 'rational' or 'professional' image of the housewife's work was presented, paving the way for the rationalization movement of the 1920s onwards.[3]

Despite these changes in emphasis, the rhetoric of the texts is remarkably consistent over the 1870 to 1920 period. My analysis will concentrate on the recurring motifs; the problems, rules and solutions identified to gain control and to create a hygienic and harmonious home. In the texts we find, first, practical instructions for how to e.g. wash a floor, eradicate moths, polish silver, or make the bedroom really fresh and clean. To this *descriptive*, handbook-style rhetoric was often added an *affirmative* one: the gendered division of power in which these practices occurred was presented as proper, desirable, or inescapable. Last, but not least, there was an *invocative* rhetoric: cleaning the home, it was argued, would keep chaos at bay, give worth to woman's state of subjugation, and create order and harmony in an otherwise disorderly world.

THE SIGNIFICANCE OF ORDER

> A home without *order* is like a ship without a helmsman, driven hither and thither by the waves, sooner or later running aground on the dangerous skerry of disorder, ruin. No, the helmsman of the home must not let her hand fall from the helm, but hold a true and uncompromising course towards the constant goal, the harbour of order that safeguards comfort and continued existence.

Thus wrote I. Zethelius in *Rådgifvare för hemmet* [Guide for the Home] in 1910.[4] Keeping things clean and orderly was a matter of avoiding social and moral ruin. It was a social and symbolical as well as a practical task.

Cleaning—anthropologists remind us—is a matter of maintaining meaningful patterns in existence.[5] In clearing and cleaning we separate wanted from unwanted material, 'junk' from that which is to be saved, 'dirt' from acceptable wear and tear. This activity is symbolically charged. What is seen as dirt varies from one time period to another, between classes and individuals. The same variations apply to definitions of what constitutes disorder.[6] The period we will discuss here was one in which order and cleanliness were raised to important social goals. Before the eighteenth century, or even later, cleaning had not been seen as a particularly central task, neither in the home nor elsewhere.[7] People, especially in the countryside, lived their lives in what we would consider rather filthy surroundings. During the working week no one had time to clean. Some cleaning was done for Sunday. It was mainly the Christmas and Midsummer cleanings that marked the annual rhythm and the difference between everyday and holiday.[8] Thus, cleaning had a definite ritual significance. It was linked to religious calendars rather than to hygiene requirements or social prestige. As the English historian Leonore Davidoff has pointed out, up to the seventeenth and eighteenth centuries lifestyle differences between rich and poor were expressed by the quantity and type of food they consumed, what gold and silver ornaments they owned, and what clothes they wore, but not in how neat, clean, and polished things were in their homes.[9] Only later, did this become an almost manic preoccupation.

The Home as Social Project

During the seventeenth century a striking and pedantic cleanliness with strong ritualistic features developed in the Netherlands. Houses and their immediate vicinity were to be kept spotless and impeccably clean. Everything was scrubbed, varnished, and polished. Keeping the home clean was not a struggle for health and hygiene. Rather, argues historian Simon Schama, it was a question of symbolically upgrading the primary cell of society—the home. The family household formed the foundation of the Dutch Republic's structure. Spotless homes were a formal prerequisite for a properly run society. At the same time, a strong symbolic and practical line of demarcation was drawn between the home and public life. The home was a separate and special place. Here, intimacy and pedantic cleanliness expressed a moral contrast to the crass materialism that permeated society in this colonialist and mercantile state. High morality and materialism could, thus, coexist in society in a way that satisfied a strong Calvinistic faith.[10]

The cultural line of demarcation between the dirty world out there and the pure world of the Home, which the Dutch drew at an early stage, also became increasingly prominent during the nineteenth century in Western societies without Holland's specific history or form of government.[11] The well-ordered Home was seen as a sanctuary from the disorderly world of commerce, politics and competition between men. Previously deprecated tasks of removing filth and keeping things in order took on new stature. Women were given a new, civilizing mission. To be sure, the man was the provider of the family, but the wife's efforts and concerns were, as one Swedish writer put it, the 'home's *perpetuum mobile*, the unremitting, quiet, invisible driving force' that created order and morality.[12] With a steady hand on the helm and with constant vigilance against the storms of indifference and the skerries of negligence, the housewife was expected to guide the domestic ship into the harbour of harmony.

In the Swedish texts, the maintenance of order was seen to involve a constant effort of control and self-control. Otherwise, ruin and spiritual misery were imminent: 'Every day in an untidy home is one rung downward on the ladder to the heaven of happiness,' Agathon Burman warned in 1875.[13] An ideal of perfect order was presented where rooms were clean and tidy, clothes and bedclothes were clean and aired, and everything was in its proper place. No disruptions due to the unpredictability of things or material wear or the capricious nature of people were to be allowed.[14]

Ambitions and Distinctions

We may wonder at this constant preoccupation with order. Clearly it had a great deal to do with the middle class family's desire for social distinction. In the dynamic world of the market economy associated with industrialization and urbanisation, the success or failure of the members of a broad new middle class of professionals, salaried employees, merchants, entrepreneurs and civil servants was, to a great extent, their own doing. Their own abilities, efforts and knowledge determined their social placement, not social connections or inherited wealth. In this context, the material order of the home became important as a social marker. A clean and well-ordered home became a sign of competence and social respectability: 'The difference between civilized and uncivilized people is seen more clearly in their cleanliness, just as the degree of refinement in various homes can readily be measured by this yardstick,' argued one Swedish handbook writer in 1904.[15] Poverty and low moral standards were blamed on the housewife's ignorance. Here is Ebba Rodhe, founder of the Home Economics Training School in Gothenburg, in 1893:

> Housewives' lack of knowledge in household matters is one cause of the working class' poor status... Their homes are messy and poorly kept, their clothes dirty and tattered, so that the sense of order and cleanliness disappears among both young and old. The husband is driven out of the house to the pub and the children into the streets. Many a man who did not have bad tendencies from the beginning thus became a drunkard and many a child a good-for-nothing.[16]

'[W]omen of the lower classes do not understand the art or are incapable of keeping filth away and *therein* lies the uglification of their homes... They never properly learned or have forgotten the art of straightening and cleaning,' maintained Elsa Tenow in her influential book *Solidar* in 1905.[17] A beautiful, well-kept home was a sign of breeding and social standing. To maintain the desired order cost time and knowledge—but the effort was necessary to prevent any association with the classes below.

The Order of Things and Family Harmony

Even more important, perhaps, was a positive and self-assured ideal of the desired state of things. The new middle classses wanted to create a new kind of *Home* as a model for family happiness and domestic harmony.

Swedish housing researchers have shown how a preoccupation with one's own home, its planning, construction, and furnishing, underwent an expansive growth around the turn of the century.[18] There was an important reaction against the upper-class style of home-making, with its heavy furniture, thick carpets and curtains, ornamental palm trees, gold frames and dark walls. Light, air, and simplicity should instead characterize the home. To the handbooks' standard requirements for order, cleanliness, precision, and thrift was added a call for 'good taste.' Simple beauty in the home would create people in harmony and contribute to serene social development. This ideal of simplicity was seen as better suited to the upwardly striving middle class, with its often modest income, than the overburdened furniture style advocated previously. Middle class gaze may have been upward towards the consumption habits of the upper class, but its member's financial means were not up to their ambitions. Again and again the importance of thrift and caution was pointed out to readers, of 'living in accordance with one's position,' and not giving in to social pressure or the temptation of advertising. The things that one owned should have a 'genuine' value, and not be used as decoration or showpieces. 'Finery and ostentation' were to be avoided.[19]

Purchasing, caring for, and preserving this carefully acquired property was the special task of the woman. 'Waste not,' it was said over and over again with emphasis. This required discipline and vigilance. 'Disorder and indifference gnaw like a worm at our fortunes, great and small,' Mrs. Laura G. warned in 1902 in *Oumbärlig rådgifvare för hvarje hem* [Essential Guide for Every Home, a best-seller in five editions since 1888].[20] The housewife constantly had to 'let her vigilant eye peruse the stores of the home. For most of us the chattels and belongings in the home are the only capital we can store up and it is a capital that requires our constant and vigilant attention. We must keep it all in good condition and not neglect its daily care in any manner,' argued an anonymous writer in *Svenska Husmodern* [Swedish Housewife] in 1877.[21]

Objects were endowed with a moral function. They were to witness unpretentiously of a family life of simple habits and togetherness, depicted for example in the writings of Mathilda Langlet, the most well-known adviser to housewives of late nineteenth century:

[In the study] the family gathers in the evening about the pleasant, round table in front of the sofa when mama's lamp has been lit. Here, the family father unabashedly smokes his cigar with coffee and his evening pipe. Here, tales are told and yarn wound, elephants and horses cut from paper, patterns are drawn for use with the fret saw, handkerchiefs are hemmed, and homework done. The chairs are a bit worn, true, the rocker somewhat rickety, and the sofa has lost some of its youthful freshness, but nowhere is there such cheer as here.[22]

Harmony and Male Power

This idyllic picture conceals a family pattern built on a fundamental asymmetry of power. According to the marriage laws in effect during the period in question, the wife was under the guardianship and mastery of her husband. The wife's personal rights were summarized in her matronage, meaning that she was in charge of the 'inner housekeeping.' It was her formal duty to foster and care for the children and to perform or organize the housework. Thus, the housekeeping work was the wife's obligation and the counterpart to the husband's duty to provide.[23]

That the wife was the one provided for, that 'her place was second,' was something that was pointed out again and again. The husband had financial power. His slightest wish must be fulfilled, 'since it is... he who, by his labour, bears the cost of supporting the entire family and since, without him, its members could not exist,' Langlet emphasized in 1884.[24] On the other hand, the struggles of public life often left the husband 'dejected and irritated by adversity, full of righteous indignation' that could not be vented there. The home was to be his sanctuary. There he should get the obedience and admiration so often denied at work, enjoy 'rest and invigoration,' happiness and delight.[25]

Increasingly, in the nineteenth century, the men turned away from the public arena towards family life as a source of personal and social pleasures.[26] But the change was partial and precarious. 'Man's calling, his pride, ambition, and longing for temporal benefits' always threaten to take him away from the home, as do his patriotism and lust for profits, the handbooks said.[27] In the final analysis, the order and harmony of the home was the wife's only insurance against being abandoned. The husband always had the 'option of fleeing from an unpleasant home,' 'to the bar, the pub, the gaming table, the theatre...'[28] The wife, on the other hand, had no real means of providing for herself. Therefore, the man's needs must be law in the home.

In the accounts I have read, there is strong tension between the competent woman who controls and runs the home and her subservience to her husband, between her technical and organizational expertise in the minutest domestic detail and her being declared incompetent in the world of men. The woman was supposed to be the serving, calming creature with intuitive understanding of the man's slightest wish. At the same time, she was to be the determined helmsman, with a firm hand on the rudder. To be sure, the husband was legal master of the house. He was to be served and looked after. But the texts also depict the woman as the *de facto* ruler of the house. She was, in the words of I. Zethelius in 1910, the 'actual governor' of this 'state in miniature.'[29] She was the legislator and police of the home. She developed the home's 'unwritten law' of punctuality, duties, and rights, a 'model order' leading to cleanliness and love of labour, an order that protected and prepared for battle in the world outside the home. As we shall see, the threats to this order, and the means of restoring it, were defined in this field of tension between subservience and authority, submission and control.

THE THREATS TO ORDER

Nature and the Vulnerability of Things

Constant and vigilant work was required, first of all, to hold the destructive forces of *nature* at bay. 'After all, cleaning... is a rather negative task by which we combat the processes of destruction and dissolution with which nature implacably threatens our possessions: rust, dust, mould, moths, etc.,' Gertrud Norden pointed out, in 1924.[30] Agents against fleas, roaches, rats, insects, flies, gnats, ticks, mosquitoes, ants, vermin, lice, moths, and other natural enemies were listed, for example, in Hagdahl's handbook of 1885 with practical hints for everything in the home.[31]

The texts vacillate between a soberly practical and a morally outraged tone. Despair over this constant and defensive cultural battle against invisible enemies is often expressed. Order in the home was seen as a kind of 'Penelopean web that the wife sits down to continue each morning and which, a constant source of anxiety, unravels each day anew,' the journal *Svenska Husmodern* wrote in 1877. After all, nature's threats to things were *everywhere*:

> There are yet other enemies in the house, quiet, but ever active demons that seek unremittingly to destroy all therein. A grain of sand or dust is caught in the curtain and wears away a thread; smoke gets in and blackens the glass; moths get in the upholstery; the gilding is darkened by moisture; the meat is spoiled and the butter is made rancid by the high heat; a great, ugly spot has appeared on the new tablecloth; that hateful nail there ripped a hole in your dress; there, the lock is broken on the door, the bell cord pulled off, etc. Not today, tomorrow or the day after, but day after day, unceasing, your entire life; at first it is unnoticeable, not at all worth the effort, but tomorrow it is worse than today and the day after the damage is already impossible to calculate.[32]

Nature's gnawing at beloved things led to indignant outbursts in the midst of otherwise quite detached descriptions of cleaning techniques: 'It feels indescribably degrading and helpless that the most treasured thing we have, our home, our memories, the things to which we are attached and the clothes in which we are comfortable, become repulsive and unpleasant as soon as our care for them wavers in the least,' wrote Célie Brunius in 1917.[33]

The threat was not just against things and possessions. More dangerous still was nature's attack on the health and working capacity of family members.

Dust

Towards the late nineteenth century a new kind of sensitivity arose towards the dangers discovered by natural science. Pasteur's theory of bacteria gave rise to a general distrust of everything: water, air, the things themselves. 'Uncleanliness arises in the air, everywhere we live and move,' Lotten Lagerstedt pointed out in 1894. 'Some is in gaseous form, some hovers in the air as a fine dust.'[34] Only science knew how bad the situation was: '[T]he microscope has... shown us how our daily dealings are subject to and surrounded by deadly microbes, the cause of all manner of disease,' *Mitt hem* wrote in 1905.[35] Two years later, the journal listed with delighted indignation what had been found 'in a microscopic study of dust between a pair of double windows': 'Bits of quartz and coal, grains of lime, wool, cotton, and linen threads, grains of starch, fly wings, bird feathers, vegetable hair, twelve different kinds of fungal spores, pollen from various types of plants, two kinds of infusoria, a number of algae and bacteria...'[36]

The distrust focused on *dust*, the visible sign of invisible dangers, of decay and ill health. 'Dust is one of the home's most dangerous enemies and it is up to each and every one to try and eradicate it or at least to prevent it from collecting in rooms,' Lagerstedt admonished.[37] Again and again the texts described the numerous 'seeds' whirling around in the dust. 'Dust should be despised, for dust always contains bacteria,' which could penetrate the body and cause discomfort and disease, wrote Möller in 1917.

> [Bacteria] are found everywhere and in all places and no one can avoid coming into contact with them. They hover in the air we breathe, fall into our food, attach themselves firmly and deposit themselves on everything, where there is room only for a grain of dust. *But they are primarily in dust and dirt.* Everywhere there is dirt, there too lurk the foes of our health, and perhaps of our lives... In crevices in the floor, in dark, dank corners, in cellars and such places, they may lie dormant if the dirt is long left untouched, only to revive once again after entering the human body in some manner and reproduce with dizzying speed.[38]

The very comfort of the middle class home contained these seeds of destruction. Plush covers, heavy curtains and rugs collected 'seeds of disease, all kinds of wandering and flying bacteria, bacillas, microbes, and what else they are called, these invisible entities, that in recent times have begun to embitter our existence,' Langlet complained in 1891.[39]

The *watchful eye* on everything was the housewife's ally in the constant battle against this creeping menace. Kamke, for example, told her reader in 1910 that you must 'be observant and often look under bureaus, chairs, and tables and let your finger test for danger on objects seldom dusted.'[40] 'Remember to seek out... [dust] in the most inaccessible places, under and behind heavy pieces of furniture that are seldom moved, on top of the tiled stove—everywhere, just where you least expect it!' *Hemtrefnad* [Home Comfort] wrote in 1899.[41] The threat was everywhere, but possibly most, I. Norden warned, in '"dead corners", where the air is stagnant and where trash and dirt accumulate': behind furniture, in cellars, and on staircases.[42]

In 1895 the journal *Idun* announced a contest for the best way to 'destroy dust in our homes.' The prize-winning essay, appearing under the name 'Dixi,' gave detailed advice for a 'rational cleaning' of rooms. Here, as in many other texts, we will find a condemnation of the 'errors' commonly made by others in the battle against dust, and a passionate defence of one's own method for correct dirt removal.[43] If we are to believe the handbook writers, most people during this period lacked the 'common horse sense' which made them dry sweep the floors instead of using the 'modern' style, where dust should be *bound*, with the help of wet rags:

> A worse abomination [than dry sweeping] cannot be imagined and this is a worthy counterpart to dry sweeping the streets. The dust swirls up to high heaven, but only to be deposited again after a time on all objects in the room [Ingrid Norden complained

in 1913]. Better than sweeping in this manner would be to let the dust lie where it lies. In that case, at least, it would cause less harm and aggravation and not just change places once a day.[44]

Hygienism became an important influence in turn-of-the century Sweden. Stockholm in the mid nineteenth century had been one of the filthiest cities in Europe, with a mortality rate that was higher than in other comparable cities.[45] Medical doctors presented arguments against dust and dirt, advocated improved building standards and increased personal and domestic cleanliness. Dust and dirt were a 'sanitary danger,' they warned the housewives, and had to be transported away. Behind their admonitions lay a concern for the spread of tuberculosis and other diseases, especially in low-income families in unhealthy and crowded dwellings.[46]

In household manuals for middle class women, the battle against dust and bacteria became part of an aesthetic/moral/hygienic program. Cleanliness was to permeate not only the housewife's habits and thoughts, but also the entire house, its planning, and possessions. Dust-collecting beds, curtains, and upholstered furniture should be replaced to facilitate work and promote health. Daily dusting and cleaning of the entire apartment or house was recommended. In addition, there should be weekly cleanings and two or more major cleanings a year. This was the norm that was advocated in greater or lesser detail long into the twentieth century.[47]

Bad Air

To the 'dust neurosis' and 'bacilla fright' brought on by Pasteur's and others' discoveries, was coupled a strong aversion to bad air.

Since the eighteenth century a highly developed sense of smell had become an ally in the modernizing professionals' battle against disease and poor health. As historian Alain Corbin has shown, it was used to demonstrate the need for city planning, drainage ditches, and other urban infrastructure. During the nineteenth century the sense of smell also became an important tool for guaranteeing household cleanliness.[48] Disgusted descriptions abound in the household manuals to drive home the need for *fresh air*. Nature's decomposition products create stinking uncleanliness, housewives were told. Ferment and decay were everywhere. 'Sweat, fat, and toxic varieties of air... *that are constantly excreted through the skin and in other ways*' make the air impure and foul smelling, Ingrid Möller pointed out in 1917.[49]

Such accounts were, more or less consciously, based on the so-called miasma theory for infectuous diseases. This theory saw health risks as originating in the spontaneous combustion that was assumed to occur in bad, stagnant air. Well into the twentieth century it coexisted with bacteriological arguments for a clean and healthy home, both in medical and lay accounts.[50] Bacteria thrive in dark, dirty, stuffy rooms, Dr Berg wrote in 1924: 'Untidiness and bad air breed disease, first by direct contamination of the blood when unhealthy vapours are inhaled and, secondly, by the increased chance of infection.' The results are 'rickets, anaemia, consumption, and numerous other sickly states.'[51] Colds and flu result from 'homes that had become imbued with bacteria and then not sufficiently aired,' Elise A. maintained in 1914.[52] Other writers pointed to 'poor complexion, headache, and weak nerves.. [as] an almost unavoidable punishment' for not following the 'law' of fresh air.[53]

In this context, the *plain board floor* became a sensitive, class-related issue. The smell of scrubbing inexorably revealed a lack of hygiene in the house and, thus, a lack of respectability. The dirty scrub water ran down between the boards and rotted in the double flooring, becoming 'fertile soil for rot and disease-causing fungi,' Norden stated in 1913. An unpleasant, musty smell distinguished such apartments, indicating ferment in the dirt and the emanation of gaseous poisons.[54]

An ideological battle for and against the traditional plain board floors was waged in the columns. The architect Lars I. Wahlman gave his opinion in 1902:

> I have seen a good old floor, scrubbed hundreds of times, broken up; I have seen the double flooring below billowing with worms, worms that had been my neighbours and fellow builders for years—and I drew my conclusions as to the suitability of unpainted floors and scrubbing.

Oiling and varnishing the floor was recommended, by him and others, in order to get rid of 'the scrubbing brush, the floods of dirt, the stench of filth, the recurring up-and-down-turning of the home.'[55] Around the turn of the century, those who could afford it replaced the naked planks with linoleum floors and the scrub brush with a vacuum cleaner.[56]

Above all, however, *airing* was advocated. It was presented as a moral imperative based on the 'scientific' estimates of physicians and sanitary engineers of how much fresh air was needed per minute and cubic meter.[57] 'Open the windows' was the motto in *Hemtrefnad* in 1914. 'Let in air, air, and more air,' *Mitt hem* wrote in 1909. Incessant airing of rooms, bedding, and clothing was recommended in text after text. 'Air in the rooms shall *be renewed several times a day*,' I. Möller insisted in 1917. 'It is not enough, as some people believe, to let the window remain open long enough for the *foul smelling* air to leave. For the air is not always clean just because the uncleanliness cannot be seen or perceived by the olfactory organs.'[58] The texts give precise rules for airing: so and so many minutes' draft during cleaning, airing after every meal, open windows before going to bed, at night, in the morning, airing schedules for the summer, the winter, for the children's room, the bedroom, and so on.

Thus, there was a persistent plea for new ideals and new habits in the home. The vigilant housewife could identify the hidden dangers of the home. She knew how to remove dust, dirt, and impure air. This incessant struggle against nature's threats can be seen as the domestic counterpart to the activities of professional doctors, sanitary engineers and building inspectors in public life. Sewage and city planning, on the one hand, airing and dusting, on the other. Each in his or her own sphere was expected to contribute towards the creation of a clean and healthy society, based on the new awareness that science had brought.[59]

Servants as Hygienic Hazard

The threats also had human carriers. Cleanliness was an economic and cultural property that was unevenly distributed, as was knowledge of the dangers of dirt and foul air. This is not the place to discuss the various attempts by the middle class to teach the working class self-control and cleanliness.[60] We will simply say a few words here about the threat to order that the *servants*—a necessary ally in the battle against dirt—presented with their presumed sloth, forgetfulness and indifferent attitude to the demands of cleanliness.[61]

Kitchen odors, in particular, came to symbolize lower-class presence in the middle-class home. Langlet's book is full of accounts of the unpleasantries that 'ignorant farm girls' could bring about: food had a slight flavour of containers never cleaned, 'gnawed bones, herring necks, boiled potatoes, etc.' were put aside and allowed to 'stand and sour'.[62] Norden suddenly—in the middle of an otherwise neutral description of dusting—erupts into a disgusted outburst against those 'souls serving in the kitchen [who]... sear and fry with the kitchen door wide open .. so that the smell of grease and cooked herring, rancid meat, cabbage, and burnt peas and milk reach up to the very rafters.. [and drifts] into the apartments' of other people's homes.[63]

Dishwashing was another touchy task, which according to Möller was performed in a manner that was 'negligent and filthy ... often almost unbelievably so.' She went on to present a detailed and almost delighted description of these filthy practices:

> The water is turbid and repugnant, the dishwashing implements are soiled and sour, and the work is done in a slovenly and poor manner, often with greasy and filthy hands. For example, a shiny smooth plate offers poor conditions for bacteria, while china that is rough and covered with a layer of grease—leftovers from the dingy dishwater and greasy washcloth—provide an excellent place to which bacteria may cling. The same applies to greasy and dull silver, which is often washed quite casually, even in an otherwise proper home. Other dishwashing may be judged with the eye, but poorly washed silver 'smells silver,' as they say. The metal does not smell, but the nauseating, aforementioned layer of grease that has entered in chemical reaction with the silver under the effects of the air, creates the unpleasant odour.[64]

Being able to be *private* in a fresh-smelling, well-ventilated home was the desirable arrangement, a privilege the upper and middle classes strove towards in the home of the nineteenth century.[65] Middle-class women were able to live up to the desired ideals of comfort and cleanliness, since *other* women did the heaviest and dirtiest housework, away from the sight of their masters. But the control was not perfect. When the activities of the 'serving souls' nevertheless invaded the middle class family's private life—with noise and decay, cooking odours and burning smells—the reaction of disgust and indignation was strong.

"Natural" Feminine Disorder

Chaos, however, also threatened from within their own ranks. Middle-class women had to discipline their *own* feminine nature for the sake of order and cleanliness.

Turn-of-the-century women were in a dual symbolic position. They were seen as being closer to nature, by reason of childbirth and nursing. Their spiritual state and health were, it was thought, strongly affected by their reproductive organs.[66] At the same time, it was the task of women to discipline and civilize Nature. They had to educate uncivilized children, eliminate dirt, create order among material things. This position between Nature and Culture was existentially ambiguous, and potentially both physically and morally unclean. Again and again the texts criticized women for their supposedly 'natural' tendencies towards indolence and filthiness, negligence and untidiness. As Norden said in 1913:

> When an [untidy woman] dresses or undresses and, all the more so, when she makes a grand toilet, it is not long before all the furniture in the bedroom, the dresser, chairs, sofa, tables, and even the beds are littered with discarded, more or less intimate, items of apparel. Petticoats lie on the floor where she climbed out of them, hairpins, combs, and toiletries are strewn on the floor, wisps of hair, pomade and makeup spots dirty the dresser, soap lather and water are splashed around and the entire bedroom is in a state of chaos and confusion... [and]... bears witness to untidiness and slovenliness.[67]

'Most women's' manner of caring for their personal hygiene is reprehensible, another writer stated in 1906. If we simply 'consider the manner in which they treat their hair brushes..., the many careless female hands that meet ours..., the dragging skirts, the waist of a dress in which dust and dried skin accumulate for years in the seams, the black or grey corset that the owner has proudly worn month after month...'[68] Lack of self-control could have disastreous results: 'Stained clothing, torn, poorly attached trimmings, unkempt hair, curlers, slippers down at the heel—such things can transform the loveliest little wife into a shrew'. A housewife who was not clean with her own person, was seldom clean with other things—'and then cleanliness throughout the entire house is of a highly dubious nature,' Langlet warned.[69] Moreover, the moral purity of the home was threatened if the woman did not constantly strive towards cleanliness: 'If our mind longs for cleanliness, then we will suffer no filth in any form,' [one writer exclaimed in 1909].' Clean in my mind and in my children's, clean in my home, clean in the kitchen... and everywhere,' and a 'strong hatred of all that is dirty.'[70]

The civilizing process in the home must therefore begin with the woman herself, and go on each day, incessantly. Constant vigilance, discipline, and hard work were needed to tame woman's feminine nature, to create purity and cleanliness. The housewife should rise a half hour earlier in the morning to deal with her unclean body—'to wash herself, comb her hair, ... and brush her teeth before it is time to work'—and then at all times try to discipline her disorderly nature.[71] Anything less would be unhygienic, unhealthy, and unappetising.

'HOLDING AN UNCOMPROMISING COURSE TOWARDS THE HARBOUR OF ORDER'

To avoid the chaos caused by natural enemies or by women's, children's and servants' more or less 'natural' slovenness, cleaning and airing were not sufficient. *Orderly patterns* must be created in the home. Two major ways were advocated: spatial organization and temporal organization.

Spatial Organization

'Dirt,' says the anthropologist Mary Douglas, involves dangerous mixtures of categories:

> Shoes are not dirty in themselves, but it is dirty to place them on the dining table; food is not dirty in itself, but it is dirty to leave cooking utensils in the bedroom or food spattered on clothing; similarly bathroom equipment in the drawing room; clothing lying on chairs; out-door things in-doors; upstairs things downstairs; under-clothing appearing where overclothing should be, and so on. In short, our pollution behaviour is the reaction which condemns any object or idea likely to confuse or contradict cherished classifications.[72]

'Dirt is only something in the wrong place,' said also Dr Öhrvall in 1900, with reference to some unknown authority.[73] 'Everything in its place' became a maxim that was repeated *ad nauseam* in turn-of-the-century household manuals. Confusion and chaos were imminent if distinctions were not maintained. 'Order consists not only in having a room cleaned, so that nothing 'stands out'. That which is not visible must have its specific place and *be in that place*,' Mathilda Langlet pointed out emphatically in 1884.[74] Temporary placement was not tolerated. *Everything* must be put back in its rightful place immediately after use, otherwise things would accumulate and the 'tangle become more and more difficult to unravel.'[75]

The anxiety expressed a new situation; there were indeed *more things* than before—knick-knacks, curtains, rugs, books, kitchen implements, furniture—to keep track of in the middle class home. They represented a capital and had, as we have seen, to be protected again wear and decay. In the new, larger apartments, new classifications also appeared. 'Backstage'—the kitchen, bathroom, children's rooms—was the sphere of women, servants and children, seldom visited by the men. There, nature was close by, taken in hand in its 'raw' form and civilized, to be presentable on the frontstage of the home. In the living rooms and salons, a cultivated order with everything in its place should reign. All indications of unpleasant household work—odours, cooking sounds, the confusion of housecleaning—were to be avoided, or were permitted only when the husband was not there.

The Bedroom as Border Zone

The bedroom, now often called the 'sleeping room,' underwent an interesting development. During the nineteenth century, this room passed symbolically from front stage to back stage. During the 1890s it became a room that was disengaged from the rest of the living suite, devoted entirely to sleep and rest. No company or visits were allowed there, as had been the case previously. The children were to have their own bedrooms, as would the servants. The practice of sleeping several people in a bed was condemned.

A number of ambiguities were manifested in turn-of-the century discourses about the bedroom. Supposedly this was where the middle-class body's right to a private, well-aired, and odor-free sphere was to become reality. In real life, however, the room had a 'tendency to become the most disorderly and untidiest' part of the apartment. Men's and women's possessions intermingled, waste and odors were produced. It was here that the housewife had to first prove her mastery of nature's destructive forces. 'The greater one's prosperity and culture, the greater weight one places on the bedroom and its suitable appointments,' Roswall stressed in 1904. Extra vigilance was required here to keep chaos at bay.

The bedroom was also a border zone between public and private. It must be put in order every day, 'so that one need not be ashamed if some outsider should happen to cast an eye therein', Roswall said.[76] Social respectability must be maintained. 'Among good housewives, nothing is condemned so much as the *contrast* between that which is intended for outside eyes and that which is meant for the family itself.'[77] 'Behind the scenes *first*, on the stage *second*, not vice versa,' Reimer admonished in 1909 with reference to the bedroom, 'the true hearth of hygiene in the home.'[78]

This civilizing mission must begin first thing in the morning. 'The bedroom must always be in total order before cooking begins in the kitchen,' Mrs. Laura G. stressed in 1902.[79] All bedding must be removed, immediately aired out for at least an hour before cleaning begins—but first, Norden pointed out, 'all things used, especially for the lady's toilet, must be put away, after first removing hair, which any person laying claim to being tidy and orderly must never leave in these toilet articles.'[80]

Medical advice books and household manuals advocated sun, fresh air, and cleanliness in order to get rid of the air spoilage caused by human 'expiration and transpiration' in the bedroom.[81] 'Soiled clothes, unclean linen, and dirty shoes must never be allowed to stay in the bedroom.'[82] Nor were '… cigar remains, leftover food, and the like' allowed to remain there.[83] A white bed-making apron, sometimes white gloves, were recommended for bed-making. An environment similar to that of a hospital was the ideal—or at least something like an 'English bedroom', with its iron bedsteads, Spartan furniture, and bright range of colours.[84]

A lively discussion arose around the turn-of-the-century concerning the most hygienic way to arrange the bedroom. This debate reflects a shift in power over home planning. Those whom housing researcher Greger Paulsson calls 'engineering architects'—highly trained male engineers—had devoted themselves mainly to the larger, impersonal units as they planned their ideal homes: city planning, boulevards, the organization of sewage and water works. The new hygienism beginning in the 1890s, on the other hand, dealt with the most private and intimate: clothing, the bedroom, personal cleanliness. To a certain extent planning then became a question for a 'middle-class intelligentsia' of women.[85] These women were the handbook writers, philantropists and emerging household experts. They were backed in their endeavours by medical men, and they acted within a sphere between the public and the private, bringing ideals of hygiene and rationality to the home.[86]

Towards the end of the century, however, a reaction appeared against the strict ideals of hygienism. In an *Idun* article in 1898, E. G. Folcker took up the gauntlet for the canopied bed—a dust collecting abomination, according to the advocates of hygienism.[87] Author Laura Fittinghoff attacked the hygienist ideal in that same magazine in 1901. When it took over a home, its pleasantness was lost, she wrote. Instead of the fine, warm, cozy bedroom where the family and guests felt at home, where mother lovingly worked at her sewing table and knitting basket, where flowers and pictures brightened the room, there was barrenness and coldness. Iron bedsteads, marble shelves and washstands replaced flower pots and grandmother's white comforters—and the 'windows are open around the clock.' The bedroom was now to be used only for sleeping. Socializing and work were to be done elsewhere. Thus, Fittinghoff complained, feminine Harmony disappeared when Hygiene took over in the home.[88]

Temporal Order

The bedroom discussion leads us to another key dilemma. It is not possible always to have everything maximally clean. Cleaning was a never-ending task. Mrs. Laura G. spoke of the 'sickly nervousness' born of the fundamentally unending task of taming and ordering nature. It could be cured, she said, only by a 'moral compulsion placed on oneself to *always be guided by the clock.*'[89]

A plea for an almost manic control over one's time use runs through the period examined here. In other words, this was not something that belongs only to a later debate, inspired by Taylorism.[90] Also the late nineteenth century home had to function like clockwork. 'The clock must rule over everyone, over man and wife, for the first paragraph in the law of order commands us to utilise time properly. Waste destroys resources, disorder destroys half one's life,' Burman warned already in 1875.[91] Without discipline in matters of time, the housewife would succumb and order never be achieved: '[W]ith each lost hour a part of life is lost,' Veritas wrote in 1900.[92] Cleaning, in particular, involved the constant risk of temporal collapse:

Of the multitude of small tasks, first one, then another will take on gigantic proportions [Célie Brunius wrote in 1917] and the more their execution is postponed the more impossible it seems to become to catch up. A definite order for the tasks of the day and week are both easier to implement and more useful here than in kitchen chores, but once you lag behind it becomes exceedingly difficult to restore that order once again.[93]

The remedy was a careful division of the housewife's time and assignment of each task to its proper time. Detailed daily timetables were listed in the texts: for the housewife herself, for a middle class home, for a rural household, for the week, for Sundays and holidays, for daily cleaning, for weekly cleaning, for major housecleaning, for cleaning days, and occasional tasks...[94] Time was divided up into what seems to be more ritual than strictly needed cleaning sessions. In addition, all tasks had to be done in a certain sequence and at a specified time, which could be altered only in case of emergency—and everything, as Langlet pointed out, 'that was to be done in the house, the housewife must keep in her head like a timetable.'[95]

The Master's Order and the Ruling of the Home

The control of time and space was intended to defend the home against nature's threats to health and possessions. But the body of laws governing everyday life was not gender-neutral in its application. The father of the house had 'sole... right to be waited on.'[96]

Every meal had to be served at precisely the proper time, so the husband would never have to wait and perhaps be prevented from enjoying the time of rest after supper to which he has a right [Veritas pointed out in *Hemmets rådgifvare* in 1900].[97]

A housewife who did not subtly create the order of the house in accordance with the husband's more or less explicitly stated needs, risked his irritation, and the dreaded loss of household harmony:

Many an otherwise good man becomes exacting and irritable if the wife fails to heed some of his habits, some of his whims, perchance. If he must wait for his supper, his morning paper; if his overcoat is not brushed, his match box not in its place; if the curtains are drawn crooked, if the piano is open, if the maid is wearing shoes that are down at the heel...[98]

Ultimately, as we have seen above, the entire home was threatened with collapse if the husband did not receive the service he demanded. He would abandon it for the temptations of the gaming table, the theatre, or the pub.

In the texts, the man appears as a large, moody child, 'seldom capable of producing thorough cleanliness in his external appearance.'[99] He was a source of filth and disorder:

> With what dissatisfaction [the housewife] furrows her brow when her thoughtless lord and husband, who like all other men values his comfort more than his furniture, *sans façon* throws himself onto the sofa, wrinkles the tablecloth, or even leans his pomaded head against the wallpaper, a sofa cushion, or such [wrote Mrs. Laura G. in 1902].[100]

The point, however, is that the husband (in the home) *did not need* to exercise restraint and control. That was the privilege of patriarchal power: to avoid the responsibility of maintaining the boundaries between order and chaos, to be spared dealing with dirt, waste, stench and pollution... That was completely the task of the wife. 'This constant work to maintain [order]... is *her* task, which the husband *cannot* understand or perform at all,' one writer wrote in *Svenska husmodern* in 1877.[101] In the nineteenth century these degrading tasks were ideologically and practically converted into their opposite: a civilizing mission and a savoir-faire that gave the woman the authority she lacked in public life.

Gender Power and the Contradictions of Cleaning

Cleaning, in this context, created a genuine dilemma. On the one hand, it was meant to create the order and harmony that the husband expected in his home. On the other, this messy and essentially endless job was supposed to be done without his even noticing it. The wife 'should make sure the home is as harmonious as possible at all times, so that everything related to scrubbing and cleaning is done while [the husband]... is away.'[102] She also had to 'take care that no soapy odour or trace of moisture remain in the room and, above all, protect all manuscripts, papers, books and tools from contact by profane hands...[103]

The matter of the *'master's room'* was particularly complicated. 'Most middle-class heads of household spent most of their time at home here,' historian Birgit Gejvall wrote in her study of homes in nineteenth century Stockholm. Its doors were most often closed, something which meant there was a private space inside, the father's own territory, separate from the rest of the apartment. The children were seldom allowed in and cleaning had to be done there with great care.[104]

Two types of logic seemed to clash in the master's room—the feminine and the masculine type of order:

> To men [Folcker wrote in *Idun*, 1898], order consists in having all materials for the work with which he is involved, all books, notes, and the like 'at hand.' The feminine sense of order, on the other hand, requires that every object be in its place, books on the shelf, and all pieces of paper in various stacks, according to size.[105]

It was, however, the husband's order that ruled. Woe to the wife who tried to create her type of order in her husband's room. 'A foolish ordering here can cause incalculable damage,' Kamke warned in 1910.[106] 'Putting his papers 'in order' is ordinarily to do him the greatest disservice. He always wants to do that himself,' Roswall insisted. 'Each scrap of paper must lie in the same place, an open book must not be closed, and even the least little trifle belonging to his work must remain undisturbed...'[107]

Responsibility for seeing that the husband's order prevailed rested with the wife and with none other: 'No foreign hand may touch this,' Roswall pointed out.[108] There was a constant risk that the uncultured servants would clear away the husband's books; that letters, notes and papers, 'that in the eyes of the servant are of no import,' would find their way to the waste basket or be used to light the tiled stove. With a shudder, the tale was told of John Stuart Mill's servant girl who thoughtlessly burned an irreplaceable manuscript by Carlyle, or of the cleaning ladies whose zealous scrubbing destroyed priceless musical instruments.[109]

During *major cleanings*, however, feminine subservience to the master's time and space was broken. In came the cleaning ladies and, like at a medieval carnival, the hierarchy was turned upside down. The women took over, their time schedule prevailed, nothing was in its right place—and husbands fled their homes, where their convenience was no longer law!

'Old time housecleaning', a 1941 text remembered, meant 'a veritable state of siege':

> Least of all could the master of the house feel calm at such a time. No consideration was given to the sanctity of his private life or his ingrained habits when the furies of major housecleaning drove him from room to room like hunted quarry and if he was wise, he preferred to flee and be done with it, rather than stay and fight in vain against the enemy hordes who occupied his home.
>
> Everything was turned upside down, furniture and carpets were removed, if possible into the open air, lamps and chandeliers were taken down, pictures moved, and book collections removed from their shelves, only to be returned in what for their owner was the most incredible jumble, seldom in their proper place, but helter-skelter, costing him many hours of effort before he was able to restore his precious clarity and order.[110]

The upheaval created by major housecleaning underscored the latent power that was built into the woman's ordering skills. Control over cleaning gave power—and thus the possibility of gender-based and generational conflict. The wife's role included not only 'tender, loving, care' but also the right to organize things for others, to steer their behavior and even to force them to work. A glimpse of this power is sometimes seen in the household literature. The housewife was, called, as already noted, the helmsman, legislator, and policeman of the home.[111] Moreover, not all husbands were depicted as thoughtless patriarchs. In fiction we sometimes see anxiety-ridden male helplessness in the face of women's ruling and bossing, memories of childhood and adult inferiority. The home, its order and priorities, as well as the husband's and children's well-being, were in the women's guiding hands.[112]

THE COMPETENT FEMALE HAND

Keeping things clean and keeping them in order was *work*. It required knowledge and self-control.

Practical Competence

Unlike the middle-class husband, who had been through a long period of general and professional training for his adult occupation, the wife was often unprepared for her complex task—at least according to the housekeeping books. Her studies had left her with a disdain for practical work. She presumably often went directly from obeying to organizing, 'from dolls' houses to duties.'[113] Mothers, according to the literature, could not always be relied upon to transmit the appropriate knowledge of correct cleaning, of household chemistry or hygiene, and the servants often kept their knowledge to themselves. Therefore, the manuals were necessary, and ideally also a year or two at the newly founded household schools.[114]

The texts often include the assumption of, at times, ridiculous incompetence in practical matters. Book after book point out that dust rags were to be shaken *outside* the window and not inside it, from an upwind, not downwind, position; that dust and dirt should be swept up onto a dustpan and not into the air; that it was not necessary to stoop down and get water on your knees and in your lap when you scrubbed the steps and floor; that you could sweep with a long-handled broom, instead of constantly bending over with a short-handled one, and so on.

On the other hand, the experienced housewife—and the one who followed the advice of the handbooks—could proudly point to her extensive competence. She was knowledgeable in domestic chemistry and biology. She could stuff mattresses, combat pests, and furnish a home. She mastered the many practical aspects of daily cleaning: its intensity, frequency, scope, order, methods, and organization. She was a good domestic engineer.

The Control of Others

This is not to say that she herself had to exercise all this expertise. Handbooks from the turn of the century often assumed that *someone else* performed the lion's share of the practical drudgery. The middle-class housewife was expected to dust and perhaps to clean her own bedroom. The rest was to be delegated to one, two, or three 'maidservants.'[115]

The housewife nevertheless had to be a competent supervisor and manager. Now and again she might also have to demonstrate her practical competence. The maidservant should be 'aware that the housewife herself understands how to do chores and that she can do without the maid,' Veritas wrote in 1900.[116] Mathilda Langlet admitted that she had never personally scrubbed a floor. She nevertheless held very definite opinions—which differed from those of her maids—as to how this work was best done.[117] The role of the housewife was presented as that of an educator and organizer—not always an easy task. In 1907 in *Hemtrefnad*, a certain A. Lundberg called for courses in the art of cleaning for maidservants. They would include 'all existing details, such as sweeping, washing, scrubbing, window washing, care for cork rugs, cleaning carafes, and especially dusting.'[118] During the new century, such maidservants became more and more uncommon, less and less competent, and more and more expensive, it was claimed. New, less authoritarian relationships had to be developed towards the servants in a changing society, where young women were more attracted to work in factories and offices than in someone else's home.

Thus, the housewife's competence must include a friendly, but firm way of putting others to work. This also applied to other female family members of the household. Considerably less frequent was the need for male assistance. It did happen, however, more so in the twentieth century than before: 'That a gentleman himself can brush his clothes and shoes and, in case of emergency, sew a button [should] not in the least [mar] his manly dignity,' Christine Reimer stated in 1909.[119] A man could perhaps even clean his own desk if, for nothing else, then for the sake of his own peace of mind: 'This cleaning... is not as difficult as one might think and is easily enough done,' Ingrid Norden said encouragingly in 1913.[120]

Moral Competence and Control

Still, overall responsibility for order was in the hands of the women. Cleaning, in particular, required a 'truly strong and morally complete personality,' Brunius noted in 1917. The work demanded 'new heroism [every day]... much more will power, much more character and persistence than other domestic chores':

> Wipe, rub, sweep, brush, beat eternally and then enjoy a moment when all is in its place—then a few hours later it is time to start again, this endlessly replacing a thousand little things...
>
> The temptation to neglect one thing or another, from turning the mattresses and shaking the pillows, to wiping the windows, changing the water in the carafe, the precise arrangement of the room after each change of clothing, comes anew each day.[121]

The housewife was strongly admonished not to cheat with the cleaning, take shortcuts and breaks. 'It hardly need be pointed out that the time for cleaning should be used for that purpose and not for reading magazines and books or for standing and gazing out of the window,' *Handledning i husliga göromål* [Handbook of Household Tasks] nevertheless pointed out in 1912.[122] A lack of perseverance was immediately punished. 'What sad consequences have we not seen because a little shirt button was missing, a key misplaced, a single minute lost,' Burman told his readers in 1875.[123] Again and again the housewife was reminded that she should never 'compromise with herself or her own convenience.'[124] Nothing must be neglected, everything should be observed and taken care of—at once.

Knowing Moderation

There was a key paradox in the imperative of order. Cleaning was necessary—but *too much* cleaning could destroy the material and moral fabric of the home. The competent housewife also knew how to exercise restraint and avoid excesses of cleaning zeal.

'Of all domestic concepts, neatness and order have a great penchant for growing into pedantry and thus wasted time, wear on furnishings, and a generally unpleasant atmosphere,' Reimer complained in 1909.[125] It is possible to destroy *everything* in the home with too much assiduous cleaning, Möller pointed out in 1917: Oil paint disappears, wallpaper gets streaks, textiles and carpets become worn, polish is rubbed off...[126]

> Many homes are cleaned constantly and still nothing is ever in order and, despite all this effort and trouble, the furniture is in abysmal condition. Trifles and ornaments are broken, chairs and tables given all manner of flaws, tables and bureaus are full of scratches, and within not too long a time a beautiful new home has become a ruin.[127]

More dangerous than the loss of capital, however, was the loss of the *comfort of home*. 'Scrubbing and washing too often bring constant disquietude and make the home a veritable house of torture for the family,' wrote Mrs. Laura G. in 1902.[128] Warnings against the 'tidiness passion' and 'cleaning mania' became more and more common after the turn of the century. Christine Reimer spoke in 1909 of the need to '*dampen* the passion for cleanliness that is running rampant through some social strata... [and which] in some middle-class families has degenerated into a kind of mania.'[129] 'Far too many women exaggerate their duties [Holm pointed out in 1921]... and in their misguided zeal they are gripped by a kind of cleaning frenzy that, despite the greatest cleanliness and order, destroys all the comfort of home instead of creating it.'[130] And in a fascinating turnabout of the period's predominant message, Zethelius blamed the downfall of the family on the cleaning mania of the housewife: 'Husband and children shun the home where the pedantic housewife becomes engrossed only in her own petty ideas', she, and others, maintained.[131]

The time-honoured nineteenth century preoccupation with major housecleaning several times a year was called into question in the early twentieth century. *Mitt hem* (1906) rejected the very idea of such events 'where the whole house is seized by spasms and the housewife herself, in the name of domestic comfort, is in such a tizzy that she becomes an affliction to those around her.'[132] 'Bacteria fright' was scorned by doctors,[133] and a new, more 'rational' approach to hygiene was called for. 'Major cleaning *should be abolished*, as it is out of step with the times,' Elna Tenow maintained in 1905:

> It should be replaced with an effective daily cleaning, rendering all the fuss and to-do unnecessary, since nothing remains for major housecleaning and since it is barbaric to tear one's home apart and chase one's husband and children out of the house. Although not a man, I hate this ruffian and foe of domestic comfort as intensely as does any man and I was never able to appreciate that nirvana of powerlessness that our housewives say reconciles them to the trials and tribulations of major housecleaning.[134]

A number of proposals were made as to how the social effects of major housecleaning could be mitigated. As far as I can see, most entailed an *increase* in cleaning work.[135] Daily care could not be neglected.[136] A scaled-down major housecleaning was required each week,[137] or spring and fall housecleaning were to be extended over several weeks instead of several days, to reduce the inconvenience to the family.[138] In other words, with the imperative to reconcile hygiene and harmony the housewife had to clean *all* the time—but discreetly and with a new kind of restraint.

The 'cleaning craze' was also defended by some. 'No household passion is more readily understood than the cleaning mania, and as unpleasant as it may be as it runs rampant and free, its results are a delightful reward for the effort,' Brunius argued in 1917. The ends justified the means. Once the toil was over, the family could gather in a clean, orderly, and fragrant home, the symbol of Homeyness in modern times:

> It is always with a feeling of respect, almost admiration, that I enter a room that has just undergone a thorough cleaning. I need not see it—I feel the cleanliness in the air upon my first breath and the impression is confirmed by the faint reflection of the chair legs in the floor, the clear and sharp contours of objects, the entire atmosphere of peace and tranquillity that engulfs you and gives you an unconscious feeling of well-being.

After all, Brunius concluded, it is on this 'practical and unimaginative' orderliness that the 'comfort of home, its fostering power, and its character as a haven from troubles stand or fall.'[139] In this ode to harmony in the well-ordered home, she was backed by a chorus of domestic writers from the 1870s onwards. With a variety of voices they sang the praises of the housewife, her constant vigilance, organizational skills, and practical domestic competence.

TECHNOLOGY AND CONSCIOUSNESS

The writers of household manuals and articles helped create a new definition of reality for women in turn-of-the century Sweden. Women were made conscious of new problems in the home, new ideals of hygiene and harmony were defined, as well as the right and proper means to reach them. Technology was not yet an important ally in the struggle for control of domestic disorder. Gradually, however, this began to change.

First, there were the iron bedsteads, new building materials and improved, smoke-less stoves, advocated by engineers and public health doctors. Then, urban homes were plugged into the public networks of power, water and waste disposal emerging from the late nineteenth century onwards.[140] By 1920, 70% of city households and about 50% of rural and small town ones had access to electricity. Around 70% of city households (but virtually no rural and small town ones) had piped water in their homes.[141] The stage was set for the introduction of domestic appliances. And the cultural insistence on order and cleanliness, analyzed here, had made households sensitive to the dangers of dust and dirt and responsive to the solutions that technology promised to give. New chemical wonders were introduced to 'control germs' and make the house spic and span. And many eagerly greeted the vacuum cleaner once it was made available in the early nineteenth century:

> There is a saviour from all this nuisance [of major housecleaning][Dr Henrik Berg claimed in a public lecture in 1905], and it is a company called *'The Vacuum cleaner'*. Ladies and gentlemen, you must have seen that steam-engine standing outside houses, huffing and puffing, with a long hose up to the apartment which sucks out the dust. I believe that 'The Vacuum cleaner'-method will replace major housecleaning in the future. I have made a firm decision to use one myself as soon as possible.[142]

A few years later the portable electric vacuum cleaner was invented and the cleaning task returned from the professionals of the vacuum cleaning companies to the women in the home. New ways of cleaning with the aid of machine technology had to be learnt:

> [It is] no easy task [Mrs Hein maintained in 1923] to handle a vacuum cleaner. It is much more convenient to whisk around a bit with a rag. But the utility of the vacuum cleaner is not that it is so easy, but that it is so good. One must exert oneself greatly and work quite slowly to utilize it to the fullest, but if this is done, the vacuum cleaner is effective beyond all praise... It is possible, if so desired for the sake of superstition, to wipe [the walls] afterward with a cloth.[143]

In the early 1940's, 89% of middle class households and 31% of working class households in the cities owned a vacuum cleaner; the figures for small towns and rural areas were much lower.[144] But the cost of electricity and appliances was prohibitive until the late 1950s. The vacuum cleaner was still not considered a tool 'of primary use' by some household writers and many people could not afford it.[145] In the boom years of the 1960s and early 70s, however, adoption of vacuum cleaners and washing machines became almost universal.[146]

But already in the 1940s and 50s, the *ideal* of the well-ordered home with the competent housewife in control had spread from the middle to the working class. Hygiene and comfort, rational housekeeping and technical competence was to characterize the 'modern housewife' in all kinds of homes.[147] Moreover, this was no longer only a private ideal for the woman in the home. It had become a public and political project of modernization, promoted, first, by women's organizations, then also by public authorities and industrial enterprises. Together, they made the home into a 'consumption junction'[148] for goods and appliances, advertisement and advise. What had started as the moral ambitions of doctors, engineers and 'experienced housewifes' in the late nineteenth century to change individual habits and concerns had, thus, gradually evolved into a cultural agenda for modern, technologized living in a consumption society.

NOTES

1. The article is based on B. Berner, *Sakernas tillstånd. Kön, klass, teknisk expertis* (Stockholm: Carlssons), chapter 2.
2. See for ex. K. Johannisson, *Den mörka kontinenten. Kvinnan, medicinen och fin-de-siècle* (Stockholm: Norstedts, 1994), and the discussion in P. Gay, *The Cultivation of Hatred* (London: Fontana Press, 1995).
3. This later story is told in J-E. Hagberg, *Tekniken i kvinnornas händer* (Linköping, 1986); Berner. *Sakernas*. Ch.4; B. Lövgren *Hemarbete som politik* (Stockholm: Almqvist & Wiksell International, 1993).
4. I. Zethelius, *Rådgifvare för hemmet* (Stockholm, 1910), p. 5. Italics in original.
5. M. Douglas, *Purity and Danger* (London: Routledge and Kegan Paul, 1966).
6. For individual differences, see B. Martin, 'Mother wouldn't like it!; Housework as Magic' *Theory, Culture & Society* (1984) 2:19–36; M. Horsfield, *Biting the Dust. The Joys of Housework* (London: Fourth Estate, 1997).
7. C. Davidson, *A Woman's Work is Never Done. A History of housework in the British Isles 1650–1950* (London: Chatto & Windus, 1982).
8. J. Frykman, & O. Löfgren, *Den kultiverade människan* (Lund: Liber, 1979).
9. L. Davidoff, 'The rationalization of housework' in D. Leonard Barker & S. Allen (eds), *Dependence and Exploitation in Work and Marriage* (London & New York: Longman, 1976), p.127f.
10. S. Schama, *The Embarrassment of Riches* (London: Harper Collins, 1987), chapter 6.
11. Discussions of this development can be found, for Sweden in Frykman & Löfgren, *Den kultiverade;* for Australia, in K. M. Reiger, *The Disenchantment of the home. Modernizing the Australian family 1880–1940,* (Melbourne: Oxford University Press, 1985); for Denmark: L-H. Schmidt & J. E. Kristensen, *Lys, luft og renlighed. Den moderne social-hygiejnes fdsel* (Copenhagen: Akademisk forlag, 1986) and T. Vammen, Tinne, *Rent og urent, Hovedstadens piger og fruer 1880–1920* (Copenhagen, 1986); for England L. Davidoff & C. Hall, *Family Fortunes. Men and women of*

the English middle class 1780–1850, (London: Hutchinson, 1987); for United States: B. Ehrenreich & D. English *For Her Own Good. 150 Years of the Experts' Advice to Women* (London: Pluto Press, 1979). See also W. Rybczynski, *Home: A Short History of an Idea* (New York: Viking Penguin, 1986).

12   C. Reimer, *Hemmets bok* (Stockholm, 1909), p. 38.

13   A. Burman, *Illustrerad Hushålls-kalender. Handbok i praktisk hushållning* (Stockholm, 1875), p. 5.

14   Cf. Davidoff. 'The rationalisation'. p.130.

15   A. Roswall (ed) *Fråga mig! Handbok för hemmet* (Stockholm, 1904), p. 21.

16   E. Rodhe, *Undervisning i hushållsgöromål för skolbarn* (Gothenburg, 1894), p.14.

17   E. Tenow, Elsa, *Solidar. En lifsfråga för hemmen, I-III* (Stockholm, 1905), part II, p. 110.

18   G. Paulsson (and collaborators), *Svensk stad Del 1. Liv och stil i svenska städer under 1800–talet* (Stockholm: Bonniers, 1950); E. Stavenow-Hidemark, *Villabebyggelse i Sverige 1900–1925* (Lund: Berlingska, 1971); E. Stavenow-Hidemark, 'Hemmet som konstverk. Heminredning i teori och praktik på 1870- och 80–talen' *Fataburen* (1984), 129–148; K.Thörn, *En bostad för hemmet* (Umeå: Department of History of ideas, 1997). For international discussions, see A. Forty, Adrian, *Objects of Desire. Design and Society 1750–1980* (London: Thames and Hudson, 1986); Rybczynski. *Hemmet.*; P. Sparke, *As Long as it is Pink: The Sexual Politics of Taste* (London: Pandora, 1995).

19   See, for example, Veritas, *Hemmets rådgifvare* (Stockholm, 1900), p. 6; Reimer. *Hemmets bok,* p. 197. See Thörn. *En bostad.* pp. 153ff. for a discussion.

20   Mrs Laura G., *Oumbärlig rådgifvare för hvarje Hem* (Malmö, 5th ed. 1902), p. 3.

21   Anonymous, 'Ett kapitel som borde intressera varje kvinna' *Svenska husmodern* (1877), 1:2.

22   M. Langlet, *Husmodern i staden och på landet* (Stockholm, 1884), p. 821.

23   G. Kyle, 'Genrebilder av kvinnor. En studie i sekelskiftets borgerliga familjehierarkier', *Historisk Tidskrift* (1987), pp 39, 45.

24   Langlet. *Husmodern.* p. 15.

25   Mrs. Laura G. *Oumbärlig.* pp. 10ff.

26   A. Briggs, *Victorian Things* (London: Penguin, 1988), p. 220.

27   Mrs. Laura G. *Oumbärlig.* pp. 10ff; A. de Frese, *Den ordnande handen i hemmet,* (Stockholm, 1870), p. 8.

28   Langlet. *Husmodern.* pp. 13, 16. See also L. von Lagerström, *Hemmets rådgivare, Praktiska råd för hemmet och hushållet* (Gothenburg, 1924), p. 8.

29 Zethelius. *Rådgifvare.* p. 5; see also, for example, Veritas. *Hemmets.* p. 3.
30 G. Norden, Gertrud, *Styra och ställa* (Stockholm, 1924), p. 87.
31 H. Hagdahl, *Det bästa af allt! En nödvändig bok för hvarje hem* (Stockholm, 1885), passim.
32 Anonymous. 'Ett kapitel'. p. 3.
33 C. Brunius, *Sin egen tjänare. Husliga studier sommaren 1917* (Stockholm, 1917), p. 72.
34 L. Lagerstedt, *Kokbok för skolkök och enkla hem jämte korta anvisningar i huslig ekonomi* (Stockholm, 1894), p. 187.
35 E. Fletcher, 'Hygieniska råd och anvisningar', *Mitt hem*, (1905), p. 91.
36 R. D. in *Mitt hem* (1907), p. 33. The original information may have come from H. Öhrvall, *Renlighet och frisk luft* (Stockholm, 1900), p.8.
37 Lagerstedt. *Kokbok.* p. 186.
38 Möller. *Konsten.* p. 18.
39 M. Langlet, *Ett eget hem* (Stockholm, 1891), p. 108.
40 H. Kamke, Hanna, 'Städning', *Hemtrefnad* (1910), p. 301.
41 Anonymous, 'Rengöring', *Hemtrefnad* (1899), p. 247.
42 Norden. *Illustrerad.* p. 58.
43 Dixi, 'En rationel städning af våra boningsrum' *Idun* (1895), 84–85.
44 Norden. *Illustrerad.* p. 53; See also S. Nilsson, *Hushållslärans första grunder för skola och hem* (Stockholm, 1904), p. 65; Öhrwall. *Renlighet.* p. 18.
45 See the discusssion in Thörn. *En bostad.* pp. 37ff.
46 See e.g. E. Heyman, *Om luften i våra bostäder* (Stockholm, 1881); C. Wallis, *Hemmets hälsolära* (Stockholm, 1906); N. Lundberg, *Vägledning vid tillsyn över bostäders sundhet* (Stockholm, 1912); A. Christer-Nilsson, *Bostadens Hygien* (Karlskrona, 1926).
47 Langlet. *Husmodern.* p. 626; see also, for example, T. Holm, *Rengöring i hemmet* (Stockholm, 1921).
48 A. Corbin, *Le Miasme et la jonquille* (Paris: Flammarion, 1986).
49 Möller. *Konsten.* p. 21. Italics in original.
50 U. Graninger, *Från osynligt till synligt* (Stockholm: Carlssons, 1997).
51 H. Berg, 'Renlighet', in L. von Lagerström (ed), *Hemmets rådgivare. Praktiska råd för hemmet och hushållet* (Stockholm, 1924), p. 49f.
52 Elise A., 'Öppna fönstren!' *Hemtrefnad*, (1914), p. 11.
53 Fletcher. 'Hygieniska'. p. 91.
54 Norden. *Illustrerad.* p. 53.
55 L. I. Wahlman, 'En gård och dess trefnad', *Ord & Bild* (1902), p. 32.
56 See e.g. T. Thunberg, *Hälsolärans grunder* (Uppsala, 1893), pp. 6f; Öhrvall. *Renlighet.* p. 16; H. Berg, *Hygieniska strövtåg* (Stockholm, 1906), p. 127.
57 Rybczynski. *Home.* See also *Mitt hem* (1905), p. 91.
58 Möller. *Konsten.* p. 38.

59 Most clearly stated in Tenow. *Solidar.* For later analyses, see E. Palmblad, *Medicinen som samhällslära* (Gothenburg: Daidalos); K. Johannisson, Karin, 'Folkhälsa. Det svenska projektet från 1900 till 2:a världskriget' *Lychnos*, (1991), 139–195. For sanitary engineers in Sweden, see B. Sundin, *Den kupade handen. Människan och tekniken* (Stockholm: Carlssons, 1991), pp. 250ff; for the relationship between sanitary experts and housewives in England, see P. Williams, 'The Laws of Health: Women, Medicine and Sanitary Reform, 1850–1890,' in M. Benjamin (ed), *Science and Sensibility. Gender and Scientific Enquiry 1780–1945* (Basil Blackwell, 1991), 60–88.

60 See P. Aléx, *Den rationella konsumenten. KF som folkuppfostrare 1899–1939* (Stockholm: Symposion, 1994); Frykman & Löfgren. *Den kultiverade.* (1979); Johannisson. 'Folkhälsa'; Palmblad. *Medicinen.* 1989; Åkerman et al. *Den okända vardagen.*

61 de Frese. *Den ordnande.* p. 4.
62 Langlet. *Husmodern.* pp. 81ff.
63 Norden. *Illustrerad.* p. 72.
64 Möller. *Konsten.* p. 66.
65 Corbin. *Le miasme.* pp 189ff; C. Dyhouse, 'Mothers and Daughters in the Middle-Class Home, c. 1870–1914,' in J. Lewis (ed), *Labour and Love. Women's Experience of Home and Family, 1850–1914* (Oxford: Basil Blackwell, 1986), 27–47; P. Palmer, *Domesticity and Dirt. Housewives and Domestic Servants in the United States, 1920–1945* (Philadelphia: Temple University Press, 1989).
66 Johannisson. *Den mörka kontinenten.*
67 Norden. *Illustrerad.* p. 66.
68 Anonymous in *Mitt hem.* p. 149. The writer then added: 'If we consider, however, the unhealthy building system, the absence of bathrooms in our apartments which men have built, the untidy bakeries and slaughterhouses that men run and care for, all the dirt that escapes the eye of men's sanitation police, all the bans against spitting and pollution that apply only to men, well, then you have to stop and think again [whether men are much cleaner].'
69 Langlet. *Husmodern.* p. 27.
70 S. Melander, 'Renhet', *Hemtrefnad*, (1909), p. 95.
71 Möller. *Konsten.* p. 33.
72 Douglas. *Purity and Danger.* p. 48.
73 Öhrvall. *Renlighet.* p.16.
74 Langlet. *Husmodern.* p. 9.
75 de Frese. *Den ordnande.* p. 2.
76 Roswall. *Fråga mig!* p. 60.
77 Ibid. p. 63. Italics in original.
78 Reimer. *Hemmets bok.* p. 194. Italics in original.
79 Mrs. Laura G. *Oumbärlig.* p. 4.

80 Norden. *Illustrerad.* p. 63.
81 See, e.g. T. Kjellberg, 'Våra sofrum', *Tidskrift för hemmet*, (1907), 43–45; Melander. 'Ett sofrum'.
82 Norden. *Illustrerad.* p. 63.
83 Elise A. 'Öppna fönstren'. p. 118.
84 E. Stavenow-Hidemark, 'Hygienismen kring sekelskiftet', *Fataburen* (1970), 47–54; Paulsson. *Svensk stad.*
85 Paulsson. *Svensk stad.* p. 28f.
86 Berner. *Sakernas.* chapter 4; Thörn. *En bostad.* passim.
87 E. G. Folcker, 'Sofrummet' *Idun* (1898), 388–389; *Mitt hem* (1909), 127–128.
88 L. Fittinghoff, 'Sängkammarinteriörer', *Idun* (1901), 299–302.
89 Mrs. Laura G. *Oumbärlig.* p. 7. Italics mine.
90 See note 3.
91 Burman. *Illustrerad.* p. 8.
92 Veritas. *Hemmets.* p. 6.
93 Brunius. *Sin egen.* pp. 35ff.
94 See e.g. Roswall. *Fråga mig!* pp. 33ff; Möller. *Konsten.* pp. 34ff.
95 Langlet. *Husmodern.* p. 9.
96 Ibid. p. 78.
97 Veritas. *Hemmets.* p. 3.
98 Langlet. *Husmodern.* p. 15.
99 Roswall. *Fråga mig!* p. 22.
100 Mrs. Laura G. *Oumbärlig.* p. 12.
101 Anonymous in *Svenska husmodern* (1877), p. 3. Italics in original.
102 Veritas. *Hemmets.* p. 7.
103 J. Burow, *En moders ord till fosterlandets döttrar*, (Stockholm, 1885), p. 148.
104 B. Gejvall, *1800–talets Stockholmsbostad* (City of Stockholm, 1967 (1954)), p. 228.
105 Folcker. 'Sofrummet'. p. 388f.
106 Kamke. 'Städning'. p. 302.
107 Roswall. *Fråga mig!* pp. 60, 68.
108 Ibid.
109 Norden. *Illustrerad.* p. 69; The musical instrument example is from Anonymous, 'Stor rengöring', *Hemmet. Läsning för ung och gammal* (1890), pp. 506, 508.
110 I. Malmström (ed), *Hem och hushåll* (Malmö, 1941), p. 58.
111 Zethelius. *Rådgifvare.* p. 5; Veritas. *Hemmets.* p. 3.
112 Literary examples can be found in C. Hardyment, *From Mangle to Microwave. The Mechanization of Household Work* (Cambridge: Polity Press, 1988); Martin. 'Mother wouldn't like it'; Horsfield. *Biting the Dust.*
113 de Frese. *Den ordnande.* p. 1; Tant Malla, *Den unga frun. Handledning vid de första osäkra stegen som husmoder* (Helsingfors, 1898), p. 1.

114 H. Haglund, 'En tidsenlig uppfostran av husmödrar,' *Dagny* (1888), 95–107.
115 See e.g. Adelsköld et al. *Hemmets.* p. 198 and E. Kleen, *Gwens bok för hemmet* (Stockholm, 1907), pp. 58ff, for detailed instructions on the proper division of labor among various numbers and types of servants.
116 Veritas. *Hemmets.* p. 8.
117 Langlet. *Husmodern.* p. 626f.
118 A. Lundberg, 'Några ord om behofvet af kurser i städningskonst,' *Hemtrefnad* (1908), p. 120.
119 Reimer. *Hemmets bok.*, p. 227.
120 Norden. *Illustrerad.*, pp. 69f.
121 Brunius. *Sin egen.*, pp. 34f.
122 *Handledning i husliga göromål* (Stockholm, 1912), p. 29.
123 Burman. *Illustrerad.*, p. 7.
124 Anonymous, *Vi och vårt* (1912), p. 215.
125 Reimer. *Hemmets.*, p. 52.
126 Möller. *Konsten.* p. 88.
127 Norden. *Illustrerad.*, pp. 51f.
128 Mrs. Laura G. *Oumbärlig.*, p. 7.
129 Reimer. *Hemmets.*, p. 183. Italics in original.
130 Holm. *Rengöring.*, p. 8.
131 I. Zethelius, *Tidskrift för hemmet, dess sysslor och intressen*, (1907), 1:2.
132 *Mitt hem* (1906), p. 4.
133 See H. Berg, *Hygieniska ströftåg* (Stockholm, 1910), pp. 29ff.
134 Tenow. *Solidar.* Part II, pp. 147f.
135 There were also pleas for simplified cleaning; see for example the articles on 'Domesticity in Step with the Times' in *Mitt hem* (1906), pp. 3–4, and 'Unnecessary Dusting' in *Vi och Vårt* (1911), p. 402.
136 Mrs. Hein, *Sin egen jungfru* (Stockholm, 1923), p. 93.
137 Langlet. *Husmodern.*, p. 626; Kleen. *Gwens bok.* p. 60.
138 Holm. *Rengöring.*, pp. 45f.
139 Brunius. *Sin egen.*, pp. 72f.
140 T. P. Hughes, *Networks of Power. Electrification in Western Society, 1880–1930* (Baltimore: Johns Hopkins University Press, 1983); A. Kaijser, *Stadens ljus* (Malmö: Liber förlag, 1986); A Kaijser, *I fädrens spår...* (Stockholm: Carlsson, 1994).
141 A. Nyberg, *Tekniken—kvinnornas befriare?* (Linköping: Tema T, 1989).
142 Berg. *Hygieniska* (1906), p. 128.
143 *Sin egen jungfru.* p. 95.
144 Nyberg. *Tekniken.*, p. 90.
145 B. Holme *Städning* (Stockholm, 1956), p. 11; K. Henrikson, *Glatt och lätt hemarbete* (Stockholm, 1957), p. 35. The domestic pride and competence that the handbook writers helped to instore was, paradoxically, also an obstacle to the wide-spread adoption of some household tech-

nology, notably the washing machine. Its mechanical motion was, into the late 1940s, thought to destroy the fabric of the clothes. See G. Kjellman. *Från bykbalja till tvättmaskin* (Linköping: Tema T. Arbetsnotat, 1989), pp 40ff.
146 Nyberg. *Tekniken.*
147 See the analysis in Berner. *Sakernas.* chapter 4.
148 R. Schwarz Cowan, 'The Consumption Junction: A Proposal for Research Strategies in the Sociology of Technology', in W. E. Bijker, T. P. Hughes, and T. Pinch (eds), *The Social Construction of Technological Systems* (Cambridge & London: The MIT Press, 1987), pp. 261–280.

CHAPTER 5
How the Motor Car Conquered the Road

Catherine Bertho Lavenir

When the automobile appeared in Europe after 1895, and when tourists left the railroad and began to drive to small villages and remote areas, a new culture of control had to be built. As a new means of transportation, the automobile had to find a place within the existing networks. The use of the road had to be renegotiated between its former users, such as pedestrians, cartmen and riders. The road network itself had to be thought anew and constraints such as speed limits had to be fixed. This could be done in different ways. Laws and regulations could set official standards. Technical devices such as a speed indicator, a horn or lights could be incorporated in the car itself. But rules of conduct also had to be incorporated by drivers themselves in order to structure their behavior. Cultural attitudes, which differed according to the social position of drivers, influenced what one could and should do.

A culture of control had, in this case, many meanings. Regulations represented not only an expression of the power of the State, an external constraint which imposed itself on drivers, they also expressed political and cultural agreements which were not fully settled before the 1920s. The adoption of legal rules for car traffic was accompanied by a process of negotiation which was characteristic of the political culture of the country. Elected bodies, public administrators and justice officials competed for the elaboration of formal legal decisions. Lobbies and social groups strove hard to make their interests respected, especially within the new set of rules concerning the use of the road. Drivers, too, had to incorporate in their behavior such new obligations. The corpus of regulations was not only an abstract set of texts but also a set of manners, values, and moral constraints, all of them infiltrating into the intimate life and private sphere of contemporaries' existence. Thus, the culture of control connected with the rise of 'automobility' was determined by political choices as well as by social arbitration, and was related to the habitus of members of different social groups,

114 *Cultures of Control*

the representation they had of themselves, their gender status, and other elements intimately connected with cultural behavior.

With the development of automobile tourism, another dimension of social control appeared. Manufacturers had to create a market for the use of cars for leisure. When traveling, drivers and tourists wanted to find good inns, clean rooms, 'typical' cooking, and also museums, historical buildings, and untouched landscapes. Automobile clubs and tourist associations wanted to adapt rural life to their needs: they set to convince inn-keepers to share their hygienic concerns and went to war against peasants and entrepreneurs who threatened to spoil their favorite scenery when they built factories or pulled down old houses. Therefore, they exercised a symbolic form of violence on social groups who didn't share their aims and values – such as an interest in old buildings or a taste for cleanliness. This symbolic violence, which led to forms of social control, has to be present in the study of the construction of the whole culture of control linked to the diffusion of the automobile.

This paper shows how the appearance of the automobile produced a discourse in which technical necessities, moral standards and prescriptions for 'correct' use were strongly intertwined. We will first see how technical standards for a road adapted to the automobile were elaborated; how the urban bourgeoisie imposed the idea that the correct way of driving was primarily a question of personal behavior and moral values, not of technical choices or public regulation. We will then examine how cultural standards prescribed by the tourist industry led to a soft but real control of the old way of life in rural areas. In the conclusion we will try to asses the exact weight of the State in this process. In France the political tradition is that government, through its administration, should have a decisive bearing upon technical and legal issues. In the development of the culture of control linked to the automobile, it seems that other forces were directly at play. Users, through their associations, and the manufacturers themselves were all important. They invented new technical devices, and they devised a novel pattern for cultural control.

BEGINNINGS

The first Paris-Rouen-Paris race inaugurated the automobile era in 1894. At the time, practically everything concerning the modern motorcar had yet to be invented. The very concept was uncertain. It is not surprising then that although Count de Dion crossed the finishing line first, he wasn't granted the prize, his vehicle being the bizarre combination of a sturdy steam-tractor and a light, four-wheeled carriage. This car was thus fell outside the specifications set down for the race. The prize was in fact divided between Messieurs Panhard and Levassor on the one hand, and the Renault brothers on the other. The two teams arrived driving vehicles that were duly 'horseless, safe, easily managed by the travelers, and of little cost on the road.'[1] This was not an overly strict definition for a car, but it was much simpler than any of the design innovations being made in the automobile world at the time.[2] This was an immensely complex affair; turning out new models, and launching them abroad was by no means a straight forward issue. One must not imagine lines of lovely cars rolling smoothly out of the workshop on to a market eager to receive them. The automobile was an entirely new technical object, a new means of transportation, with unprecedented requirements, and as such it found itself at the core of an emerging social-technical system that can be imagined as a set of concentric rings.

The car, as an object, was a world in itself. It wasn't only an engine. There were the chassis, the body, the steering, braking, lighting systems to be worked upon. And the inside space, seats, windows, trunks to fit to all purposes. The automobile was a complex challenge. Adapting the technical system of the road to the newcomer was just as exacting. This process included improving the roads themselves, the structure and surface, but also the network, and signals and other devices contrived to regulate the intersection with other systems, notably the tramway and railway networks. Some satisfactory agreement also had to be devised for all the road users, an agreement acceptable to all. To that effect, the happy owners of a De Dion light car or a Darracq Tonneau were to be presented with a code of road manners. In their day, the railway companies had found it efficient to inspire their customers with a longing to travel and the positive knowledge they would come back home safe and sound. Car manufacturers followed the same trend: French motorists must be taught how to drive and how to behave. They must learn that frightening horses is dangerous, and running over pedestrians simply not done. On top of all that, a system of references for car travelers had to be evolved. Stray drivers ought to be able to rely upon road maps and sign-posts, hotels and inns must be conveniently listed up with their deserved number of stars, according to a system borrowed from the nineteenth century Swiss guides.[3] In this way, today's travelers would be sure to enjoy board and lodging up to the standards of urban comfort.

THE WRITTEN WORD AND THE AUTOMOBILE

The written word found its inescapable place within such changes. Racing competitions, and notably the long mileage races along open roads played a prominent role in the settling of technical choices, but the results had to be relayed by print of all sorts. Quite early in the process, specialized papers and magazines came to light, with two outstanding titles. One was *L'Auto*, a daily paper founded in 1900, which claimed it wrote 'in defense of the interests of the car and cycle.' The other was *La vie automobile (Automobile Life)*, a weekly paper supported by Dunod publishers, which had a definite technical bent and was always on the lookout for novelty. Its leading man was the indomitable Baudry de Saunier, who stood as the beacon of the car world at the turn of the century. Newspapers with large audiences were keen on feeding their readers with the spectacular, and they organized competitions they could turn into big shows. *Le Journal,* a newspaper founded by Le Tellier, who had been a civil engineer and who was of Belgian origin, gave extensive publicity to such competitions. Le Tellier himself organized the car races from 1897 to 1899. At about the same time, associations brought car owners together, and quickly became enthusiastic partners of the car manufacturers, over whom they kept a watchful eye.

The Automobile Club of France was founded as early as 1895. Sister associations soon appeared in the province. The Touring Club, which had originally been devoted to the bicycle and cycling, switched over to the car and motoring even before 1900. *The Ligue des chauffeurs* (Drivers League) supported their lobbying action. Industrialists cleverly supported the associations. The ongoing debate between manufacturers and consumers would reach a peak on the occasion of the Salon de l'Automobile in Paris. Devised on the model of painting salons held periodically, the Salon de l'Automobile also linked the car industry to the glamour and marketing methods associated with the established couturier fashion shows. The larger associations would at times join together in order to promote some topical issue—the 'Antidust League' is an example. But their main concern was elsewhere. They functioned as pressure groups on the political process, with a view to obtaining technical policies and regulations suited to their own choices and needs.

Indeed, the development of the automobile was a process full of conflict. Before 1914, in France, owning a car remained the privilege of men with a noble family background, or belonging to the uppermost layers of the urban 'bourgeoisie,' or professional men bent on novelty who used their cars as tools. For example, together with true blood noblemen, doctors, pharmacists and traders made up the bulk of the Armorican Automobile Club at the turn of the century.[4] The strategy of French industrialists then consisted in maintaining the automobile within the high luxury goods category. They made cars ever more powerful and heavy, the price of which didn't come down in any significant way. They developed a habit of selling the chassis separately, and of having it equipped by a body builder, which entailed higher cost. The price of an automobile was still up 13,000–15,000 Francs in 1914. There were under 100,000 cars in 1913 in France, versus over a million in the U.S. Their owners were rich, and they were men of power. This, paradoxically, did not help them in their conquest of the road.

CONFLICTS ON THE ROADWAY

In this context, the road, and even more the right to use the road at will, were to become hotly disputed issues. Contemporaries felt that with the development of the railways, the road network had been progressively abandoned, and consequently gradually restored to its former rural users. Galloping carters and mail-coaches had vanished. Vanquished by its opponent, the railways, the last relay had closed down. Since 1879 and the Freycinet Plan, the authorities of the Republic had been intent on multiplying local railways lines, in direct competition with coaching entrepreneurs. Only small-scale traffic was left: carters on their way to the neighboring market town, or cart-drivers making for the fields. This type of traffic grew during the 1880s, as did road repair efforts. Roadside people had regained possession of the way. In villages, children and dogs could be seen anew playing in the dust. Cattle and pigs enjoyed a right of pasture along the verge of the quiet roads, while workmen in search of jobs, tinkers and vagrants could be seen going their way along the road.[5] This balance was only temporary. The arrival of the automobile disturbed the situation.

Paradoxically, conflict arose first over the bicycle, not the automobile. This machine, which we now perceive as so harmless, truly caused people to panic when it was introduced in the 1880s. The 'vélocemen'—that is how they called themselves in French—initially all came from the upper bourgeoisie. They congregated into pressure groups throughout Europe, aiming to take possession of the road and submit it to their wills and whims. Acting together with the Touring Clubs that multiplied all over the old continent after 1890, they put all their social talent at the service of their ambitions. They first aimed at adapting the techniques of the road to their specific needs, denouncing devices then in current use for road-repair. The jagged stones laid out in sheets would rip up their tires, loose chippings would cause terrific falls. Thus, those highly qualified engineers, barristers or attorneys at law, together with men of letters, who were typical members of the Cycling Associations, produced innumerable technical articles to prescribe new devices. They had no qualms in making known the names of the civil engineers reluctant to follow suit. The Belgian Touring Club turned itself into a survey organization. Its members were to report to the central office their observations identifying road stretches in need of repair. They also financed, with their own money, the making of cycling lanes, and sued the villains guilty of ruining them. Better still, the cyclists from Brussels bought steam-rollers and presented the Administration with them, along with the firm advice to put them to use.[6] In Italy, the dynamic 'bourgeois' of Milan and Turin, who were the majority in the first cycling associations, went to war against any regulation limiting their freedom. They set on quibbling about any point, like the shape of the lamp they were compelled to carry, or the size of the duty plate for their machine, or against the rules applying to carts or tramcars.[7] In France, The Touring Club, which numbered 40,000 members as early as 1896, behaved similarly. With a famous lawyer as their chairman, the club's members strove to control the emerging jurisprudence pertaining to road matters. All subscribers were requested to forward to the legal matters committee any judgement passed throughout the country about so-called infractions of the road regulations, with a view to annulling any unfavorable decisions. Besides, fashionable Parisian velocemen were all for putting to good use the stylistic resources they mastered as men of education. They delivered papers meant to disqualify the rural mayors

118 *Cultures of Control*

or thick-headed chiefs of police daring enough to stand in their way. Incredible venom and contempt were in order. In 1898, Léon Petit, a prominent member of the French Touring Club wrote:

> A greasy, bloated, pot-bellied 'gentleman', the mayor of E. will not like a bicycle for the reason an elephant will scorn a harp. A stubborn and slow-witted tardigrad*, the said mayor will hate cyclists for the very reason toads hate butterflies they can't catch. His thick rustic mind has long been planning cruel retaliation upon those city men daring enough to cycle through his estate without asking for his permission.[8]

(*a kind of animal. From the Latin tardis – slow – and gradi – to walk.)

ADAPTING THE ROADS TO THE AUTOMOBILE

When the motorcar entered the picture, the struggle for the road followed exactly the same pattern, involving the same contestants. Car owners were even more wealthy than the cyclists of early days, they were more than well connected, and they knew how to use their social weight to turn the scale and secure the road for themselves. The first conflicts were technical. The old road had been adapted to the horse. Distributed according to an administrative organization which was typical of the French, the road network followed a hierarchy with the national roads at the top—their very name a symbol of their importance—down to the local small roads which were within the jurisdiction of mayors, assisted by rural policemen who were constables of a sort. From a technical point of view, roads were in the care of the Ponts et Chaussees (Bridges and Roads) civil engineers, most of them from the most famous scientific school in France. These gentlemen all shared a remarkable corporate feeling. They ruled over a host of minor agents, road surveyors and menders, and they elaborated a technical policy in agreement with a comprehensive social and political vision. The doctrine of Saint Simon was influential in these circles, and there was a widespread notion that multiplying roads was bound to favor communication between men, and lend coherence to the social body.[9] Quite naturally, such men were convinced that decisions in their field of action must be taken by themselves as they were far better qualified than political men, not to mention ordinary citizen.

At the time, the structure of the road was ill-adapted to the car, even when 30 km/h was about the limit. Small unstable cars and long chassis with uncertain steering would only too often be drawn to the side ditches by over cambered roads. Downhill slopes were too steep for the braking capacity of cars, and they would frequently come to a right-angle turning or a narrow bridge. Tunnels plunging suddenly under railway lines were another type of mortal trap. By law, level-crossings were permanently closed on the smaller roads, and they would come into view like so many obstacles around the corner on country by-roads. Cartmen had suffered from the very same inconvenience, but they did not go so fast, and above all they could hardly make their protest heard, socially speaking.

Not so with the first motorists who engaged in a methodical battle to get the road technically adapted to their needs. The two higher bodies of civil engineers entered a grim contest. Old boys from L'Ecole centrale were very numerous in the automobile industry. They stood against old boys from Polytechnique, overwhelmingly present at the control posts of the Road Network Administration. They argued about the gradient of the road cross-section, about possible ways to suppress the dust, or about the durability of bituminous or other surfaces. Technical reviews and specialized papers were forums for furious exchanges, which found an echo in popular newspapers. Motorists Associations with Automobile Clubs and Touring Clubs in the lead followed a double policy. For one thing, they financed a handsome part of the necessary changes. The Touring Club had the four basic signals designed and installed. They stood for 'open gutter, crossing, bend and level crossing.' They were officially adopted in 1908. Subscribers also financed special equipment, like the net destined to hold back cars which had made a habit of jumping over the parapet bordering the La Turbie road. More generally the associations, with the help of industrialists like Michelin undertook the education of the Bridges and Roads engineers. Both the AC and the TC pursued a systematic policy of cooperation with the Road Network Administration. They had tests for bituminous and other surfaces made, studies of the gradient of curves sketched out, standard processes for resurfacing the road devised. All this was duly addressed to the Minister of Public Work. It was difficult for any minister to ignore an association like the Touring Club, which now numbered about 100,000, or a pressure group as powerful as the Automobile Club. Thus, in 1908 Barthou, the minister in office, informed his interlocutors that their message had been fully taken into consideration.

The question of markers and signals was also a bone of contention. The old milestones erected on the side of the road used to bear references to the path as it was listed by the administrative system. This practice was perhaps useful for road repairers, but totally useless for stray motorists. All the more so as the numbering system was different for every *département*, i.e., local administrative geographic area, in total indifference to the needs of drivers travelling long distances. The pressure groups insisted that the kilometer stones should face the oncoming traffic and they were satisfied when the stones were turned about, and the number system modified.[10] The administration could not but accept that the Michelin firm, together with the motorists' associations, should finance and install road-signs adapted to automobile traffic. The Paris-Trouville road was the first to be fully equipped. Two important congresses were held in Paris in 1908 and 1909. They saw the consecration of the joint technical management of the road and they gave it an international dimension. The concerned administrations adopted a number of innovating devices proposed by the associations. With a keen understanding of things, these associations also took care of winning over to their side the more modest agents of the Bridge and Roads administration. In Italy as well as in France significant sums were given to road-menders through

120 *Cultures of Control*

relief funds, and the French Touring Club made a point of presenting each and every road-mender with an oil-skin jacket.

FROM DEAD DOGS TO A HIGHWAY CODE

The cohabitation of the car drivers with other users of the road brought about more dramatic problems. Motorcars would suddenly loom amid cloud of dust, to the terrifying blasting of the horn, this striking universal panic; they would run over children and dogs, upturn carts, and squash cyclists flat. To be accurate, the number of accidents caused by cars was relatively small (119 casualties in 1901). It was far under the number caused by horses or trains. Yet car crashes made a vivid impression on the collective mind, and the tension grew between the supporters of the drivers' freedom and the leagues intent on the protection of pedestrians and the like.[11] The early 'society motorists' ostentatiously ignored their opponents. These pioneers were fully aware of the pleasure they were taking in driving powerful cars. The feeling of power and freedom associated with motoring was in accordance with the values of the new ruling classes and it was too precious a pleasure for them to renounce it readily. That is why car owners in 1900 called up all the resources accruing to their social superiority to try and deprive all other users of any legitimacy.

Figure 3 Here de Saunier (the author of the driving lessons book here quoted) attributes responsibility for a dangerous situation to the steeply graded road surface. The carter's refusal to move over is excused. If there is an accident, the road administration will be held responsible because it is their engineers who constructed a road unsuited to its users. The author, moreover, invites the administration to rectify its error. Private Collection. Photograph: André.

How the Motor Car Conquered the Road 121

Figure 4 The car driver is endangered as a result of the railroad employee's actions, not his own. The dramatic composition recalls newspaper illustrations of front page stories at the turn of the century. Private Collection. Photograph: André.

Figure 5 The reader, conceived as a spectator, is confronted with a situation leading directly to an accident. Drivers are made personally responsible for their fate to the degree that they fail to observe the prescribed rules of self discipline: slow down and signal. Private Collection. Photograph: André.

122 *Cultures of Control*

Figure 6 Other users of the road create dangerous situations by failing to take into account the special characteristics of the automobile. In the commentary, the cart, the peasant and the animal are equally responsible for the drama that is unfolding. Private Collection. Photograph: André.

Figure 7 Other road users (pedestrians, cyclists, carters, animals and children) are shown here as the source of potential danger to themselves and to the car driver from the point of view of a spectator aware of the entire situation. Here, the caption places the entire responsibility on the car driver, asking that he regulate his conduct in the interests of all because other users neither know how nor wish to comport themselves in a reasonable way. Private Collection. Photograph: André.

Their authority was strongly opposed by other members of the social body. Public opinion was ill-disposed towards the automobile. Certain events supported their views. For example, in 1903 during the Paris-Madrid race, accidents occurred in which drivers and onlookers, among them children, were killed when cars went out of control and plunged into the crowd. Such events strengthened a feeling—largely inspired by the popular press—that motorists were uncaring 'bourgeois', dangerous to their fellow citizens. Such hostility engendered direct consequences. In the U.S., countrymen would dig up road gutters to trap cars.[12] In France, a local justice had to be persuaded that firing at cars to frighten travelers was indeed an illegitimate pastime and an offence. Associations in the defense of pedestrians, with lawyers and other persons of standing, were founded, proving that hostile feelings were developing within the very ranks of the urban bourgeoisie. Motorists now felt they were besieged and persecuted, the victims of small-time policemen, aggressive dogs and slow-moving pedestrians.

Things were soon to calm down. The representatives of the automobile associations realized that if they wanted the car to assert its place in social practice, a more conciliatory attitude was in order, along with negotiation. They successfully imposed their will upon public authorities who had neither the political ability nor the strategic will necessary to counter the claims for freedom expressed by the motorists, with the solid backing of industrialists. The conflicts coalesced around the speed limit issue, and worse still around the legal capacity of rural mayors. A national decision, passed in 1893, mentioned 30 km/h in open country, 20 km/h in built-up areas. And it gave the local elected bodies (the mayors) full powers to lay down the rules regulating traffic throughout their territory, that is of the commune. Initially, some really overdid it. Automobiles were forced to crawl at a pedestrian pace through the villages, with an obligation to hoot to claim the right of way at every cross-road; or, on the contrary, they were strictly forbidden to use the horn at all. Prefects, representing the central power of the state, were called to the rescue to nullify decrees inhibiting the flow of traffic. Ruling from their provincial bench, police court magistrates were the ultimate arbiters. Policemen's reports all landed on their desk, and a pattern of jurisprudence began to emerge in their decisions. The motorists' associations then decided on systematic action. Their disputed claims committees set about collecting each and every judgement passed throughout France, pointing out to discrepancies and absurd conclusions. When necessary, the committees pleaded directly for their members. The issue was such a tangle that non-parliamentary committees busily began to write up a highway code. Neither the 'official' nor the 'non official' groups succeeded in elaborating a single set of rules acceptable by all concerned. Nor were they able to have their decisions implemented. Motorists wouldn't hear of a general speed limit, mayors continued to turn out innumerable local decisions and decrees. There was endless wrangling over such fine points as the definition of a 'built up area', or could a constable legally report on a motorist caught 'on the wing' as it were.

Behind these conflicts there was another question: who had a right to be on the road, and to what purpose? The speed issue revealed deep differences in the perception of time and space according to social groups, but the pressure motorists exerted was such that within a few years, all uses and ways that were not their own were ruled illegitimate. The 'killed dogs' column grew larger. A judgement passed in Sorgues was quite definite:

> Citizens are fully entitled to let their dogs wander about roads, but this is at their own risk, dogs being extremely unpredictable animals. As such, their presence on the road endangers traffic.[13]

The right of cattle lumbering along or of herds grazing on the verges was also questioned. A decision from the prefect of Morbihan permitted pigs to graze on the edge of by-roads. This proved to be dynamite. The TCF disputed claims department immediately engaged in collecting all the various texts produced about 'free grazing' along highways. Decisions dating back to Turgot (eighteenth century) were quoted. This was the beginning of a new offensive aimed at cattle and cattle drivers. The latter were to be subjected to entirely new restraints, such as carrying a lamp at night. The very presence of agricultural carts was questioned, as well as the driving methods of cartmen on their way to the next village. These hitherto undisputed uses of the rural road were now branded as 'dangerous'. Their existence was made doubly illegitimate. First, they were a material hindrance to those going far and fast. Second, they were a moral nuisance to the same. The joint circulation of horse-drawn carts and motorcars along the lesser highways took on a cultural as well as a technical aspect. For ages, cart drivers had been used to driving right in the middle of the road, thus barring the way. Very often, such drivers would lack vigilance. They left everything to the horse, which was used to the routine.

It is true that the reports written by pioneers in motoring all mention the decided lack of goodwill shown by the cart owners. They would not allow overtaking, and they would stick to their ingrained habit of making unexpected turns or surging forth onto the road. These and similar reactions were the symptoms of a perception of time and space proper to a rural way of life. In the country, they circulated at a pace in accordance with that of agricultural activities. These called for continuity and endurance, they rarely called for speed. The distance covered rarely exceeded that of familiar territory, so that the desires or projects of car drivers motoring through France from Paris to Biarritz or Nice were totally inconceivable to most country people. Clashes were all the more violent for it. The automobile reporters of the time offered accounts of the lack of understanding that from their point of view opposed two systems of moral values: one that of the future; the other that of the cart drivers, entirely archaic in character.

> The automobile can be seen speeding along, robust, ardent, full of vigour, full of health, alive with the laughter of strong beautiful young people ... But fate lies in wait at the cross-road. Plunged deep in a brutish sleep, covered by a tarpaulin, here comes

the miller, unconscious and unaware. Here comes death himself, stupid death across the way...[14]

In their fight against horse-drawn carts, the Automobile Associations scored some nice points. They were successful in having carts and herds circulating at night compelled to carry lamps. Yet the same attempt aimed at agricultural carts proved a total failure.

SELF CONTROL AND GOOD MANNERS

Having won the battle to have carts and herds held in check, motorists had to fight for their right to drive without restrictions. A temporary compromise was found. It has already been said that public authorities were unable to get a general limitation on speed accepted. On the other hand, the Associations realized they must now negotiate. They succeeded in obtaining minimal regulation where driving was concerned, stressing the fact that safe driving depended on a sound 'education.' And indeed, official control bore but lightly on new motorists. An administrative decision, taken in 1896, made it obligator for drivers to have a 'grey card' (connected with duties—meaning taxes) and a 'pink card', that is the driving licence. The issue of learning was long left unsettled, however. Car traders generally took on this responsibility up until 1917, and the creation of the first driving schools. It was freely admitted that the 'fresher' had no experience or skill whatever.[15] The automobile lobby managed to have the issue suitably minimised, to the greater benefit of the car trade. That is how the new drivers' ability came to be assessed from a cultural standpoint. According to an intellectual construction in complete coherence with the values of the French republican bourgeoisie, the Automobile Club linked together the concepts of freedom and responsibility. Any driver was deemed responsible for both his speed and the accidents that might happen through his driving. So it was imperative that he learn to control himself. The automobile magazines were now rivaling each other to define a code of manners for motoring. The general idea was that drivers must be willing to accept limits to their own personal urges and desires. They laid the key to their own safety and possible coexistence with other road users. Baudry de Saunier, the pope of automobile philosophy, had a new edition of his book *The Art of Good Car Driving* printed. The magazine *Automobile Practice* (*La pratique automobile*) dated July 7, 1907 published an article attacking the 'road-dogs' whose horns, wildly honked to intimidate all around them, would give rise to feelings of 'fright and hatred among lesser people.' The title of an article by Baudry de Saunier boldly warned 'One must not squash them flat' or in the new polite terms: 'Thou shall not run them down' ('On n'écrase pas!').

In a more general way, there developed a system in which road safety was not enacted through regulation but through cultural attitudes. This supposed the constraints were turned into accepted attitudes, though this kind of cultural control

126 *Cultures of Control*

also had clear social dimensions, as we can read in a book written by Doctor Bommier, who made dubious comparisons between sexual behavior and 'the way of driving'. He wrote in the Touring Club Review that men issued from the ranks of ordinary people were as brutal with their car as with their women. On the contrary, well-educated men were delicate. Which was the good driver? 'It is the man with the discreet foot, who smoothly releases the pedal, who changes his speed at the very psychological moment after he has got the motor up and running...'[16]

Trusting in the efficiency of the cultural control of behavior, in the bourgeois—and male—sphere, the authorities accepted not to try and lay down rules based on technical specifications. For instance, maximum speed was not assessed in relation to the roads and the cars' characteristics. In the highway code that was established in 1921 and 1922, no speed limit was specified along French roads. This went for England too. It was the driver's task to decide upon the speed of his car. Conversely, he was held responsible for accidents. This proceeded from a decision initiated by a member of Parliament in 1901. As a consequence of this, no unfortunate event happening on the road would be traced to technical and rational causes. They had to be the result of some indefinite malfunctioning, some mistake or error of judgement. In other words, it was just fate.

CULTURAL CONTROL

The culture of control associated with the development of the automobile also had a cultural dimension. Car travelers soon discovered that the countryside was not adapted to their needs and desires. The inns were dirty, the cooking disappointing, and beautiful landscapes were endangered by uncontrolled modernization. So they tried to change this and to adapt reality to their desire, imposing a subtle but effective control on the rural way of life.

In the first year of the new century, the French rubber manufacturer Michelin printed the first *Guide Michelin*. The text was devoted helping the poor car driver, trying to make his way through a hostile countryside lacking gas stations, garages, comfortable inns, welcoming hotels or helpful natives. Reading the guide, one can see that the chief problem seemed to lie in the inability of the small country inns and hotels to meet the needs, ways of life and hygienic concerns of the urban bourgeoisie. This situation, however, was not new: in 1861, the *Guide Joanne* warned its readers:

> In some country towns, radical reforms are necessary. The case is pressing. Because of a lack of suitable hotels, the better part of France is closed off to women who wish to travel the country. It would be necessary to demolish everything and rebuild it, and even then the result would not begin to be satisfactory.[17]

The problem lay in the difference between the traveler's expectations and the ordinary behavior of pub and inn owners: 'It would not be enough to substitute an inn with a [modern] hotel or a pub with a [clean] restaurant, if you did not, at the same time, change the habits, morals and manners of the people in charge of these new hotels. How can you make people clean when they don't even know what cleanliness is?' In 1896, the French Touring Club wrote about a small inn in Corsica: 'Marinca... Miss Anne Marie Paoli offers bed and breakfast. House clean but room small and inhospitable, breakfast small, awful and very expensive (for the country)... Impossible to get water, no lavatories'.[18]

The *Guide Michelin* and the Touring Club began to try to adapt these deficient inns and hotels to the needs of tourists travelling by car. The reason for the involvement of auto manufacturers like the Clermont-Ferrand firm is clear: if you want people to buy and use cars, then they will need nice places to go, monuments to visit, agreeable inns for breakfast, comfortable beds to sleep in, as well as intelligible road signs and maps for finding their way. These things were vital for the new car industry. Public officials were also interested in the economic development of the countryside. Rural areas had a surplus of peasants, numbering in excess of what was needed for the new conditions of agricultural labor. National and local authorities, however, did not wish to increase employment through the development of large industrial areas. Instead, they sought the expansion of small local activities, considered to be less socially disruptive. Tourism was seen as an alternative for development, and the Touring Club was pronounced to be 'a public utility.' The Club exercised its influence by organizing competitions for the best inn, the best inn-keeper, the village or the railroad station with most flowers. It heavily promoted the ideal 'Touring Club room,' which consisted of a modern hotel bedroom, easy to clean and graced with modern furniture. The Club launched a campaign against traditional hotel decor, the heavy curtains and eider-down that retained dust. It wanted to see light, clean surfaces, and open windows, viewed as more hygienic. The Touring Club also urged the adoption of English-style lavatories, which implied the use of running water. If a hotel offered such accommodation, the Club gave it a sign reading 'Approved by the French Touring Club,' which worked as an advertisement as well as a means of guaranteeing the hotel the patronage of Club members. The whole system was a subtle mix of incentive to virtue and constraint. The first edition of the *Guide Michelin*, in 1900 asked the drivers to rely to a set of questions about the hotels visited during their trip. The text is as follows:

> We promise to cancel mercilessly every hotel whose cooking, beds, lavatories, waiters are identified as faulty as well as every badly-stocked fuel deposit and ill managed Michelin license-holders.[19]

The Club was also interested in restaurants and dining. It promoted the development of genuine regional 'cuisines,' alternatives to the 'eternal trout' served in the famous establishments of the seaside resorts. In 1906, Henri Bolland, one of the leaders of the association, travelling through the Basque Country wrote:

We cannot emphasize enough to provincial inn-keepers the advisability of keeping alive traditional country dishes and local condiments, which are more appealing to the majority of tourists than the Bechamel, Colbert and Sevigné sauces whose second-rate, bland aromas follow travellers from one end of the inhabited world to the other.[20]

A systematic policy in favor of regional cooking would only begin during the 1920s. At this time an American traveler wrote to the Touring Club asking for 'cleanliness and local color' in cooking. He explained that local inn-keepers were unable to provide the tourist with edible local cooking and unwilling to learn how to improve it. Support of the policy was difficult to sustain at times because of the characteristics of local cuisine—the gruel, porridges, cakes and breads prepared with traditional ingredients—were often quite heavy and unpleasant for urban patrons. For example, in 1921, a Limousin woman wrote to the club magazine, responding to a public appeal for checking off the 'richesses gastronomiques' of the country. She explained that the Limousin *clafoutis* (a cake made with cherries) could not be offered to travelers because the cooks were not in the habit of pitting the cherries[21]. Nevertheless, the 1920s witnessed the renaissance of 'traditional' cooking. 'Tourist offices' such as the one in the Jura, appealed to the public for help, assisted by local members of Parliament. They rediscovered peasant plates and adapted them to foreign customers. Local fraternal organizations were created for the promotion and sale of wine or local specialities such as 'Burgundy mustard'.[22] In 1934, the Knights of Tastevin met for the first time in Nuits-Saint-Georges hosting a medieval-style ceremony dedicated to the celebration of commercial production of local items. Inns themselves began to change, taking on a more medieval decor and renaming themselves 'auberges.' The Touring Club attempted to persuade inns to redecorate; the feature article of the July-September 1920 issue of the Club magazine is titled 'Local color in hostelry'. The author wrote:

[If I were an inn-keeper], my dining room would be like a trophy room, decorated with ten point antlers and other hunting trophies; we would serve bunches of grapes from the king's vine, real brie from Brie, and melt-in-your-mouth pears from Ile-de-France orchards.[23]

'TIDY VILLAGES'

The Club also tried to shape rural and local landscapes to its specifications. French travelers returning from Switzerland or Netherlands were appalled by the sight of French villages; they found them dirty and unpleasant. Here one finds two contradictory conceptions of rural space. For rural inhabitants, especially peasants, who were frequently poor, the countryside was a space for production and hard work. They had neither time nor money for flowers or other embellishments and thus had no interest in them. Flowers were too obviously a sign of wealth, wealth that should be hidden. Travellers would change this frame of mind. Before 1914, they organized competitions for railroad stations with the co-operation of railway companies. Those decorated with the most flowers received awards, in some cases with cash prizes for the station manager and his wife. The winning station's photographs appeared in local newspapers and the Touring Club magazine. In 1935, the first 'Charming Village' contest was organized by the 'Club for flowered villages' in the Massif Central.[24] The manufacturer Michelin donated a car for contest judges to travel in. The pragmatic judges were not even hoping to find flowers, simply clean and tidy spaces. The local newspaper *Le Cantal républicain* agreed. Cleanliness was indispensable if the villages were to attract the tourist trade; more importantly, hygienic reforms would benefit peasants themselves. Insidiously, new standards of behavior were proposed to rural populations under the covering of tourism development and aesthetic acculturation.

Protection of the landscape was nevertheless a top priority. Beginning in the 1850s, all the major European nations provided for the preservation of buildings considered to be of historical interest. Historians and local specialists wrote features on the innumerable medieval castles, 'burgs', towers, churches, chapels and other ruins located all over the continent. Ruins from antiquity abounded in Italy. In France Napoleon III, a devotee of the Gallic past, paid for the excavations of Alesia in Alise-Sainte-Reine. The remains of the national past were literally sacred to the new nation-states. The landscape gradually came to join them in this quasi-religious status. Landscapes, in their specificity and diversities, were considered to be the manifestation and essence of a nation's spirit. Visiting the countryside, appreciating the scenery was a patriotic duty as well as a pleasure. In addition to its contribution to the patrimony, landscape preservation was also desirable for another reason. If 'seeing the sights' and the beauty of the scenery was the motive for an automobile journey, tourist associations and automobile manufacturers had a vested interest in protecting the countryside from alteration or destruction. The first large European association for landscape preservation was created in Switzerland. At the end of the 1896 international fair, conservative policy makers, artists and writers joined together for the protection of traditional Alpine spaces disfigured by railroads, tramways and restaurants.[25] The mission of the 'Heimatschutz' would be to preserve a certain idea of Switzerland, as clean as the snow, as pure as the crystalline water of mountain streams. Imitator associations quickly sprang up in Germany, Belgium, Italy and France.[26] 1901 saw the founding of the Association for the Protection of French Landscapes. The French Touring Club organized a Site and Monuments Committee, and, after 1908, a version of this committee existed in almost every department alongside an official departmental committee. Committee members were very active, denouncing every degradation committed by rural inhabitants, provincial mayors and local firms and writing to the authorities, taking photographs of the damages, organizing a vast citizen's network that obliged the central administration to act, classify and protect. Members were relatively well-to-do and educated; they had business experience and represented a new collec-

130 *Cultures of Control*

tive approach to action.[27] The landscape -urban or rural- and the monuments that they protected were their leisure area. They wanted it to be 'as pretty as a picture' and especially lobbied against the innovations of the industrial age, such as tramways (in Belgium), electric power plants and dams (in France) and billboards.

This strategy can be considered a act of violence one social group made against other social groups. Weren't the specialized articles devoted to the protection of monuments and landscapes in the Touring Club review titled 'pillory!'? Nothing daunted the friends of peaceful landscapes and old churches when their treasures were threatened. They denounced by name the perpetrators of 'horrors' and tried to intimidate locals through their social superiority. Nevertheless, members of associations devoted to the protection of the landscape and the embellishment of villages progressively changed their strategy, exactly as the car drivers did. During the first decades of the twentieth century, tourist spokesmen became less and less aggressive and tried to build a community of interest with other social groups instead of engaging in ferocious confrontations. They systematically rewarded those who shared their choices and ceased to threaten those who continued to rebel. The pressure they brought became less open but remained perhaps more effective. Asking people to share their values and aims, they never tried to understand their own culture. They did not think they were compelling people or violating their privacy when they offered lavatories to a country hotel which accepted a partnership, and explained to the maids how to manage and clean them, or when they asked inn-keepers to redecorate their rooms, change their cooking, or their wine, or even when they financially encouraged peasants to clean the area around their house or to place flowers in front of their windows. The control they exerted was soft but effective.

CULTURE OF CONTROL AND CULTURAL CONTROL

Both the examples of the regulation of car traffic and the cultural propaganda of the tourist trade question the role of the state. The question has to be carefully considered. Traditionally speaking, studies of the French system emphasize the role of the State. But, in the special case of automobile development and the tourist business, public services or civil servants didn't take the initiative. The initial impulse came from an active urban bourgeoisie, closely connected to the main manufacturers of the car industry. Regarding the technical fittings of the road, state services, and especially the engineers in charge of the road network, were slow in understanding the necessity to change technical specifications. For instance they didn't initiate such decisive modifications as installing new traffic signals. They were also unable to have important decisions, such as limiting the speed to 30 km/h, fully accepted. Similarly, during the entire period considered, no state services, but rather private 'schools,' controlled by automobile sellers and associations, organized exams for giving driving licenses. One can remark that few purely technical devices contributed to the 'culture of control' of the car during these initial years. Traffic signals were few. Identity plates were the main technical artifacts installed on the car itself, for the purpose of control. Other transformations of the car, such as the internal organization of the motor, or viability of various devices are discussed in the technical reviews between manufacturers and users, with little interference coming from the State administration.

Therefore, the word 'state,' must be used with caution here. Where administration is weak, and parliament subjected to the action of lobbies, jurisprudence becomes influential. Important decisions, such as those concerning the use of the roads, the responsibility of car drivers and cattle drivers, dog owners and... parents, were not made by civil servants but by local judges (and contested by influential lawyers) in a purely legal context. Some legal regulations emerged from the various conflicts which took place at the beginning of the century. The most important—the 'code de la route'—was adopted very late (1921–1922), because no consensus could emerge before. But these regulations were perhaps less important than the control imposed by social uses: as we have seen, the main issue seems to be the etiquette which made a difference between good and bad drivers.

Because of this lack of involvement on the part of the State, competing interests expressed themselves mainly in the cultural field. Drivers, lawyers, defenders of pedestrians or protectors of peasants had to say what the suitable uses of the road were, which were permitted and which were not. The model at work was a negotiation between actors who were powerful in different ways. On the one side, one finds automobile manufacturers and their associates, such as Michelin, and also bourgeois and aristocratic drivers.[28] They had the cultural ability to control representations: they created reviews, owned newspapers, some had influence over the popular press and all of them were highly experienced in lobbying. They were also able to change their policy and adapt their discourse when necessary. For example, during the 1890s, authors writing in automobile magazines freely showed that they despised ordinary people. After 1905, lobbyists understood that this was counterproductive and authors wishing to publish in influential automobile magazines had to be more careful and consensual. Then they completely controlled the production of the set of regulations concerning car traffic: until 1970 there was to be no speed limit in France.

The weakness of the State can also be seen in the management of tourism requirements. Associations had great influence in the protection of landscapes, the embellishment of the country, the conservation of ancient buildings and the modernization of the hotel trade. They progressively extended their requirements to encompass more and more dimensions of collective life, bringing cultural pressure on traditional owners or users of country houses, fields and forests. When taking pictures of historical sites in decay or rewarding clean villages, their members exerted a soft but effective control over other citizens. In conclusion, we can say that, in the process of building a culture of control around the car, state regulations were paradoxically weak in a country where the State is overwhelmingly present. Some, but not all, technical issues were indeed state-controlled. They were by no means the only decisive ones. Associations, lobbyists, lawyers, judges, and engineers were the most important actors in a process which was largely negotiated in cultural terms. The culture of control was on the whole, cultural control.

ACKNOWLEDGEMENTS

Research for this essay is part of a larger project for the Mission du Patrimoine Ethnologique (Ministère de la Culture), in Paris on 'Ordinary Writings.' Earlier versions were presented at the seminar 'On the Road' organized by the Department of French Studies, New York University and at the Department of History, Case Western Reserve University. I thank my friends and colleagues Susan Rogers, Vera Mark, Monique Yaari and Miriam Levin for their comments and suggestions and Danièle Leclerc for his assistance with the translation.

NOTES

1 J. A Grégoire, *L'aventure automobile*, Flammarion, 1953, 275 p., p. 208. See also Patrick Fridenson, 'Les inventeurs de l'automobile', in Jacques Marseille (ed.) *Puissance et faiblesses de la France industrielle*, Seuil, 1997, pp. 391–408.
2 For an analysis of the american model of development of automobile industry, see Rudi Volti, 'A century of automobility', *Technology and Culture*, 1996, pp. 663–685.
3 Laurent Tissot, 'How did the British conquer Switzerland ? Guidebooks, railways, travel agencies, 1850–1914', in *The Journal of Transport History* n°1, March 1995.
4 Patrick Harismendy, 'La difficile histoire de la 'revolution automobile' dans l'Ouest (1897–1914)', a paraitre dans *La Bretagne des savants et des ingenieurs.Le Xe siecle*, 1998.
5 Catherine Bertho Lavenir, 'Luttes de classes et d'influence,' in *Cahiers de mediologie, n. 3, Qu'est-ce qu'une route?*, 1997 pp. 131–140 and *la roue et le stylo. Comment nous sommes devenus touristes*, O. Jacob, 1999, 444 p.
6 Catherine Bertho Lavenir, 'Normes de comportement et contrôle de l'espace: le Touring Club de Belgique avant 1914', *Le mouvement social*, n° 178, janvier-mars 1997, Ed. De l'Atelier/Editions ouvrières, pp. 69–87.
7 When Italian cyclist were fined, they could write to the Italian Cyclist and Touring Club which centralized the legal cases. *Touring Cycling Club Italiano, Bolletino*, January-February 1896, p. 100.
8 Léon Petit, 'Au pays des croquants. Récit véridique et moral dédié aux membres du Touring Club', in *Touring Club de France, Revue mensuelle*.

October 1898, pp. 362–364. ('In clod-hopper's land. A true and moral tale dedicated to the members of the Touring Club')

9 Pierre Musso, *Telecommunications et philosophie des réseaux. La postérité paradoxale de Saint-Simon*, PUF, 1997, 395 p.

10 André Ribeill 'Du pneumatique à la logistique routière. André Michelin promoteur de la 'révolution automobile'', *Culture technique*, n°19, 1989, pp. 362–364.

11 Patrick Fridenson, 'La société française et les accidents de la route', dans *Ethnologie francaise*, 1991-3, pp. 306–313.

12 Ronald Kline and Trevor Pinch, 'Users as agents of Technological Change. The social Construction of the Automobile in the Rural United States', *Technology and Culture*, 1996, pp. 763–795.

13 *Touring Club de France, Revue mensuelle*, Décembre 1905, p. 567.

14 Mortimer Megret, 'Les moeurs de la route', *Touring Club de France, Revue mensuelle*, avril 1907, pp. 205–207 et sq.

15 Pierre Mayet 'Chronique d'une politique de la sécurité routière', dans *Routes, espaces incertains. Cahiers de la sécurité routière*, n° 25, 1996, pp. 118–128. A decree in March 1893 instituted the fundaments of control of circulation for automobiles. Speed limited at 30 km/h on the country roads, 20 km/h in villages and towns; grey card for the owner of the car and pink card for the driver. Motorcar didn't still actually exist.

16 Docteur Bommier, 'Le doigté,' *Tourina Club de France, Revue mensuelle*, aôut 1906, pp. 88–389. One must be conscious that 'ordinary people' are here drivers employed by wealthy people who don't like to drive themselves. True ordinary people were out of the game because of the marketing policy of French manufacturers who were not interested in manufacturing and selling cheap cars (they were probably right, considering the configuration of the market in Europe at that time).

17 Adolphe Joanne, *Itinéraire général de la France, Réseau de Paris à Lyon et la Méditerranée*, Paris, Hachette, 1861, p. XIII.

18 *Touring Club. Revue mensuelle*, 1898, p. 395.

19 Guide Michelin, 1900, The use of the word 'faulty' shows that the rubber manufacturer consider the whole economy of tourism as a part of a large socio-technical system which has to be as well managed as a factory.

20 'Un voyage par mois', in *Touring Club de France, Revue mensuelle*, October 1906, pp. 506–507.

21 *Touring Club de France, Revue mensuelle*, juillet-aout 1920.

22 Marie Thérèse Berthier John Thomas Sweeney, *Les confréries en Bourgogne*, La Manufacture, 1982, 519 p., p. 192.

23 *Touring Club de France. Revue mensuelle*, 'La couleur locale en matière d'hôtellerie', juillet-sept 1920.

24 'Concours du village coquet', in *Touring Club de France. Revue mensuelle*, janv. 1936.

25 Diana Le Dihn, *Le Heimatschutz. Une ligue pour la beauté*. Lausanne, 1992, 152 p.

26 Dany Trom, 'Natur und Nationale Identität. Der Streit um des Schutz des Natur um die Jahrhundert in Deutschland und Frankreich,' aus Etienne François Hannes Sigrist und Jacob Vogel (ed), *Nation und Emotion. Deutschland und Frankreich im Vergleich, 19–20 Jahrhundert*, Vandenhoeck und Ruprecht, Goettingen, s.d., pp. 147–167.

27 The statistics of membership in 1907 shows that the French Touring Club is mainly constituted by members of a modern elite recruited among the ranks of businessmen, engineers, tradesmen, employees (30%); professors, doctors or civil servants represent 30% from the members, women 5% students 5% too (other are unidentified). See Catherine Bertho Lavenir, 'Le voyage,une expérience d'écriture. La revue du touring Club de France', dans Daniel Fabre (ed), *Par Ecrit, Ethnologie des écritures quotidiennes*, Ed. De la maison des Sciences de l'homme, Paris, 1997, pp. 273–297. The Belgian Touring Club recruits mainly among the members of the trade and business bourgeoisie in Bruxelles.

28 Insurance companies also played a key role.

CHAPTER 6

Culture, Technology and Constructed Memory in Disney's New Town: Techno-nostalgia in Historical Perspective

Robert H. Kargon and Arthur P. Molella

In 1996, the Walt Disney World Company opened for settlement a new town in Florida and bestowed upon it the up-beat name of 'Celebration.' The origins and character of this new place are especially interesting because they shed considerable light on ideas of progress, of urban design and of technology's role in society that are held by important elements of American culture. Celebration began as Walt Disney's utopian dream, born of the technological optimism of the 1920s and 1930s, the optimism culminating in the representations of the future city at the 1939 World's Fair in New York. This potent vision took concrete form in Disney's theme parks and, especially, in his plan for a real urban development that was to be EPCOT. This vision drew broadly upon American technical enthusiasms in urban design exemplified in Disney's older contemporaries such as Henry Ford's seventy-five mile city, Le Corbusier's Radial City and City of Towers, or even in Hugo Gernsbach's *Amazing Stories*. It is a tradition that persists to this very day in the "futuropolis" ideas of Paolo Soleri's Arcosanti.[1] These exemplars of rational planning married to technical progress were intended, through architecture, spatial configurations, transportation systems and other infrastructures, to shape the behaviors of the inhabitants for the better.

This essay will trace the evolution of Walt Disney's new town idea from his original Experimental Prototype Community of Tomorrow through the succeeding generation of Disney planners who brought to the enterprise new ideas of urban design and future-oriented technologies appropriate to the 1990s. We will argue that, along with the 'New Urbanism' and emphasis on the new information technologies, there is, in Celebration, a continued belief in technology's power to shape human behavior and forms of social organization, which moreover fits very comfortably in the well-defined Disney corporate formula for success. The ironies of this attempt to mold society via today's high technology will be evident in the history of the development of Disney's new town.

ROOTS: FROM EPCOT TO EPCOT CENTER

After 1955, when Disneyland opened at Anaheim, California, the Disney Company found itself in an entirely new situation. It had gained a great deal of experience in matters that were far from its original business of animation: transportation, electronic systems, crowd manipulation, efficient handling of huge numbers of people, etc. In short, Disney was facing problems that cities faced daily. It had also established an international reputation for ingenuity and innovation.

In 1963, before a Harvard University audience, the developer James Rouse, the creator of Columbia, Maryland, and pioneer of 'recreational shopping' venues such as Faneuil Hall in Boston and Harborplace in Baltimore, asserted that 'the greatest piece of urban design in the United States today is Disneyland.'[2] Walt Disney was pleased with this attention and liked to think optimistically of his role in molding the future. He conceived the idea of master-planning a city which would be a living experiment to light the way for others.

Disney died on December 15, 1966, before Walt Disney World in Florida was built and before his Experimental Prototype Community of Tomorrow could be established. But he left a film, through which he 'testified' posthumously before the Florida legislature, outlining his ideas. The futuristic city EPCOT would be laid out like a wheel, the hub containing a downtown under an air-conditioned dome. The 50-acre downtown would contain a 30-storey hotel, close to a convention center, theaters, restaurants, shops and office buildings. There would be a minimum of traffic; monorails would connect the center with the residential areas—high-density apartments and green-belt areas for low-density housing to serve a population of 20,000.[3] Before the legislature, Disney posthumously asked for exemptions from existing building codes and regulations so that the Disney Company would have the 'freedom to work with American industry' and the 'flexibility...to keep pace with tomorrow's world.'[4] His wish was granted in the form of the Reedy Creek Improvement District, which has surprising—even startling—autonomy.[5] Popular on the west coast, 'special districts' are legally empowered special agents elected by landowners to levy taxes and to assume some of the functions of government.[6] The Reedy Creek Improvement District plan empowered the one owner (the Disney Corporation) and from Disney's viewpoint drastically reduced the prospect of messy politics interfering with the plan. It was absolutely necessary, Disney believed, both to protect the company's interests and the technocratic version of utopia—an experimental city—that he was proferring.

Walt Disney's plans were a variation on optimistic, consumption-oriented utopian dreams like those of Edward Bellamy's *Looking Backward* of the late-nineteenth century on which there is a substantial literature.[7] Disney's successors at the Disney Corporation soon realized that Walt's utopian plans were gargantuan: if carried through, the company would be faced with enormous development costs, problematic financial return, and perhaps the political problems that go along with a real population. The Experimental Prototype Community of Tomorrow evolved into Epcot Center, an amusement park with ambitions—some would say pretensions—to educate. Epcot in Florida is a huge tract of land to the south and east of the Magic Kingdom and the Disney Resort Hotels. It comprises two distinct areas: The World Showcase (in which various 'nations' are represented) and Future World. The latter has pavilions, each with a corporate sponsor, 'celebrating the limitless potential of science, industry and technology in creating a better tomorrow.'[8] Instead of Disney's original idea of a living 'experimental community,' the new version is a perpetual, and vastly updated 1939 World's Fair. It was a harbinger of what two decades later would become a relatively big business: *edu-tainment*.

In the 1990s, a generation has passed and the Disney Company is in the hands of new people under different corporate circumstances, but with a great deal of continuity in corporate culture. The old corporate culture was an extension of Walt Disney himself a business selling the American idea of progress, driven by private initiative, know-how, up-to-dateness and small town values. The new culture of the 1990s has commodified these themes and wrapped them in new packages appropriate to the times: community, environmental sensitivity and social and ethnic diversity. It is important to note that in the world of Disney the amusement parks are no longer a mere addendum to a film empire; they are the driving force. Synergism between parks, films and merchandising is a powerful Disney strategy. At Florida's Walt Disney World there are a ring of new 'themed' hotels and resorts displaying the synergism. It is in this context that the CEO of the Disney Corporation, Michael Eisner, began in the early 1990s to see a future for the undeveloped land west and south of Epcot.

Eisner is a man fascinated by architecture, building design and development. Under his leadership, the Company began a greatly accelerated development strategy, the crown jewel of which was sometimes termed 'Dream City,' an update of Walt Disney's original concept of a high-technology experimental prototype city.[9] 'Dream City' was renamed 'Celebration,' the choice of Eisner and his wife. Celebration is a town for a projected population of 20,000 (Walt's 1960s size as well) in about 8,000 housing units. It comprises a shade under 5,000 acres, plus a greenbelt of comparable size to 'protect' it from the perceived cluttered fate of the Anaheim park.

Not only had the Company evolved from Walt Disney's ideas under Eisner, but prevailing ideas about architecture and urban design had changed as well. In 1966 the artist's rendition of Disney's EPCOT depicted structures drawn from 1930s 'futuristic' ideas and displayed a modernist city with greenbelt protected suburbs. The original EPCOT city is strongly reminiscent of Le Corbusier's ideas for his *ville radieuse* and even more strikingly of Norman Bel Geddes's city of 1960, a model constructed for the 1939 World's Fair.

The new town of Celebration on the other hand draws heavily on post-modern architecture (including one of the Disney Company's favored architects, Michael Graves) and upon an urban design movement popular in the 1990s known as the New Urbanism. In fact, the pioneers of the New Urbanism, Andres Duany and Elizabeth Plater-Zyberk, were brought early on into the planning of Celebration. Their Florida town, Seaside, is a New Urbanist icon, and one of the models for Celebration. The concept of 'New Urbanism' is somewhat loose, but most agree that ingredients include higher density than most suburbs allow, a disdain for the automobile and a preference for the pedestrian, a positive attitude towards transit, public open spaces, mixed land usage, and easy access from residential areas to shops, schools and even work. All of these together are intended to encourage a closer sense of 'community'—a notion often invoked but rarely defined. Here Celebration drew on the experiences of new towns like Seaside, a resort town of about 80 acres on the Gulf Coast.[10] It should be remembered that some of these ideas were congenial to Walt Disney—for example, his 'pedestrianism' and the notion of 'community'—but most have been developed in reaction to suburban sprawl and the rise of Edge Cities that have arisen in the past twenty years.

THE NATURE OF CELEBRATION

With appropriate fanfare, The Celebration Company opened phase one of its master plan on July 4, 1996. Celebration now consists of some 1500 residents in several hundred homes and apartments, a downtown and business park. This model utopian community is projected to take some fifteen years to complete. Celebration is a complex, evolving phenomenon, with many interesting facets: political, cultural, aesthetic, social, and technological. It has already received a great deal of attention from journalists, scholars, and cultural critics. What concerns us here is the cultural role of technology, both actual and ideological, in Disney's new town.[11]

Celebration makes a bold and colorful first impression, but it seems at first glance far from the futuristic spectacle originally envisioned by Walt. A dramatic tension between the past and future has always been the hallmark of the Disney theme park. It is the tension between the nostalgia of Main Street USA and the futuristic wonder of Tomorrowland. However, when it came to building Celebration—not a fantasy world, but a real community for real people—the past seemed to emerge triumphant. Celebration's planners apparently believed that only a return to the past could truly serve the imperatives of the new urbanism.

Culture, Technology and Constructed Memory in Disney's New Town 139

Figure 8 Westinghouse Singing Cascades, New York World's Fair, 1939–40. Photograph: Conrad H. Ruppert.

At first sight then, futuristic technology—in fact, technology of any sort—seems to have receded into the background at Celebration. Although only the first stages of Celebration are open, one can already see the outlines of a Norman Rockwell scene in the making: tree-lined streets, old-fashioned architectural styles, picket fences, a golf course, a town square facing on a small manmade lake. And, everywhere there are front porches, which Celebration's

140 *Cultures of Control*

Figure 9 Front porches abound on Celebration streets. Copyright: The Walt Disney Company.

planners see as key to neighborhood intimacy. Conspicuously absent were any signs of advanced forms of public transportation (such as the old Disney standby, the monorail), TV and radio towers, or other familiar accoutrements of modern technological society.

As we shall see, Celebration suggests a new model for the future of technological society, a radical departure from the sort of futuristic thinking that animated Walt Disney and his generation. It is fixated less on physical transformation than on the essentially invisible but no less powerful applications of the telecommunications revolution—a revolution designed to give us both our past and our future at the same time.

Despite its invisibility, cutting edge technology plays a prominent role in the Disney scheme for Celebration. Disney's promotional brochure proudly lists technology as one of the five cornerstones anchoring the new community—the others being place, health, education, and community. Information and communications are the featured technologies. The sales pitch is made in a techno-jargon designed to appeal to hard-core enthusiasts of the Information Revolution. AT&T, which at the time had the contract to build the local communications network for Celebration, advertised its fiber-optic systems in breathless terms:

Figure 10 Celebration's business district, billed as a 'pedestrian friendly downtown' with lake and boat rentals. Copyright: The Walt Disney Company.

When completed, the Celebration Network is envisioned to be a comprehensive, digital, fiber-to-the-curb network providing voice, data and video communication throughout the community. Plans call for the centerpiece of this infrastructure to be an AT&T SLC -2000 Access System with FLX Switched Digital Video. The System, a joint development between AT&T Network Systems and BroadBand Technologies Inc., is planned to enable Celebration subscribers to obtain services ranging from standard telephone service to full motion high-definition digital interactive multimedia services.[12]

To implement the system, AT&T planned to invite the first 300 families to participate in a 'living laboratory' in order to 'analyze and evaluate consumer attitudes and behavioral data concerning proposed products and services....'[13]

Some of Celebration's critics have decried its unbridled technological enthusiasm. Yet what is remarkable about Celebration, given Disney's record of technological flamboyance, is the attempt to temper this enthusiasm with humanistic concern. According to Eisner, 'the real magic is not in the building, physical structures, or even in the technology...We are interested in the civic infrastructure, because it is the human element that will make the community great.'[14]

Technology, in short, would seem closely connected to the primary cornerstone, that of *community*. Joe Barnes, architectural manager and arbiter of Celebration's rigidly controlled architectural standards—and our host during our

Figure 11 A view of Celebration's Mulberry Avenue, lined with approved Charleston siderows, modeled after the 'single house' found in Charleston, South Carolina. Copyright: The Walt Disney Company.

visit to the new town—pays very little attention to the technological infrastructure; it is someone else's department. His eye is rather on aesthetic matters, on the community values that are to be the bedrock of the new town and, above all, on tradition. In architectural terms, this means conformity to the six architectural styles deemed acceptable for Celebration: Colonial Revival, Coastal, Classical, Victorian, Mediterranean, and French. Customers can buy ready-made models or use their own architects, who are required to abide by the detailed specifications contained in a giant 'pattern book,' developed by Pittsburgh-based UDA Architects.

The styles adopted for Celebration's apartments and homes, according to Barnes, reflect the demands of the market. Extensive surveys and focus groups determined that the public preferred the psychological comforts of pre-War architectural styles. Armed with this mandate, Celebration's planners researched Southern vernacular and domestic building sites, deriving from them their own architectural palette.

Where there is a departure from these traditions is in Celebration's downtown, a showcase for the work of star architects. Buildings include a post office by Michael Graves, a bank by Robert Venturi and Denise Scott Brown, a town hall

Figure 12 Celebration bank by Robert Venturi and Denis Scott Brown. Photograph: Arthur Molella.

by Philip Johnson. Cast in colorful pastels, the downtown is a study in a kind of conservative post-modernism. The only hint of futuristic fantasy is in Cesar Pelli's impressively spired two-theater cinema. Although designed as part of the master plan, Celebration's downtown looks less like a planned ensemble than a melange of show pieces—a kind of architectural theme park.

In sharp contrast to the old-style futurism, where modernist skyscrapers and futuristic domes, soaring highways, and wheeling aircraft dominated, even defined imagined urban scenes, technology is discreetly hidden in Celebration. Its technologies are underground and within the walls, or, in the case of cars, tucked in the garage behind the house. Disney's planners have clearly sensed that old-style versions of technological progress no longer sell the way they used to—at least in terms of how people want to live. They have worked very hard to give the impression of subordinating technology to human needs, rather than vice-versa.

In terms of technical expectations, the visit to model homes in Celebration is almost anti-climactic. The only visible signs of technology are base-board plug-ins for a central vacuum cleaner system and a slightly enhanced home-entertainment center, housing CD-players, computer station, radio, TV, and VCR. Of course, the TV set or computer does not have to look spectacular to dominate a room or an entire home. It is what is on it and how it is used that count.

144 *Cultures of Control*

Figure 13 Two-theater cinema by Cesar Pelli & Associates. Photograph: Arthur Molella.

Celebration does not claim that its technology *per se* is anything beyond state-of-the-art. What is new, Disney claims, is the application of a cutting-edge fiber optics network to the total community environment. They are particularly proud in this regard of the Celebration School, where students 'have access to the latest technology, and work in flexible "neighborhood" classrooms with a team of teachers.' The network is designed to connect students' homes with school and with such services as an on-line electronic school library. The Celebration Health Center, designed by Robert Stern Architects and emphasizing health maintenance, is to be similarly high-tech. Plans include telemedicine facilities, allowing doctors, for example, to monitor a patient's heart rate and blood pressure via the network. Among other future benefits of Celebration's information network are interactive banking, voting from home, virtual office at home, home energy management, instant communication among the residences and between residents and community facilities and retail establishments, and, not least, home security linking each resident to a central monitoring point.[15]

The information network is still in its early stages. Clearly, however, Eisner and company have grand designs for this infrastructure. It is key to their plan. Disney's faith in information technology as a cornerstone of community may indeed seem, to some, utopian fantasy, but it is nothing new; it is typical of the high—some would say exaggerated—hopes frequently associated with the telecommunications revolution. It is often cited as a promoter of egalitarianism, of a shift of power from large cities to towns and villages, as a preserver of individual, person-to-person communication in mass society—all key values of the new urbanism.[16]

Perhaps we should see Celebration's communications system as not only a thrust to the future, but also a technological fix. Today, it is recognized that some of the early models of the new urbanism are losing their original community spirit. According to recent reports, citizens in Rouse's Columbia, Maryland, show signs of wavering on these basic values, putting self-interest before the welfare of the community. Evidence ranges from preference for private over public schools to residents' avoiding neighbors at their mail boxes, purposely clustered to promote citizen encounters.[17] In the face of these anti-communitarian trends, information networks offer both physical privacy *and* personal communication. In a way, the new digital technologies offer a virtual community should the physical community flag. Even if people decline to sit on all those front porches in Celebration, they can at least meet on the Internet. Frankly, even allowing for the fact that Celebration is still under construction, there were remarkably few people on the street, and almost none on their porches. Perhaps, they were all inside logging on. Whether the new information networks will work for or against the communitarian ethos so many desire is yet to be worked out on larger stages than Celebration; perhaps the Celebration case study will ultimately illuminate the issue.[18]

Moreover, whether the new information networks ultimately empower the individual and the village or, in fact, foster just the reverse—greatly enhancing the power of central authorities—is one of the great conundrums of modern technological society. The reaction to Celebration has been ambivalent in this respect. Those skeptical of Disney's utopian aspirations emphasize the disturbing aspects of communications technologies, while the believers naturally emphasize the positive.[19] Critical reaction to Celebration is in great part a reflection of our general ambivalence as a society to the information revolution.

THE EARLY REACTIONS

The market reaction to the opening of Celebration was phenomenal. So many customers lined up to purchase homesites in Celebration that the Corporation had to institute a lottery. Twelve hundred people paid $1,000 deposits just to make an appointment to discuss buying a lot.[20] The winners were first in line to build residences in the new town, billed as 'a 19th-century town for the late 20th century.' Not only potential residents, but also journalists, academics, architects, builders and sightseers patrolled the streets of downtown and the unfinished neighborhoods. The early appraisals, as expected, were mixed. John Kasarda of the Kenan Institute of Private Enterprise at the University of North Carolina was positive: 'Disney again has its thumb on the pulse of the American public,' he says, stressing the notions of community, neighborhood, and control. Patrick Burke of Michael Graves Architects declares that 'I think they've done the right thing...I just wish it had a different name.' Peter Miller, an urban planner at the University of Miami, opines that the Disney brand name is key: people want to be part of the Disney 'name.' Visitors had some mixed reactions: a retiree insisted that 'Anything Disney touches, everything they do is first class.' A psychologist from a nearby town says 'This really scares me. I just asked the girl if she heard of the movie 'The Stepford Wives' and she hadn't. I'm not sure I believe that.' Some visitors reject the insistent emphasis on community: 'I want my privacy.'[21] Some early residents, expecting *McGuffey's Reader*, were dismayed by the experimental ideas for schooling.

Some architectural and planning professionals, too, are dissenters. John Henry, an Orlando architect, calls Celebration 'a subdivision on steroids' and says that it 'there is nothing cultural there, nothing scenic there....A nucleus for a community it is not.' Aesthetically, he maintains, it reeks of 'pastel banality with homogeneous finish' (due to single developer build-out of the entire ensemble and too much stucco finish) and is a Disney trademark.[22] A Florida developer who chose to remain anonymous adds that 'it looks more like an amusement park to me. Everything is so cutesy and looks so artificial like out of a Disney movie.'[23]

On the whole, however, the Celebration idea has found its market. Many people find the protective umbrella of the Disney Corporation comforting. For example, the Erharts of Rockville, Maryland, are selling their suburban home and moving to Celebration. After pulling a low number in the lottery, they will be spending about $550,000 on their new home. 'If it wasn't Disney,' says Joe Erhart, 'we wouldn't be moving there.'[24] Many willingly are trading political responsiblity for what they perceive to be a benevolent and protective corporate authority associated with the Disney image. Ironically, recently, however, this has led to the first signs of political dissent within Celebration.[25]

TECHNO-NOSTALGIA—BUT, YOU CAN'T GO HOME AGAIN

On the surface it may seem that Celebration is not the high-tech utopia that Walt Disney envisioned shortly before his death in 1966. Witold Rybczynski in an article for the New Yorker titled 'Tomorrowland' writes that:

> Michael Eisner's Celebration is actually the opposite of Walt Disney's urban vision. Walt Disney imagined a world in which problems would be solved by science and technology. Celebration puts technology in the background and concentrates on putting in place the less tangible civic infrastructure that is a prerequisite for community.[26]

But is it really the opposite? We strongly disagree with Rybczynski. Celebration is a 90s version of the blend of technology and old-fashioned values that Walt Disney held dear. Most journalists and commentators when discussing Celebration invoke the name of the magazine cover illustrator and painter Norman Rockwell, conjuring up images of small-town America of the past. But though the skin may be Norman Rockwell, the muscle beneath is more like Rockwell International, the aerospace company.

The advertising for Celebration invokes the small-town, homespun image, but underlines the high-technology backbone as well. Early on, prospective residents were often reminded of the 'communications-rich communications tapestry' provided by AT&T's Advanced Technology Panel for Celebration.[27] The journal *The Economist* noted that 'what people want in their homes...are all the conveniences and technology of modern life, but hidden in "timeless" architecture.'[28] Inside the small-town skin is a fiber optic network to link every resident to the internet and to Celebration's own network. A brochure on Celebration's 'strategic alliances' boasts an impressive list of high-tech companies: AT & T, Honeywell, GE, and others.

In fact, we would argue, contra Rybczynski, that Celebration is a better thought-out combination of Walt Disney's Main Street nostalgia with his love for the 'futuristic' than his original idea for EPCOT in 1966. Celebration is built upon a shrewd 'techno-nostalgia' that combines a yearning for a mythical 'way it used to be' with a profound admiration for technical progress.

In a deep sense, Celebration remains true to Disney's formula for success: draw out from within each of us our images and stereotypes and through clever technique make them 'real'. Whether by its fairy-tale films, Epcot pavilions, the

148 *Cultures of Control*

Magic Kingdom or even its resort hotels with their simulated 'histories' distributed to each guest, the Disney visitor is furnished with memories of a simulated past. These 'memories' reinforce the images or stereotypes which the visitor already possesses upon arrival. For example, in Epcot's France, one finds men with berets, striped shirts and neckerchiefs; in Japan, women with kimonos; in Germany, men in lederhosen. One guest excitedly exclaimed that he 'saw more here in two hours than [he] saw in two weeks in Europe.'

What Disney does, and does expertly, is draw out these constructed memories from each individual and make them tangible with flair, style and technical wizardry. The simulation becomes, in a sense, more real than the original. While crowds are gaping at artificial alligators in some pirate's lagoon in the Magic Kingdom, few are meeting real alligators a few short miles away in a Florida river.

Celebration does precisely the same. Celebration invokes a simpler, more neighborly life, one that draws more from Andy Hardy movies than from historical research or living memory. One of its brochures declaims:

> Before World War II, many Americans lived in small towns, enjoying a convivial and comparatively simple existence. The intimacy of small town life has vanished over the past five decades. As cities grew and suburbs sprawled, neighborliness became little more than a memory. Celebration is designed to offer a return to a more sociable and civic-minded way of life.[29]

As one commentator has put it, 'Celebration promises to enact memories that most Americans have never experienced but desperately desire.'[30] However, Celebration may present itself as old-fashioned in the best sense, yet it insists on its futuristic credentials. Its brochure 'Celebration Network' maintains that 'Every apartment and home in Celebration will be linked by a fiber optic network that will carry telephone, video and all data services. The idea is old-fashioned one-on-one communication, but with sophisticated technology.'[31]

Russ Rymer in his perceptive article 'Back to the Future: Disney reinvents the company town,' writes that:

> Walt Disney was the Louis Pasteur of history, who perfected ways to protect from the viral effects of memory by injecting it back into them in a denatured form. As long as the technique was used in the service of amusement, it could be very amusing. Applied to public life, as it is in Celebration, it becomes something more grave.[32]

Savoring constructed memory is one thing; actually trying to 'relive' it is another. Celebration's planners understood very well that few citizens were literally prepared to return to a small town past (even if it were possible)—however inviting the 'memory.' Twentieth-century Americans left the security of small towns for the cities, and then the suburbs, for a variety of compelling reasons: jobs, schools, health care, consumer goods, cultural life, and life-styles. To be sure, there is now a reverse trend under way, but toward small towns

Culture, Technology and Constructed Memory in Disney's New Town 149

with a difference. And, as we have seen, even carefully planned experimental communities like Columbia are having their problems.

No small part of the Disney genius in selling Celebration was to combine the inducements of a rose-tinted past with those of a fabricated future—a sort of reverse nostalgia. They advertise a new brand of futurism based on the promises of the telecommunications revolution. Since the full implications of that revolution are still only dimly understood, what Celebration offers is as much a partially drafted hope as a reality. Whether or not these technological advantages can make the urban village work for a new generation of Americans remains to be seen. But, as Disney theme parks have proven beyond a doubt, simulated past and future fantasies make a potent blend—a blend that few consumers can resist.[33] And, Celebration's planners are banking on it.

NOTES

1 Little McClung, 'The Seventy-five Mile City: What Henry Ford Wants to Do with Muscle Shoals,' *Scientific American* (September, 1922), pp. 156–157; Robert Scheckley, *Futuropolis* (New York, Bergstrom, 1978), *passim*.
2 Bob Thomas, *Walt Disney: an American Original*, (New York, Simon and Schuster, 1976) p. 387.
3 'Disneyworld Amusement Center with Domed City Set for Florida,' *New York Times*, Feb. 3, 1967.
4 Quoted in John Findlay, *Magic Lands* (Berkeley, U. California, 1992), p. 111.
5 See Stephen Fjellman, *Vinyl Leaves* (Boulder, Westview, 1992), p. 121. See also Joshua W. Shenk, 'Hidden Kingdom: Disney's Polticial Blueprint,' *The American Prospect* no. 21 (1995), pp. 80–84.
6 Richard O. Brooks, *New Towns and Communal Values* (New York, Praeger, 1974), p. 61.
7 See Krishan Kumar, *Utopianism and Anti-utopianism in Modern Times* (Oxford, Blackwell,1987); Howard Segal, *Technological Utopianism in American Culture* (Chicago, U. Chicago, 1985); T. J. Jackson Lears and R. W. Fox, eds. *Cultures of Consumption* (New York, Pantheon, 1983).
8 Epcot brochure.
9 Ron Grover, *The Disney Touch* (Burr Ridge, Il, Irwin Professional Publications, 1991), *passim*.
10 James Howard Kunstler, *Home from Nowhere* (New York, Simon and Schuster, 1996), pp. 150–152.
11 Given the incipient state of the enterprise, particularly in its technological features, the description that follows relies on both promotional literature outlining the overall Disney concept and our personal observations.

12 AT&T, 'AT&T and Disney to Build High Tech Community of the Future,' AT&T news release, July 26, 1995. VISTA United now owns and maintains the community network.
13 *Ibid.*
14 The Celebration Company, 'Celebration: American Town Taking Shape in Central Florida,' Disney brochure, p. 1.
15 AT&T, *op. cit.*
16 For example, Mike Mills, 'Orbit Wars,' *The Washington Post Magazine*, August 3, 1997, pp. 8–13.
17 Katherine Shaver, 'Columbia's Community Values,' *Washington Post*, August 24, 1997, B1 and B8.
18 See Steven Lubar, *InfoCulture: the Smithsonian Book of Information Age Inventions* (Boston, Houghton Mifflin, 1993). This theme is widely discussed in popular periodicals such as *Wired*.
19 Mark Slouka, 'The Illusion of Life,' *New Statesman and Society*, January 12,1996, reprinted in *World Press Review,* May 1996, pp. 28–29.
20 Caroline Mayer, 'Disney's Latest—a Town,' *Washington Post*, Nov. 15, 1996, p. A1.
21 Craig Wilson, 'Celebration Puts Disney in Reality's Realm', *USA Today* (October 18, 1995), p. 1A.
22 Mayer, *Washington Post, op. cit.;* John Henry, 'Is Celebration Mayberry or a Stepford Village?' *ProBuilder Magazine News*, September, 1996.
23 Mayer, *Washington Post, op. cit.*
24 Jean Marbella, 'Mickey House,' *Baltimore Sun*, May 20, 1996, p. D2.
25 Michael Pollan, 'Town Building is No Mickey Mouse Operation,' *New York Times Magazine,* December 14, 1997, pp. 56ff.
26 Witold Rybczynski, 'Tomorrowland', *New Yorker*, July 22, 1996, p. 38.
27 Sullivan, "Virtual House People," *Home Office Computing*, October 1995, p. 136.
28 *Economist*, Dec. 1995, p. 27.
29 *Dowtown Celebration Walking Tour*, p. 2.
30 Andy Wood, "Arcadian Spaces" IDS (on-line journal), http: socrates.berkeley.edu:4050/ids.celebration.html.
31 *Celebration Network*, p. 1.
32 Russ Rymer, 'Back to the Future: Disney reinvents the company town,' *Harper's* (October, 1996) p. 77.
33 Many American cities (like Baltimore, Boston and others) are attempting to implement 'the city as stage,' a reproduction of the city resembling real life 'but not urban life as it ever actually was: the model is Main Street America...at Disneyland.' Peter Hall, *Cities of Tomorrow* (Oxford, Blackwell, 1988) p. 351.

Part 2

Managing Machines

CHAPTER 7

How to Make Chance Manageable: Statistical Thinking and Cognitive Devices in Manufacturing Control

Denis Bayart

The industrial enterprise is an excellent place to view a great diversity of forms of control: control of finances and accounts, controls on the material operations of fabrication, of logistics, and control of people at every level. Managerial knowledge seeks very explicit control objectives and their study is thus particularly fruitful for one interested in the history of techniques and in the sociological aspects of control. These modes of control are often embodied in very complex plans and devices which exist, at one and the same time, as ideas (they have been conceived by humans, they are founded on certain bodies of knowledge), and in material form (written texts, numbers, graphs, measuring instruments, software applications, etc...).

We propose to address the problem of the birth and diffusion of management knowledge from another perspective: that of the *devices* or *objects* through which this knowledge acquires a certain materiality. Without wishing to deny the importance of ideas, it is useful to examine as well the role played by material objects in the construction of ideas and in their diffusion and application in the world of business.

By 'devices' or 'objects,' we mean all material or graphic concrete forms which are produced in support of specific knowledge, and which might be used as an illustration, an argument, a proof, or means of application. Graphical representations hold an important place among these objects, as one can observe in leafing through any management manual. The *manual*, a particular type of book, is also a specific object which plays a certain role in the diffusion of knowledge, a role generally little studied in the domain of management (while the history and sociology of science and technology is interested in this sort of object). The application of knowledge, as in 'time and motion studies', requires particular instruments (special timepieces, equipment for recording the scene, data sheets..). One can also think of software objects (for example, the packages of statistical tools for the control of quality).

We will examine the role played by objects in the construction of managerial knowledge from three points of view:

- in the construction of management theories, both as elements of the development of knowledge and as means of support of the rhetoric of their promoters (these two aspects being difficult to dissociate in practice);
- in the application of knowledge, as mediators with respect to action; the properties of knowledge for action are in fact tied to objects;
- in the diffusion of managerial knowledge: being the material side of knowledge, they are engaged in social life in the same manner as any other object; for example, they can have the form of merchandise and be subject to circulation, commerce, and exchange.

We adopt a constructivist's and ecologist's conception of what is usually called the 'production/ diffusion' of knowledge in management: this knowledge, constructed under certain unique conditions by a group of promoters, is put in circulation within the social domain by means of objects (mock-ups, models, texts of different kinds...); entering into the world of the firm, they are subjected to a process of selection which brings into play the properties which the objects appear to bear. This analysis puts the accent on the interactions among, on the one hand, the objects that are produced and put into circulation, and, on the other hand, the contexts which give sense to the objects and establish their properties.

Let us say a few words about the role objects play within an organization. Researchers in organizational science have frequently studied the unanticipated effects[1] of managerial tools, for example some systems of budgetary control or control by objectives: The meaning of these control systems is transformed by their users according to their local context, which sometimes leads to results which differ in quite significant ways from what was intended at the outset. Today, an interest in objects, and the way in which they are 'engaged in action', is manifest within the social and cognitive sciences[2] by those who study the role of objects and the environment in the coordination of individuals at work; we see interesting parallels developing with respect to organizational science. From this perspective, an industrial organization is not only a collection of abstract procedures (rules) which coordinate its people, but also and above all, an assemblage of material plans and devices which make the doing of real activities possible. A management or organizational method is a composite of ideas, abstract principles, and objects or methods of practical import which engage the individual bodily and mentally in the execution of certain procedures. The control of the organization over its members is effected largely through these objects and methods: its members must understand and learn to use them in a way conforming to the intentions of the management. One can thus see the objects as signs (in the sense defined by C. S. Peirce): they do not possess in themselves any literal meaning but their meaning is constructed by the members of the organization, through processes of social interaction under certain circumstances in relation with their work (or with other preoccupations). One of the challenges of managerial control is thus to frame these interpretations, to limit their reading which could turn them against the aims of the organization. We know that one of the most prevalent forms of worker resistance is to use the objects of work in illegitimate and/or unauthorized ways. We are thus led to study all the social processes which reveal, establish or identify the properties of objects.

Here we consider the case of statistical methods of quality control in industrial manufacturing. The relevant objects in this case are of a cognitive type: they give form to information (data) in a way which enables certain operations which would not be possible without them. These objects are based upon scientific knowledge, a knowledge which they bring onto the shop floor, but they shape and represent this knowledge so that it is not evident in the object's day-to-day use; it enters only in special occasions.

Desrosieres[3] has ably characterized the double aspect of statistical objects with respect to action—and this statement might hold for many cognitive artifacts:

> Statistical tools permit the discovery or the creation of things which serve as a basis for describing the world and acting upon it. One can say of these objects, at one and the same time, that they are real and that they have been constructed, from the moment they are taken up in other assemblages and circulated as things in themselves, cut from their origins, this which is, after all, the fate of all manner of products.

Statistical methods of quality control were developed in the decade of the 20s to meet the needs of the American telephone industry at AT&T, Western Electric and at Bell Laboratories; in their final form, as a standardized technique, they appear as 'control charts' which allow one to track the consistency of manufacturing performance and detect early on the deregulation of a machine.

The principle of the control chart is relatively easy to understand, but requires some explanation. In order for a manufactured object to be judged of 'good quality', a certain number of its characteristics which have been selected as critical measures of quality—for example specific geometric dimensions—ought to satisfy some specified tolerances. But the machinery of production, as precise as it may be, is incapable of producing objects *exactly* alike; in fact, the characteristics of the fabricated products are distributed according to some statistical distribution. All goes well as long as the distribution of each characteristic remains within the limits set by the tolerances; one then says that the machine is *under statistical control* or *well set*. But it always happens at some time or another that the machine goes off. When this deregulation is progressive, the machine begins to produce some bad pieces among the majority of good pieces.

The control charts are a graphical tool which, with the aid of sampling techniques, allows one to detect this deregulation very early on, before it affects a large number of fabricated parts. One can then, with confidence, interrupt the fabrication process to reset the machine and thus avoid the production of scrap as well as the cost of reworking later on—a cost which can be very high. The control charts are today a fundamental tool for tracking quality in manufacturing, principally in the context of quality assurance procedures.

THE CONTROL OF MANUFACTURE PRIOR TO THE 'PROBABILISTIC REVOLUTION'

The principal innovation introduced in the 20s in the control of fabrication was *taking chance into account* by the methods of mathematical statistics which allowed the rational definition of procedures for making sampling decisions. As we shall see, this step constitutes, in the field of industrial production, a veritable 'probabilistic revolution.'[4]

Prior to the 20s, there evidently existed some procedures for the control of quality, even some which made use of sampling, but they did not explicitly rely upon probability theory.[5] At the beginning of the twentieth century, *determinism* was the dominant notion among those engineers and scientists who turned their attention towards industrial organization. The example of Henry Le Chatelier illustrates how, in this way of thinking, it was impossible to take chance into account. Le Chatelier, an eminent chemist, member of the French Academy of Sciences, is also recognized for his role in diffusing the work of F. W. Taylor in France. He declares, in the preface to a book on scientific management:

> All phenomena are interwoven according to some inexorable laws... The belief in the necessity of laws—that is, in the non-existence of chance—leads in industry to a continual struggle against irregularity, against the wastes of fabrication and, in almost all instances, allows one to eliminate all such irregularity and waste.[6]

The opinion of Le Chatelier *vis a vis* chance is explained by his conception of shop management. Only one way appears to him to be legitimate and fertile: to know, with as great an exactitude as possible, the laws of the material put into play in the fabrication process by the machines... Neither his conception of quality nor his ideas about industrial organization have need of, one might say paraphrasing Laplace, a 'hypothesis of chance.' On the contrary, one must reject chance with the greatest vigor because it offers an easy way out for the managers of a factory who show a distaste for taking the laborious and costly path of scientific knowledge which would render a true account of phenomena. To accept the idea that chance exists is to refuse to banish disorder.[7]

Sometimes, notably when control required a destruction of the product (e.g., a rupture test), it was necessary to take a sample of the items. Le Chatelier did not consider what the size of this sample should be. Nor did he consider the validity of conclusions that one might draw from the test. He probably was incapable of treating these questions because he seemed to have ignored the theory of probability, a subject that Laplace, however, had very clearly articulated, along with a deterministic philosophy, at the beginning of the nineteenth century.[8]

The same deterministic conception seems to have reigned as well in American industry. F. W. Taylor gave an example of a perfectly deterministic organization of quality control in the manufacture of bicycle ball bearings.[9] Also in 1916, Nusbaumer[10] had followed the plan of Taylor to a 'T' in reorganizing a gun powder manufacturing plant for which he was responsible. Even with respect to those subjects which, from today's perspective, lend themselves remarkably well to a probabilistic approach, such as the preventive maintenance of power transmission belts in a shop with the aim of avoiding interruptions of the fabrication process, Taylor adopts a rigorously deterministic approach.[11]

The faith in determinism in the American industrial milieu was equally supported by research of the greatest possible precision in the mechanics of manufacture, which appeared as the only way to ensure the interchangeability of parts. The historian A. D. Chandler notes:

> The American system of manufactures can be defined as a process of high-volume production by means of the fabrication of standardized parts that assemble into finished products.[12]

For much of industry, one of the principal objectives regarding quality in mechanical processing was the interchangeability of components according to the 'equation': quality = interchangeability = precision. The non-deterministic approach to the control of fabrication developed at Bell Laboratories would overthrow this dominant scientific ideology within industry and introduce an approach based upon statistical physics into the field of engineering.

THE CONSTRUCTION OF A SOLID THEORY OF QUALITY

Why Shewhart?

If this was a purely historical approach, it would be necessary to describe and analyze a large variety of studies made in the decade between 1920 and 1930, not only in the United States, but also in France, in Germany, in Great Britain, and perhaps also in Russia. It is in fact remarkable that, in a very short period of time in these different countries, but independently, engineers had considered probabilistic approaches to the control of quality. This suggests that the emergence of the problem did not result from a train of circumstances within a single industrial sector, but is more probably tied to an historical stage in the evolution of production techniques, in the organization of the firm, and in industrial exchange. Indeed, one might imagine that what we have here is the spread of mass production ideas and techniques, recognizing that a characterization as general as this will not suffice as an explanation. To give a more satisfactory response would require extensive research beyond the scope of this text.

Among all these independent efforts, those of Bell Laboratories are worth looking at in detail for the following reasons: It is there that the most ambitious and most complete body of theory was developed; there, too, the applied method embodied in the control charts in fact moved into industry; there, too, the published materials are very numerous and rich and allow us to track the development of this innovation. In the other countries, on the contrary, the methods that were developed remained fragmentary or limited to certain firms and have been definitely superseded by the method of Bell Labs. There is one notable exception: the British were able to climb aboard the train and integrate their own efforts with the American approach, to which they have contributed considerably in the years 1930—which is not astonishing considering the impressive potential of their researchers in statistics.

In order not to complicate our exposition, we limit ourselves to the methods of the control charts; but complementary methods of quality control by acceptance sampling were also developed at Bell Laboratories during the same period.[13] W. A. Shewhart (1891–1967) is the recognized creator of control charts, as attested by a dozen articles appearing between 1924 and 1931, culminating with a treatise[14] which assembled all of his previous work in one place. From one article to another one can easily follow the evolution of his ideas and his associated tools, and it is this construction that we will analyze, addressing particularly three stages of this development: 1924, 1926 and 1929–30.[15]

Western Electric was preoccupied during the years 1922–24 within the *Engineering Department* with problems of quality control.[16] W. A. Shewhart, a physicist by training, schooled in the methods of statistical physics, was charged with the task of examining measures to apply to telephone equipment with aim of developing procedures for quality control. He was transferred to Bell Labs when they were created in 1925 and there continued his work on quality control.

The Carbon Microphone, or Setting the Stage for Randomness: 1923–24

In the first sentence of his first lengthy article, before any talk of control charts, Shewhart makes a frontal attack on the belief in determinism, redefining the significance of the measure of any physical magnitude based on the modern physics of his time. In place of exactitude, that is to say a precision as fine as one wishes, one can only hope to find some statistical entities which no longer provide certainty but only probability:

> We ordinarily think of the physical and engineering sciences as being exact. In a majority of physical measurements this is practically true ... With the introduction of the molecular theory and the theory of quanta, it has been necessary to modify some of our older conceptions. Thus, more and more we are led to consider the problem of measuring any physical quantity as that of establishing its most probable value. We are led to conceive of the physical-chemical laws as a statistical determinism to which 'the law of great numbers' imparts the appearance of infinite precision.[17]

Shewhart transfers the approach used in statistical physics to the field of engineering of the telephone. The two first paragraphs of the article are entitled in a symmetrical way: 'Statistical nature of certain physical problems' and 'Statistical nature of certain telephone problems.' In the first, he uses an account of the historical experiments of Rutherford and Geiger (1910), showing that alpha emission by a radioactive source is a random variable following a determinate statistical law. In the second, he shows, with graphical support, a telephone component whose behavior appears to be random and concludes:

> The characteristics of some telephone equipment cannot be controlled within narrow limits much better than the distribution of alpha particles could be controlled in the above experiment.

This comparison legitimates the transfer of probabilistic and statistical thinking from the field of science to the field of telephone engineering. The recourse to probabilistic models becomes even necessary on the part of scientific researchers at AT&T, whose mission is to remain in touch with the state of the art.

The object chosen by Shewhart furnishes a particularly illustrative example of why one needs a statistical approach. The carbon microphone is a key element of the telephone system and posed many problems at the time.[18] One of its important characteristics, namely its electrical resistance, shows all the appearances of random behavior, even though measured in a laboratory with all imaginable care. The impotence of a deterministic approach to this kind of object is clear: randomness is found at the heart of the manufactured object and not just in the machines which make it.

This providential object gives Shewhart the opportunity to emphasize a fundamental problem in quality control: what standards of fabrication can one establish for products whose characteristics of quality can not be fully controlled in a deterministic way? And how then does one formulate and represent these characteristics for personnel in the shop? How to establish the limits of allowable variability? Shewhart's response is 'by using statistics and only by such means.' This then provided the basis for his pursuit of research in this field in order to arrive at some operational methods for use on the shop floor. At the same time, with this example, he is able to denounce the belief of industry according to which one could indefinitely increase the precision of machines so as, *in all cases*, to resolve questions of quality.

Let us see now how this radical questioning opened up matters. Because it does not suffice, in the industrial domain, to denounce established ideas but requires proposing some useful, working methods, Shewhart proposes an approach in this article, and one which he adopts in the following, which consists of identifying, by numerical methods, the statistical distribution of the chosen characteristic of quality under stable conditions of manufacturing. The greater part of the article is thus devoted to a review of existing statistical methods with the aim of evaluating their relevance to identifying an empirical distribution. Thus it is a methodological and problem-setting article. The practical tools of work are not yet the focus, but they soon make their appearance in the course of the year, 1924.

From Statistical Distributions to Control Charts

The 1924 article concludes that the quality of an industrial product may be represented by a statistical distribution which is identified by the numerical values of the first elements, according to the mathematical theory of Karl Pearson. Starting from there, Shewhart progressively puts this into the form of graphical tools—the control charts—which will be the essential material object of this method. It is this evolution which we are going to retrace, beginning with the initial idea appearing in an internal memorandum[19] dated 16 May, 1924, and ending with the standardized form of 1935. Our analysis aims to show that the graphic form comprises a fixed element with respect to which theoretical conceptions have evolved around as a pivot. This observation supports and confirms

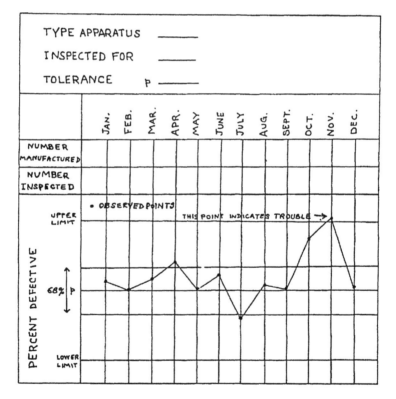

Figure 14 The original idea of a control chart (1924). Copyright © 1924 Lucent Technologies. Used by permission.

the thesis to wit: the predominant role of objects in the evolution of ideas in management.

The internal memorandum of 1924 has two components: an example of the graphical representation (Fig. 14) and a very brief text by Shewhart indicating that he is on the way to developing an operational method:

> The attached form [graphical representation] of report is designed to indicate whether or not the observed variations in the percent of defective apparatus of a given type are significant; that is, to indicate whether or not the product is satisfactory.

He adds that the underlying theory is relatively complex and that he has begun work on a memorandum which will explain it in detail. But it is clear that this graphical form is the major contribution because it permits one 'to see in a glance the most pertinent information.' Its principle is simple: a horizontal axis represents the successive dates of the measurements made, while the vertical axis shows the scale of the measured characteristic. The new element, which renders the graph valuable, is the couple of horizontal lines whose ordinate corresponds to theoretically determined values and which represent limits not to be exceeded. If the value of the measured characteristic reaches either of those lines, this indicates a problem (Shewhart has written 'this point indicates trouble' on the figure).

A posteriori, knowing the underlying theoretical developments, it is easy to understand the leading idea: given a statistical distribution, one may deduce from probability theory an interval within which the variable falls with a probability very close to 1.0. If then the result of a measurement falls outside of this interval, it is most probable that a change occurred in the statistical distribution—this being the *trouble* highlighted by Shewhart: something is out of line; it requires intervention. But before he arrived at such a clear conception, Shewhart began with the construction of a sophisticated theory closely related to the work of the British biometricians. He laid it out in 1926, employing for the first time the term *control charts*.

His objective, as in his first article, is to identify the empirical distribution of the quality characteristic. The method includes four steps: choice of a theoretical model for the distribution (normal law, Poisson law, etc.), choice of estimators, numerical estimation, test of significance. This methodology requires much too arduous calculation for it to be used on a routine basis in a shop, but it takes advantage of the graphical display principle of 1924, yet with a difference. This time, it shows a display for each of the parameters characterizing the assumed distribution (Fig. 15). The parameters are calculated for a sample of each month's production and drawn on the graph. The horizontal lines show the limits within which each parameter should stay, provided the statistical distribution remains unchanged. They also include sampling fluctuations. One clearly sees on Figure 15 important variations for some months, which correspond to variations in the fabrication process.

162 *Cultures of Control*

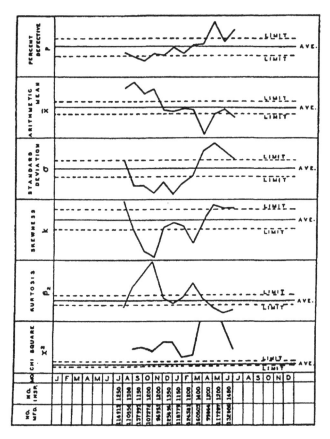

Figure 15 A control chart for the parameters of a statistical distribution (1926). Copyright © 1926 Lucent Technologies. Used by permission.

It is striking, in looking at Figure 15, how the distribution of the quality characteristic changes over the period of observation. The four first moments of the distribution displace significantly from the limits corresponding to the fluctuations of the sample. The graph thus shows clearly the existence of important causes of variations in the fabrication process.

At this stage, Shewhart has thus constructed two instruments: one simple and eloquent graphical representation and a methodology which is dense and requires much calculation. These two facets are not yet fully complementary: the graph illustrates the theory but doesn't contribute to it. The evolution of the method which ensues is very interesting in that the theory is going to be considerably simplified and the graphical tool is going to become an integral part of the whole. This evolution calls to mind that process Simondon has called *concretization*[20] of a technical object. A technical object is first thought of, then realized as a prototype, as a representation of a theoretical scheme (here, the

TABLE II.—OPERATING CHARACTERISTIC, DAILY CONTROL DATA.

SAMPLE	SAMPLE SIZE, n	AVERAGE, $\overline{X}$	STANDARD DEVIATION, σ
No. 1	50	35.7	5.35
No. 2	50	34.6	5.03
No. 3	50	32.6	3.43
No. 4	50	35.3	4.55
No. 5	50	33.4	4.10
No. 6	50	35.2	4.30
No. 7	50	33.3	5.18
No. 8	50	33.9	5.30
No. 9	50	32.3	3.09
No. 10	50	33.7	3.67

Central Lines

For $\overline{X}$: $\overline{X}' = 35.00$.
For σ: $\sigma' = 4.20$.

Control Limits

For $\overline{X}$: $\overline{X}' \pm 3\dfrac{\sigma'}{\sqrt{n}} = 35.00 \pm 1.78$,

33.22 and 36.78.

For σ: $\sigma' \pm 3\dfrac{\sigma'}{\sqrt{2n}} = 4.20 \pm 1.260$,

2.940 and 5.460.

FIG. 2.—Control Charts for $\overline{X}$ and σ. Large samples, $\overline{X}'$, σ' given.

RESULTS: Lack of control at standard level indicated on third and ninth days.

Figure 16 The standardized form of the control chart (1935). Copyright ASTM. Reprinted with permission.

Pearsonian theory). Then, with time and use, its components are redefined as a function of one another—a process which confers to the object the appearance of an autonomous life, relatively independent of the theoretical conceptions from which it emerged.

This concretization clearly shows itself in the normalized forms of the control charts (1935,[21] Fig. 16) The primitive form uses all of the first four elements of the distribution in order to avoid the hypothesis of its normality, at the cost of very considerable calculation. We see that the standardized method only uses the first two elements; what occurs then with the normality hypothesis? The manual defining the standard says quite briefly that 'in practice, the mean and the dispersion are considered sufficient.'[22] The process of concretization has thus led in the present case to a simplification of the initial object for the domain of application of the method, secured at the price of an implicitly restrictive hypothesis at the theoretical level.

Similarly, the procedure for determining the statistical distribution is also standardized, codified in an operational procedure where one aims to minimize the references to statistical theory. The very open method that Shewhart had presented in 1926 has thus, ten years later, taken the tangible form of a graphical object in accord with a *mode of application*. This object, having become to a great extent autonomous with respect to the statistical theory at its base, is now ready to be routed through the institutional channels for diffusion throughout the industrial world: for standardization, use in training. Even though these channels, as we shall see, have not been the only means of diffusion, they played an important role in identifying and making known the control chart 'product.'

From Epistemological Ambitions to Economic Advantage

But Shewhart was not content simply to propose some operational rules and tools. His ambitions went beyond those of an industrial engineer; he sought the status of *savant*. He constructed a veritable *epistemology* of statistical quality control, relating his development of ideas and methods to the grand scientific laws of nature in a form which suggests Laplace's Philosophical Essay on Probabilities.[23] In a communication of 1929 to the *American Society for the Advancement of Science,* he posed three *postulates (sic)* in order to introduce the concept of *constant system of chance causes* (in modern terms, what we call a *stationary random system*):

> Postulate 1. All chance systems of causes are not alike in the sense that they enable us to predict the future in terms of the past.
>
> Postulate 2. Constant systems of chance causes do exist in nature.
>
> Postulate 3. Assignable causes of variation may be found and eliminated.

These propositions are destined to serve as the theoretical basis of the development of statistical quality control. Each postulate is supported by several examples, some drawn from statistical physics and demography, others from the experience of engineers (which is the basis of the third postulate, a principle of action). If it had been published today, the text would probably be judged a fantasy or the dream of a megalomaniac; as a matter of fact, it connects things which appear to us disproportionate and heterogeneous: engineers seeking to regulate machinery, on the one hand, and cosmological or metaphysical principles on the other hand. Is it really necessary to invoke so general a set of propositions in order to justify a method which is by itself totally understandable? But Shewhart, in fact, exploits as far as possible his experience and knowledge as physicist to bring all of modern science to the support of his approach. The mobilization of, in the words of B. Latour, these 'alliés de poids' might explain why Shewhart's theories have never been attacked with respect to their scientific legitimacy: such a critique would have to contend with the weight of all the science with which Shewhart's texts are amply loaded.

But the weight of these scientific allies does not suffice to explain the success in practice of Shewhart's method. Yet to be demonstrated is the method's technical feasibility and economic viability—criteria which are critical and prerequisite to industrial acceptance. In an extremely dense article, Shewhart adds on some economic arguments in favor of statistical quality control: reduction of the cost of inspection, reduction of the cost of waste, maximization of the benefit of large scale production, achievement of uniform quality even in the case when one performs destructive tests, reduction of the tolerance limits when the measure of quality is indirect (then making use of correlations).

Shewhart, however, is neither a very gifted popularizer nor a great communicator. In his whole career he only published two books. The first, in 1931, pulled together all of his prior articles in an opus which was very dense and difficult to read. It constitutes a reference work, a work of legitimization, but certainly not an operator's manual... It has been enormously cited, but without doubt little studied by practitioners because it raises more questions than it answers. The second book is even more 'philosophical,' concerned as it is with the theory of scientific knowledge and operationalization of concepts.

In view of these works, it is evident that it is not the personal charisma of Shewhart (so theoretically inclined) nor his efforts at promotion which can explain the success of statistical control of fabrication. It would require the help of engineers more oriented toward practice who, coming together in committee, would produce some operational standards. A true division of roles thus appeared among the different agents intervening in the process of promotion. The weight of Bell System, of its research arm, Bell Labs, and its production division, Western Electric, is also evidently an important reason for the promotional success of the method.[24]

The very theoretical character of some of Shewhart's works ought not, however, lead us to neglect the importance of the modification he has accomplished in the domain of ideas. In fact, beyond the pure transfer of reasoning and observations drawn from the field of theoretical physics, he develops as well an approach which comes to grips with challenges specific to the world of industrial production. In the first place, he takes economic factors into account. If the best strategy when faced with the randomness of production is to eliminate the assignable causes of variability and to maintain as constant as possible the conditions of fabrication, the cost of these operations ought to remain 'reasonable' in the sense that it satisfies the judgement of the engineer. Certainly from the perspective of economic criteria, Shewhart did not take the articulation of the development of quality very far. However, his colleagues Dodge and Romig, with whom he was closely associated, published in 1929 a method of control via sampling which rests explicitly on an optimization of the costs of inspection. The preoccupation of management with costs is thus well represented in this engineering milieu and it comes to be expressed in operational terms.

Then too, Shewhart completely reformulates the notion of *control* with the aim of taking into account the indeterminism of phenomenon, notably the fundamental fact that *a controlled quality is a variable quality* and not always equal to a preestablished standard:

> For our present purpose a phenomenon will be said to be controlled when, through the use of past experience, we can predict, at least within limits, how the phenomenon will be expected to vary in the future. Here it is understood that prediction within

limits means that we can state, at least approximately, the probability that the observed phenomenon will fall within the given limits.[25]

The fundamental principles are now fully integrated into the daily practice of the quality control function.

The Mode of Engagement of Objects in Action

We have seen how in the construction of the theory, Shewhart articulates, on the one hand, science and on the other hand, some objects which link with practice: the carbon microphone demonstrates the necessity to resolve a problem of fabrication, the graphical object in the form of the control chart suggests a method which appears intuitive and easy to apply. But at this stage of the analyses we have only examined the matter from the point of view of Shewhart, the initial promoter, who expresses himself with a good dose of rhetoric. That the method appears to be useful in practice might derive from Shewhart's rhetorical abilities or from the helpful advice he received from his engineering colleagues with whom he associated and who consulted with him—since we have seen that Shewhart was more oriented toward theory. From an examination of the rhetoric of the promoters alone we can, in fact, deduce nothing about the actual conditions for applying the method.

To address this question, requires that we analyze the way in which the method was put into practice and received by users. We avail ourselves of some witnesses who, although there are lacuna, none the less allow us to draw some interesting conclusion when we place what they have to say within an appropriate conceptual framework. Let us sketch our framework for analysis. It is a question, fundamentally, of a study of *reception*. It requires radically displacing our point of view which, up until now has been that of the promoters, in order to adopt that of the users. The user confronts two types of factors: the discourse of promoters and the object for putting the proposed method into practice. The industrial user is above all anxious to know if the method works in the context of the shop floor; a priori, then, he is going to listen to the words of the promoters with distrust knowing that they contain a good dose of rhetoric. He will ask for proof, for trustworthy witnesses, for results of tests. But these are *discursive* elements which, if they can attract the attention and interest of industry, ought to be dissociated from *trying out* the method, an engagement in practical action which puts the objects to work. We will try to show that the test of the control charts was decisive, for example in the training of professionals. This will be the focus of the first point: the control chart as a new cognitive tool. We will then examine how the argument of the promoters was reinforced by recourse to objects other than the control chart, such as urns to simulate random sampling, which were utilized in the training sessions but not in the workshops. In a third part we analyze the compatibility among the ensemble of objects associated with statistical control (notably the directions for use) and the organizational structures of the enterprise, the division of competencies and tasks. Finally, we will see how the objects engage the theory in the daily life of the enterprise.

The Control Chart as New Cognitive Tool

The control chart exhibits some properties which are associated with a new way of perception: it renders visible and tangible some phenomenon which were previously hidden. A standard control chart (cf Fig. 16) allows one to follow two tendencies of the quality characteristic: its mean and its standard deviation. If the mean is a relatively intuitive notion, the standard deviation is not; one can conceive of the idea but one is at a loss to give it a precise mental representation without recourse to an image such as a histogram. Now the control chart offers the viewer, laid out on a plane sheet of paper, the concept of dispersion showing the limits that this dispersion ought not to exceed as long as the production process remains under control. It presents, in a perfectly visible and sensorial way, the variability of the fabrication process as it proceeds in time. Note that it does not use the representation of the histogram which would not be a very effective way to follow the evolution of the standard deviation in time.

If we go a little more into detail. the control chart also represents other more abstract notions: the variability of the mean and the variability of the dispersion. What is the variability of a dispersion, of the standard deviation? To understand this concept requires explaining the process of sampling. It is to understand that one estimates, with each sample made, the dispersion of the ensemble of the population, and that this estimation displays a variability due to the sampling process itself. In brief, a succession of difficult reasonings, that would be impossible to mentally deploy in the course of repetitive work. The control chart presents these not very intuitive notions, without the need for a mental representation on the part of the user. Here resides the tour de force: thanks to the control chart there is no need to rely upon the mental powers of the worker to manage dispersion. The statistical notion of dispersion, which is constructed in the theory, has thus acquired a unique, visual representation.

The control chart thus allows the transformation of a complex ensemble of abstract reasoning into a work procedure which calls upon the most general faculties of representation (vision) and on some elementary, arithmetic operations. The analysis presented above certainly does not rest on first hand empirical observation; we have constructed it from the thought process we have projected upon the user. But it is necessary to emphasize that such observations are practiced by certain researchers in the cognitive sciences and distributed cognition.[26] The principle here is to describe exactly, by means of a phenomenological observation, what the subjects do, what elementary cognitive means they put into play in the use of instruments of work, with the aim of reconstituting their 'mode of use' of the objects—and not the theory that an engineer could see behind the functioning of these objects. In the routine of the workshop, once statistical control is in place, it is not the theory that is at work, but the control chart object and the associated organizational procedure which governs its use. The procedure, applied in an automatic way, requires no reference to statistical theory. The activity of the worker can be analyzed as a succession of elementary cognitive operations: select a sample, make the measurements, then the computations, record on the graph, look at the data, conclude.

But, from another angle, it would be false to consider that the control chart object allows one to completely avoid the theory, to relegate it to the backstage. In fact, if we readily allow that the worker on the line is not concerned with the theory of statistical control in his day to day activities, it is certainly not the same for the engineers who try to understand, by means of their individual cognitive powers (rooted in the scientific concepts that they ordinarily employ), how these objects work, how they produce tangible results. It is to this audience, as well as their supervisors, that the training sessions are directed where each participant is confronted with some 'pedagogical' artifacts which generally have a very convincing effect, according to trainers' reports.[27]

Among these artifacts, we must in particular mention the *bowls*. They were filled with numbered uniform 'poker chips' in such a way so that, when one made a random selection from the lot, one simulated the random sampling of a normal distribution (following the law of Laplace-Gauss), or of a uniform distribution, or even a triangular distribution. This type of simulation has frequently been used by statisticians either to test the results obtained from analysis or to demonstrate in a vivid way the 'laws of chance.'[28] Science museums display various apparatus inspired by the same principle, and these artifacts always provoke astonishment on the part of visitors: is it not always fascinating to observe order born out of apparent disorder? We find in this an effect of the type 'to test it is to adopt it.' The astonishment that a new user experiences in observing how 'it works' is an important psychological factor, which is why partisans of the statistical method are often such militant promoters of it.

One consequence of the confrontation with this artifact, generally successful in the training sessions, is that, for the engineers so trained, the control chart object becomes an incarnation of the theory. The control charts in their routine functioning (as they work well) constitute a permanent validation of the theory; it becomes as impossible to doubt as the theory of the steam engine. We observe here a circular causal chain: the object is founded on the theory, which in turn is founded on the object's functioning, and so on. But similarly, it is necessary to recall—because this shows the multiplicity of meanings which an object can bear—that the object engages the theory in a different way according to the level of knowledge of each individual: the worker sees in it only a procedure. Ignorance of the theory does not prevent him from putting the object to use in an autonomous way. On the other hand, the theory can only make its proofs through the object on which it depends, that is the control chart.

Objects in Support of Rhetoric

The acceptance of statistical control developed along two paths: that of the persuasive rhetoric of promoters, who included the new militant users of statistics;

and that of individuals' confrontation with the objects themselves as an aspect of their experience of reality. But in all the material put to use in this historical analysis, it is impossible to separate the two types of effects: all accounts of confrontation with the objects, published in the technical journals, have rhetorical content.

In order to get beyond this difficulty, we observe the way, in the rhetoric, the objects are described and what objects are chosen and privileged in support of the argument made. We find a great number of texts which recount the application of the method. These texts lead the reader to mentally project him or herself into the situation in confrontation with the objects and to simulate this experience. The method of statistical control is put to a real test, the text exposes the conditions and renders account of the results. These represent, if not ostensibly 'advertising,' elements upon which the readers can base their opinion of the method.

A second category of texts concerns experiments with the method in a 'scientific' context, that is to say in a laboratory. Shewhart, thus, utilized the urns of normal, uniform, and triangular distributions, discussed above, in order to test the method of rational subgroups (too complex to explain here) which is the basis of the control charts. He has published the results of 4000 drawings from each of these three distributions, and his tables continued for a long time afterwards to serve as a reference since we find them still used in a manual dating from the 70s. We note that in the first French article[29] dating from 1925, we find a similar presentation which the author employs in order to confirm the results of his analysis. These explanatory objects are only engaged with by the reader, via a description in words, sketches, lists of numbers and tables of results of analysis. The reader does not have the original objects in front of him and cannot manipulate them in order to verify what he is reading. To understand the experiment, he has to do so via mental representations with full confidence in the author. None the less, these elements are taken as proofs.

Alongside these objects which engage the theory by putting the method into practice, we should also pay attention to those objects which serve to establish an argument, and which the reader encounters in the textual form of accounts of experiments. With respect to the arguments about the economic advantages of the method, we see that the objects exhibited are often less convincing. Essentially, here we find some evaluations which are not always quantitative. More than the objects, it is the effect of the example that brings acceptance: the fact that one firm as important and serious as Western Electric had undertaken between 1922 and 1924 a campaign to improve quality by applying statistical methods constitutes a powerful argument. It has behind it all the weight of the company.

But these experiences are not very numerous, so it is necessary to extrapolate. Shewhart shows, making use of graphs and series of numbers which measure the nature of quality, some situations which are not 'under statistical control' and where he must intervene to set matters right. But he takes care to state that it is necessary to 'use your good sense' and not try to improve quality if the gain in doing so is not greatly superior to the cost of doing so. This common sense approach is called 'engineering reasoning'. The engineer is brought into the picture to counter balance the scientist who might be a bit too much of an idealist; in this way the entrepreneurial reader would be reassured. We know furthermore that Bell Labs was staffed with as many engineers as academics, which would give a certain credibility to Shewhart's argument.

Objects in the Organization of the Enterprise

At work, the objects prescribed by the theory of statistical control call into question certain organizational requirements on the shop floor and its social life. The objects which might lead to the theoretically best performance are often too difficult for a manual worker to put into use, and hence might, if adopted, lead to errors. For example, the 'sequential plan' which leads, in theory, to very important gains with respect to the size of samples is little utilized because it requires too many manipulations and thus risks being applied wrongly. These plans were the work of a brilliant mathematician, Abraham Wald, who developed them under contract with the American government during the Second World War. But without doubt this mathematician did not have a sufficiently concrete sense of context, and the methods of the engineers at Bell have continued to be favored in industry. In retrospect, the efforts of Wald have had some very important consequences for decision theory and have significantly contributed to scholarly research in this area.

In the context of the shop floor, it is not good to leave to much room to chance. Hence, the methods of statistical control were rapidly standardized. This process of codifying instructions may be compared to the operation of a military organization: in the artillery, one finds a handbook for the gunner, another for the staff sergeant, another for the officer. In the firm we find a scientific treatise for the engineers, a popularized text for the directors, a technical manual for the supervisors (which does not reproduce theoretical derivations but gives examples instead), and a list of instructions for the machine attendant. Each of these texts gives some rules of conduct, but with less and less freedom to maneuver as we descend the hierarchy. The engineer can choose among different types of control charts, the supervisor among different ways to make a measurement, but the worker only has one rule to apply: to call the machine-setter if the points recorded on the chart exceed the control limits and to continue as before if they do not. Even the random selection of the members of the sample is subject to strict regulation so as to avoid the worker introducing, consciously or unconsciously, some personal strategy which would produce a bias in the control method. The experts advise as a matter of course the use of the tables of random numbers but think that these also offer too much space for maneuvering and, thus, increase the risk of error. Thus, they invented numerous ingenious apparatuses which facilitated randomly selecting a sample with a minimum of intervention by the person performing the task.

The development of statistical control is thus mixed up with the division of labor and responsibilities on the shop floor. Apparently this mixing has been carried through with success, that is to say, in a way which is acceptable to and in harmony with the social order within the enterprise. This is certainly one of the strong points of the successful development of statistical quality control, i.e. it ably lends itself to fragmentation across the hierarchial structure, which assigns to each person a task corresponding to his social rank and level of education. Not all management methods have had this capacity, which explains why a number have been rejected.

Once this fragmentation is conceived and put into practice it becomes a factor which anchors statistical quality control in the firm. It is then integrated into the organizational system, and it is no longer possible to alter one element of the system without affecting many others and the cost of changing the system becomes very great.

Objects Engage Theory in Social Process

The object, while seen as the incarnation of theory for those in the know, can find itself engaged in other relationships which were generally not foreseen by the initial promoters of the theory and which emerge as the method is applied. The theory finds itself thus tied to new objects, taking part in new relations which contribute either to consolidate or destabilize the theory according to the circumstances. We take an example: statistical control changes modes of relationships on the shop floor. One of its advantages, according to the experts, is that it in fact allows one to decide, on the basis of objective and impersonal criteria, at what moment the machine has strayed off course. This means that the worker, on the basis of his reading of the control chart, can decide to call the machine setter or to continue production as usual. The machine setter or supervisor can thus no longer reprimand the worker as freely as they choose when something goes wrong. On the contrary, the machine setter can find himself in a difficult situation if the control chart shows that he has not been able to set up the machine properly.

Statistical control can also be used to modify relations among manufacturer, inspection, and the department of product design. Control charts provide a picture of the precision that the machines are capable of attaining and it would thus seem logical that the design department should take this into account in setting the tolerances. If not, a good number of products would not be in conformity with specifications and one might need to eliminate the nonconforming ones as waste. Before the introduction of statistical quality control, the design department was rarely called into question. The responsibility for bad parts lay with people on the shop floor, and this provoked disputes between production people and inspectors. According to some witnesses, statistical quality control has made it possible to break this closed loop by implicating the design department. It has also allowed resolution of questions of this sort by providing some tangible elements for discussion (the measures of quality and their statistical distribution).

A second example shows how theory, engaged with social process via its associated objects, acquires a social image which was not foreseen at the outset. The promoters of quality control at Ford, in 1950, produced a brochure intended for personnel training. In it statistical quality control was associated with images of physical and social improvement, of medical practice, and alarm systems. The analogy between quality control and tracking the state of a patient's health was based on the similarity between the graphic form of the control chart and the sheet recording the temperature of a patient:

> Charting is a running picture which keeps us up to date on the quality of work we turn out. Some people compare it with a patient's temperature chart in a hospital. Nurses take the patient's temperature at regular intervals. They plot each reading on a sheet of graph paper and connect the points by a line. When the doctor arrives to check on the patient's progress he notes the graph, which he considers a good general sign of the patient's state of health.

By analogy with alarm systems, statistical quality control is poetically characterized as a method which signals displacements from the ideal quality:

> It would be wonderful to have a series of lights and bells hooked up to every machine and operation. Then when our work would get 'just a hair' away from perfect, the bells would ring and the lights would flash.

Through this process of association of statistical control objects with other clearly social objects, a social perception of the new theory develops and finds itself anchored in a social reality which had been foreign to it. It is important to emphasize that these associations and anchorings are made through the use of the methods and tools of statistical control by social actors; they are not inherent in the objects themselves. To take an analogy from linguistics, it is the context which weaves a sense into the message. Through this phenomenon of social anchoring, this supplementary meaning ties the message to the object in a permanent way. The theory of management then becomes much more than a body of knowledge: it is a symbol bearing a value which it is no longer possible to separate from it.

CONCLUSION

The history of statistical quality control shows that two processes of construction have occurred in parallel. On the one hand we find the scientific construction of new properties of industrial products (for example the dispersion of characteristics), on the other hand the construction of objects permitting one to see these new properties and to manage them. Correlatively, these objects have thus an essential double use: they provide faith in the solid foundation of theory, and allow one to act in the real world. Through the use of objects and the acts of training which have accompanied their use, individuals have acquired new aptitudes such as the ability to perceive a dispersion in reading a graph.

The case addressed here illustrates the relationship which exists between control theories and the objects which serve to put them to work. This relationship is both trivial and enigmatic, according to the way in which one approaches the subject. It is trivial for the statistician: the objects are only the embodiment of the theory and it is the latter which is important, not the objects. The objects are only an aid, a prosthetic device, a material extension of the mind. But seen from the workshop floor, these objects are the tools one works with; the theory is as far removed from the floor manager as from the worker and both have no means to enable them to understand that theory in the same way as the statistician. The effective factor is then the object's capacity to represent something— in this example, the world which is represented is abstract and invisible—the world of statistical parameters. But the representation itself is surely real: e.g., the designated points on a graph, some lines and axes labeled with different numbers. Metaphor allows one to give various, isomorphic, meanings to these geometric figures; e.g. it is necessary to operate in a way which keeps things on a path defined by the two limits of control; to leave the route is an accident, the cause of which it is the job of the technicians (and not the worker) to uncover.

The control chart is a good example of an object which materializes a control objective. This object is subject to human cognition, and the understanding of its functioning is thus extremely complex. One can however gain a little clarity in locating it among other modes of control according to its degree of materiality. If the spoken word is of minimal materiality, the walls of a prison are at the other extreme. When a superior gives an order to his subordinate, there remains no material trace. A higher degree of materiality is attained with a written instruction. The order or the rule is written 'somewhere' and, if need be, can be exhibited. Maximum materiality is attained by the prison where the management of bodies is inscribed, as Foucault has shown, in its architecture and spatial arrangement. Prisons, hospitals, schools of the 19th century, are good illustrations of the methods of management and control of persons. By their materiality they act directly on the body and indirectly on the mind. Think too of the example of the bridge analyzed by Langdon Winner. In public transportation today, methods for managing people use procedures similarly acting on the body, even if they are less deliberately provocative: corridors and walkways within the metro, escalators, barriers and other guides. There is a continuity among the corridors which lays out the obligatory path of the traveler and the indicator panel lights of the stations, the signal system which indicates the trajectories that the managers of the stations hope will be used by travelers.[30] All serve the same ends. The difference is that the trajectories shown by the signaling system can be modified as a function of circumstances, with the location of trains for departure, etc. These are immaterial walls which require the use of the cognitive faculties of the travelers.

These different examples of objects produced with the intention of control apparently work in different ways: Walls can't be crossed, but one can act as if one has not heard what a colleague at work has said (this is more risky if the speaker is a superior). How to envision all of them as one thing—in so far as they are objects serving to control? We propose the concept of sign in the sense of C. S. Peirce. A sign according to Peirce does not have a unique, well defined interpretation; it is not a signal. It is a point of departure of a process called 'semiosis' which might go on indefinitely. A wall of a prison can signify, for a prisoner thirsting for freedom, an aim in his life as prisoner—to escape—and this will help him to remain in a state of mental alertness (at least according to the police literature). The effectiveness of a control object is not tied so much to its materiality as to the meaning which it takes on when it is interpreted by people. The focus of research thus finds itself displaced towards the study of situations which, we hypothesize, frame and orient the individual and the collective processes of semiosis. In particular, in the context of work within an enterprise, it seems necessary to study work practice, that is the cognitive activity at work as it is continuously engaged in interaction with the work environment. Through research of this kind, we can hope to understand better how the cognitive or material objects engage the person at work and sustain this engagement through a feedback process.

ACKNOWLEDGEMENTS

I wish to thank Professor L Bucciarelli for translating this article.

NOTES

1 For example, Berry, M., 1983: 'Une technologie invisible? L'impact des instruments de gestion sur l'évolution des systèmes humains,' Centre de recherche en gestion, Paris.
2 See notably the thematic edition of the journal *Raison pratique*: 'Les objets dans l'action. De la maison au laboratoire', Ed. de l'EHESS, Paris, 1993.
 Hutchins E., 1990: 'The Technology of Team Navigation' in: Galegher J., Kraut B., Egido C. (eds): *Intellectual Teamwork: Social and Technical Bases of Collaborative Work*, NJ: Hillsdale, Lawrence Erlbaum Associates.
 Suchman L., 1987: *Plans and Situated Actions: The Problem of Human-Machine Communication*, New York: Cambridge University Press.
3 Desrosières A., 1993: *La politique des grands nombres. Histoire de la raison statistique*, La Découverte, Paris, p. 9.

4   Kruger L., Daston L., Heidelberger M. (eds), 1987: *The Probabilistic Revolution*, Cambridge: MIT Press.
5   Stigler S. M., 1977: 'Eight Centuries of Sampling Inspection: the Trial of the Pyx,' J. Am. Stat. Ass., vol. 72, pp. 493–500.
6   Nusbaumer E., 1924: *L'organisation scientifique des usines*, Nouvelle librairie nationale, Paris. Preface of H. Le Chatelier.
7   The debate is still ongoing with the partisans of 'zero default.' Certain people see the approach of statistical control of manufacturing as the institutionalization of inefficiency: the machine operators, knowing that the products are inspected at the end of the production line, do not particularly seek to correct any defaults. The primary intent of a policy of 'zero default' would be to force the workers to coordinate their activities.
8   Laplace P. S., 1986: *Essai philosophique sur les probabilités*, 1825, reed. Christian Bourgois, Paris.
9   Example to be found in: *The Principles of Scientific Management*.
10 Nusbaumer E., 1924, *op. cit.*
11 Taylor, F. W., 1907: 'L'emploi des courroies,' in: *Etudes sur l'organisation du travail dans les usines*, Dunod et Pinat, Paris.
12 In: Mayr O. and Post R. C. (eds), 1981: *Yankee Enterprise. The Rise of the American System of Manufactures*, Smithsonian Institution Press, Washington, D.C., p. 153. The whole of this book shows the primary importance of the question of interchangeability of parts and the astonishment of the industrial world when confronted with the performance of American manufacturers in this matter.
13 Bayart D., 1996: 'Savoir organisationnel, savoir théorique et situation: le contrôle statistique sur échantillon,' *Entreprises et Histoire*, No. 13, 67–81
14 Shewhart W. A., 1931: *Economic Control of Quality of Manufactured Products*, New York: Van Nostrand and MacMillan, London.
15 Shewhart W. A., 1924: 'Some Applications of Statistical Methods to the Analysis of Physical and Engineering Data,' *Bell System Technical Journal*, vol. III, No. 1, 43–87.
    Shewhart W. A., 1926: 'Quality Control Charts: a brief description of a newly developed form of control chart for detecting lack of control of manufactured products,' *Bell System Technical Journal*, vol. V (1926), 593–603.
    Shewhart W. A., 1930: 'Economic Quality Control of Manufactured Product,' communication Am. Assoc. Advancement of Science, Des Moines, Dec. 1929, published in *Bell System Technical Journal*, vol. 9 (1930), 364–389.
16 Fagen M. D. (ed), 1975: *A History of Engineering and Science in the Bell System, The Early Years (1875–1925)*, vol. 1, Bell Telephone Laboratories. Chapter 9: 'Quality Assurance'.
17 Shewhart, 1924, pp. 43–44.
18 Fagen, 1975, *op. cit.*

19   Note reproduced in Fagen, 1975, *op. cit.*
20   Simondon G., 1969: *Du mode d'evolution des objets techniques*, Paris, Aubier.
21   American Society for Testing Materials, 1935: *Manual on Presentation of Data, Supplement B*.
22   In the meantime, Shewhart had conducted many experiments with his 'bowls', and this conclusion is rather empirical.
23   A connection which is not simply due to chance, Shewhart having been introduced to this work of Laplace by E. C. Molina, of Bell Labs, a mathematician and connaisseur of the history of probabilities.
24   A French example provides an element of comparison: Maurice Dumas, an engineer who developed an accurate probabilistic thinking about acceptance sampling in 1925, did not meet with any success. He was not backed up by heavy industrial forces.
25   Shewhart 1930, *op. cit.*, p. 4.
26   see note 2.
27, For example, Grant E. L., and Leavenworth R. S., 1972: *Statistical Quality Control*, McGraw Hill, International Student Edition.
     Peach P., 1947: *An Introduction to Industrial Statistics and Quality Control*, Raleigh, N. C.: Edwards Broughton.
28   Stigler S. M., 1986: *The History of Statistics*, Harvard University Press.
29   Dumas M., 1925: 'Sur une interprétation des conditions de recette,' *Mémorial de l'artillerie française*, tome 4, fasc. 2., pp. 395–438.
30   These remarks are based on an empirical study realized within a large Parisian railway station with the support of the transport agencies.

CHAPTER 8

Ideology Counts: Controlling the Bodies of Concentration Camp Prisoners

Michael Thad Allen

NAZI RACIAL IDEOLOGY AND PRODUCTIVISM

In April of 1943, Oswald Pohl, the chief executive of SS slave-labor industries, wrote one of the dry letters emblematic of both the convoluted prose and the dispassionate, bureaucratic cruelty of Nazi Germany:

> It has been reported to me by the camp *Kommandanten* about the conditions of health in the concentration camps. This yields the following picture after the situation of 1 April, 1943.

Typically, Pohl termed ongoing attrition among prisoners a state of 'health' to be measured in numbers, which he duly gave:

> 1) 12,658 protective custody prisoners (*Sicherungsverwahrten*) were taken over by the concentration camps.
>
> 2) Of these 5,935 have died.
>
> 3) The strength at hand on 4/1/43: 6,723.[1]

These prisoners had recently transferred from the Justice Ministry's jails, and Pohl could account for them down to the last prisoner.

Such scrupulous statistical discretion over life and death has come to epitomize the banality of the Holocaust: Pohl might as easily have spoken of scrap metal or margins of profit and loss instead of the broken bodies of men and women. However, although mass killings and slave labor had begun years before, SS managers began keeping such figures in any rigorous fashion only after 1942 as Germany mobilized every last available resource for total war.

Hitler's campaign against the Soviet Union had grown increasingly bitter with each passing month, and straightened circumstances had led to key negotiations between the SS and Albert Speer, who had become Minister of Armaments and Munitions in February. March to September of 1942 marked a turning point as the SS sought to convert its prisons into labor depots for weapons plants. In order to do the bidding of modern industry, the SS needed to introduce the methods of modern control; that is, Pohl had to substitute abstract, statistical observations for personal, on-site supervision of machines and working bodies in production; his managers had to convert the masses of prisoners, whose individuality they had already stripped as much as possible, into numbers.[2] It is a grim irony that our common perception of the Nazis' demonically scrupulous accounting of industrialized killing largely stems from techniques which the SS deployed only in the depths of wartime, a full decade after Adolf Hitler had come to power.

This irony is pertinent to the study of control as well as to the history of the Holocaust. Over the past decade historians of science have argued that the Enlightenment gave issue to an 'avalanche of numbers,' an 'orgy of rationalization.'[3] Much in the tradition of Michel Foucault's *Discipline and Punish*, several recent essays on the culture of precision allude to the effort undertaken since the Enlightenment to alter entire populations through the use of statistical surveillance.[4] The Holocaust was undeniably one such experiment in the manipulation of demography on a vast and terrifying scale.[5] Social scientists, predominantly Zygmunt Bauman and Wolfgang Sofsky, have identified the Nazi's mania for social control as modern society's most evil legacy.[6] Here control implies much more than stability or steering; it means laying statistical hands on the body politic in order to 'heal' its perceived infirmities and even to create a 'new man.' And it was precisely the desire immanent within National Socialism to mold a new German spirit and form the clay of humanity into a 'Thousand Year Reich' which Bauman, following Hannah Arendt, declared totalitarian and condemned as the root of the Nazis' contempt for human life. Yet to Bauman the techniques of control born of the Enlightenment and the systems they were intended to sustain often appear as purely instrumental: the alienated tools of morally stunted citizens. As he writes, a 'meticulous functional division of labor' in modern bureaucratic states led to the 'substitution of technical for a moral responsibility.'[7] The Enlightenment led to great evil, in his estimation, because it inured ordinary Germans to moral judgment; in the name of instrumental reason, it stripped modern society of ideological discourse, including discourse about the moral good.

On this point, historians of science and technology have proven themselves a bit more analytically astute than social scientists like Bauman or Sofsky because they have delved into the moral commitments that lie sleeping in numbers. In other words, they have uncovered the historical processes in which ideology counts. Precision, as A. Norton Wise notes, 'is never the product simply of an individual using a carefully constructed instrument. It is always the accomplishment of an extended network of people.'[8] The Holocaust was managed not by a mere technocracy in Bauman's sense, that is, by experts operating in ideological alienation or ignorance of larger political agendas. SS perpetrators acted within bureaucracies, organizations that were, at one and the same time, sustained by dense webs of cultural meaning that guided SS officers' interpretations of production and information transfer among like-minded men.[9] The modern hierarchy of the corporation, as Olivier Zunz has pointed out, constrained but also empowered as it invited its initiates to unleash the heightened command and control possible through collective action.[10] Far from erasing humanity's moral impulses, bureaucracy can magnify and focus them.

As Ted Porter and Norton Wise point out, 'rigorous quantitative methods served primarily to discipline qualitative decisions, not to replace them.'[11] And for such discipline, ideological judgment is needed. In spite of the fact that many of the Enlightenment's advocates sought to eliminate ideology from their technics of control, we would err if we judged the Enlightenment to have succeeded in this. The seeming paradoxical dualism of social values and technics is and always was a false dilemma. The measurement of men necessarily contained, at its core, substantive judgments about the nature of the citizen, the state, political economy, or in the case presented here, about the nature of cancers in the body politic which the SS sought to remove surgically.[12]

The rhetoric of nature, body, and disease appears here for two reasons. First, as numerous historians of the medical profession have demonstrated, the Third Reich's doctors conceived their task in these terms.[13] Second, within the concentration camps this rhetoric structured daily practice. For example, it lead the SS to call upon doctors and medical orderlies (*Santitätsdienstgräde*) to manage its labor force (instead of, for example, engineers or industrial managers).

The SS conceived of prisoners engaged in industry as 'human material,' a designation that had distinct ideological roots and caused the SS to diverge from the way in which German industry typically evaluated and tracked civilian workers. To be sure, industrial managers also sought control over their workers' bodies through statistics, but such numbers often measured productivity, hourly wages, labor hours per unit product, and countless other specific indices of humanity, technics, and organizational structure (what Bruno Latour would call 'collectives').[14] Such managers, especially those who controlled the factory floor, were necessarily engineers or technical specialists. It would be a mistake to consider the German engineer immune to medicalized ideals of the body politic, which were as much a current topic of discussion in technical trade journals as in those of the medical profession. Yet with their expert knowledge engineers could combine accurate surveillance of both humanity and the material world of industry.[15] The SS, by contrast, assigned the key position of oversight in prison factories to medical orderlies and doctors, who managed only the material of the prisoner's body, not that of the factory as an amalgamated whole.[16] As one might expect, these unusual managers focused not on the quality of production but on prisoners' health. Thus Pohl's reference to his statistics as measures of 'health' instead of a mere body count was not a mere cynical euphemism but an accurate representation of what the SS had set out to manage.[17]

SS medicalized management, which became the norm by 1943, resulted from substantive judgments passed upon the nature of prisoners as biological degenerates, a human material that the SS as well as German industrial planners considered to be a danger, not a contribution, to national industrial output. On this point Nazi doctors, economists, and engineers largely agreed as demonstrated by the Four Year Plan, the Third Reich's blueprint for rearmament promulgated in 1936. Its offices warned of conspiracies aimed at 'economic sabotage' and demanded a law 'making all Jewry liable for any damage caused to the German economy—and thus to the German people.'[18] Beyond the intensified persecution of the Jews, the Four Year Plan also led to new categories for criminal detention such as 'asocial,' a broad term for anyone found lagging in productivity or otherwise failing to contribute adequately to industrial output.[19] Nazi 'racial hygiene' cast this term as a genetic trait: 'human beings with a hereditary or irreversible mental attitude, who, due to this nature, incline toward alcoholism and immorality, have repeatedly come into conflict with government agencies and the courts, and thus appear ... a threat to humanity.'[20] As a resulting conclusion, those whom the Nazis criminalized quickly became the subject of medical surveillance and control; they were considered 'burdensome lives,' 'life unworthy of life,' 'useless eaters'; and the Nazis deployed a host of other terms that defined individual worth by reference to use value to the body politic or *Volksgemeinschaft*.[21] Prisoners, according to this logic, had to be removed from the mainstream of social production, not analyzed as part of it.

The SS set itself the task of sending such 'ballast' into the concentration camps. Once inside the gates, no one tried to harness prisoners' agency as workers; rather, guards sought only to drive them as slaves. The SS wished to force prisoners to contribute to social production against their nature as degenerates. In this light it never occurred to SS henchmen that the camps were barbarous institutions. In the minds of their wardens, the camps were progressive, part of a moral imperative to improve the recalcitrant nature of those who would otherwise be a drag on the nation. As Heinrich Himmler put it, production had to be 'earned by putting the scum of humanity, the inmates, the habitual criminals to work,' or in another context, '... we must ... help with our energy on location to drive things forward with the bludgeon of our word.'[22] Nevertheless, as much as the SS wished to see itself as progressive, 'racial hygiene' in the absence of technical knowledge did not, at first, encourage an advanced rationalization of the concentration camps. Prisoners needed to be driven with a whip rather than subtle attention to production organization, and the SS continually overlooked the role of factory engineering and machines precisely because of this misconception. The consistent application of violence needed little statistical accounting. Up until 1942, it had been routine for SS guards to beat prisoners to death even when it ruined output. As irrational as this was, Nazi ideology was nevertheless consistent. Concentration camp personnel already believed they were doing a service to the Reich by protecting the German economy from inmates. Only the requirements of total war made this practice unacceptable, but long after the SS nevertheless remained captive to its medicalized ideology—it merely merged with rationalized industry in new forms.

When total war came and the SS had to initiate systematic management, *what was measured* was as significant as *what was not*. Instead of factoring prisoners into tables of productivity, the SS often reduced management to an analog function: either a prisoner was 'fit to work' (*arbeitsfähig*) or 'unfit to work' (*arbeitsunfähig*), and those declared unfit were subsequently returned to the stream of victims headed for immediate extermination. Camp industries treated the prisoners as a raw material in an open-loop system: the almost unceasing flow of fresh workers came freighted in cattle cars, and, as output, came the grey ashes of the dead from the crematoria. SS doctors were arbiters of both killing sites and work sites and, in balancing the two ends of this loop, they increasingly turned mass murder into a factory-like operation. As they stepped into the role of labor managers, they also began to see their job in industrial terms, a perception that was no mere analogy. Thus, to state, as I have above, that the SS neglected factory organization turns out, in the end, to be an exaggeration; the SS did concern itself with such management and control, but only in matters of murder. The SS always intended its most successful products to be corpses.

HOLOCAUST STATISTICS: THE SS'S FIRST EFFORTS AT CONTROL

Of course, the SS had been operating concentration-camp industries before total war, but these were insignificant and, for the most part, also inefficient (even in the SS's estimation) not to mention horribly wasteful of human life. When, as every other avenue to labor dried up, increasing numbers of armaments producers began to ask the SS for slaves, the SS had to make some concessions toward the war economy even as it freighted millions of Jews toward the gas chambers of Auschwitz and Majdanek.[23] For example, Heinrich Himmler promised to set aside some '100,000 male Jews and up to 50,000 Jewesses' as workers for 'great economic tasks' even as he continued to send millions to their death.[24]

Of note, however, Himmler's call for 150,000 'Jews and Jewesses' went almost entirely unheeded within the Inspector of Concentration Camps (IKL or *Inspeteur der Konzentrationslager*) which lacked any administrative system to track camp populations. In early 1942 the SS had almost no idea of the number, location, or condition of its prisoners. A top official in the IKL had already ordered the transfer of 100,000 Jews to a central labor camp at Lublin (Majdanek), but the request had almost no effect. Local concentration camp *Kommandanten* often could not tell how many prisoners they had on their hands and could not effectively control such transfers.[25] Labor camps routinely received the wrong numbers and the wrong prisoners at the wrong worksites, and throughout the years 1939–1941, as the SS had tried to organize its own modern industry (separate from German armaments production), acrimonious complaints filled internal correspondence.[26]

SS correspondence with the Armaments Ministry only underscores this point. In urgent meetings in March, the Inspector of Concentration Camps (Richard Glücks) had to admit that he did not know how many prisoners the SS could provide to Germany's war industries. Even though the SS had already agreed that slave labor should be shunted into armaments production 'as swiftly as possible,' the IKL could only offer vague promises of 5,000 prisoners at Buchenwald, 6,000 at Sachsenhausen, Auschwitz and the women's camp Ravensbrück, and 2,000 at Neuengamme. The gross statistics, rounded to the thousands, contrast markedly to Oswald Pohl's precise data less than one year later (cited at the beginning of this article), and they displayed the fact that the SS had no idea of how many working inmates it had or where. The IKL even left Majdanek out of its tallies, where the SS was supposed to have 100,000 fit, working inmates at the time.[27] To add to the confusion, an armaments official quickly found out that Neuengamme, for one, could muster only 200 workers for a pilot project, and camp administrators were interfering impossibly with his plans.[28]

The Armaments Ministry was used to getting precise information from industrial managers, and the language it spoke was the statistical language of technological management hammered out between engineers. Contemporary engineers went so far as to describe this work as the 'work community of technology,' referring to the spirit of like-mindedness among technical men.[29] One of Albert Speer's first achievements as armaments minister had been the imposition of an accurate quotas and vouchers system for raw materials allocations, which demanded not only the manipulation of existing stockpiles, but precise estimates of productivity, output, and the projected needs of future manufacture.[30] The engineers and managers of private industry were not necessarily adverse to the SS's insistence on force or 'racial hygiene.' In fact, they labored mightily throughout the war to reorganize factories for compulsory labor, and often badgered the SS to get rid of sickly prisoners who were no longer useful in their plants. Industry had great success with assembly lines that curtailed as much as possible the agency of forced laborers (achieving efficiency rates up to 98% those of civilian German workers).[31] In early 1942, the SS could not muster equivalent mechanisms of production control.

Egregious mismanagement of slave labor had not gone unnoticed or unanticipated by the SS itself. By the end of 1940, a full year before Germany's mobilization for total war, Pohl as chief executive officer of all SS companies, had already ordered the formation of a special SS Labor Action Office, whose sole duty was to track the camps' working populations. Pohl chose Wilhelm Burböck, a long-standing Nazi from Austria to lead the office, and Burböck generated a series of initiatives in a burst of activity. At least on the surface, he began creating a modern system of statistical control which, in an era before computers, meant beginning cardfiles and compiling data into charts and graphs.

Burböck split his Labor Action Office into two divisions. One, in his central office near Berlin, gathered and collated information. The second, composed of officers he sent into the field to regional labor operations, managed slave labor on-site. For this second branch, Burböck also established the Detention Camp Führer-Labor Action (*Schutzhaftlagerführer 'Einsatz'*), a new post within each *Kommandant's* staff. He made their main responsibility a quintessentially bureaucratic task: they had to develop and watch over a cross-referenced card file.³³ As banal as this seems, these banks of cards are of special importance because they formed the core information system for all subsequent SS slave-labor management. The Labor Action Führer was supposed to list working inmates by name, number, skills, and their history of labor experience within the camp system. If maintained diligently, for any given project a Labor Action Führer could tally at a glance through his cards the exact number and kind of workers available, where they were, their skills and history.

Burböck also laid down rules for the allocation of prisoners. Ostensibly, his Labor Action Führer were to intervene in production management as liaisons to factories (almost exclusively SS companies before 1941). All firms had to first apply in writing for workers. On one hand, the Labor Action Führer had to negotiate with industrial supervisors regarding the duration of labor, the feasibility of operations, and wages; on the other hand, he had to hammer out with camp *Kommandanten* the selection of prisoners and security conditions.³⁴ In theory, this system would allow the IKL to coordinate all camps with only a minimal staff in to review reports.

Of note, in its original conception, this system of control encompassed both prisoners' bodies and the skills they literally incorporated. Color labels were supposed to differentiate skills in the files. However, Burböck did not, at first, create the kind of statistical shorthand necessary for modern control: the reports he requested came embedded in prose instead of codes, and this had consequences when the SS began allocating inmates. Burböck had introduced a new information system with quantifiable standards (wages, population, guard personnel), but he did not extend statistical surveillance to the sophisticated management of machines, raw materials, and labor organization (quantified entities like productivity, unit output per unit time, depreciation, or labor hours per unit product). Perhaps because the SS charged only a pittance to its own cost accounts for leasing prisoners, Burböck did not believe such controls were important; but, in contrast, low rates for slave labor never stopped German engineers in private industry from calculating such figures. In fact, once the Third Reich's war economy came to rely on myriad forms of compulsory labor after 1941–1942 (making up over one third of laborers in some key sectors), trade journals dedicated to labor management carried articles detailing how supervisors could "count" slave laborers with precision.³⁵ Labor still entailed an expense even when no wages went to the prisoners: for example, firms often had to carry the costs of provisions and security, among other things, and usually had to reckon *decreased* productivity and *increased* depreciation of machinery into their books. Burböck's statistical control did not reference factory production; it sought only to measure crude numbers of prisoners' bodies with only a cursory acknowledgment of their skills. In time, emphasis on skills would fall away almost completely. While, in theory, the SS Labor Action set out to place dependable data at its officers' fingertips, in practice the IKL reneged on its charge to control labor and production.

Nowhere was this neglect more evident than in Burböck's gestures toward directing labor sites by using photographs. He asked his Labor Action Führer to take snapshots of work details and send the pictures to his office in Oranienburg. In theory, he would study these photos in order to determine more efficient guidelines for supervision without actually having to visit each site. In reality he allowed the Labor Action in individual camps to proceed without his direct intervention. Remaining in Oranienburg, he expected his organizational scheme to run automatically once he had dictated it to his subordinates. Completely consistent in this regard, he never imposed statistical measures upon the performance of his own managers, which left him without any feedback mechanism. On location they followed Burböck's example. Each delegated responsibility and supervision on down the camps' managerial hierarchy and failed in turn to monitor production for effective control. Regardless of Burböck's orders, the actual oversight of work details was passed on to mere SS guards or Kapos (prisoner supervisors often more renowned for brutality than the SS itself).

By late 1941, just months before Germany's transition to a total war economy, Burböck's entire operation had acquired the character of a sham. His initiatives were convincing on paper but amateurish in application. Over the course of 1941 his office became a wellspring of querulous letters and ineffective directives as Labor Action Führer routinely ignored the central directives of the IKL.[36] In February of 1942, the Inspector of Concentration Camps, Richard Glücks, liquidated Burböck's office and his regional representatives stationed at each camp. Glücks simultaneously ordered camp commanders to redouble their efforts to facilitate labor management but refused to stir a finger to insure that their directions would actually be carried out.[37] Thus on the eve of Germany's transition to total war, when modern industries quickly turned to the IKL as a slave labor lord, the SS actually moved to eliminate even these partial vestiges of statistical control over camp population.

STATISTICAL CONTROL OVER THE 'UNFIT TO WORK' AND THE ACTION 14 F 13[38]

It is important to point out the conditions this managerial turmoil created for prisoners involved in the SS Labor Action. Here one simple innovation deserves special attention because it partially enabled the SS to manage the catastrophic suffering of the camps in conjunction with the genocide. Burböck, as noted, failed at first to introduce statistical controls upon his mid-level managers, but in the late summer of 1941 he did urge the implementation of one such category: the inclusion of 'Unfit to Work' on standard reports. That prisoners were already sick and dying at this time says much for the poor organization of supplies, the atrocious conditions of shelter, and the brutality of daily life in the camps. Death had been endemic to the camps since their inception, and mortality rates had become a recognized problem leading to requests for crematoria from the IKL by mid-1938. The first crematoria did not yet directly serve in the extermination of the Jews (who were a minority of all concentration-camp prisoners at this date) but to cover up the indiscriminate flow of corpses produced by the vicious conditions prevailing in the camps.[39] Thus very early on and in advance of full-scale genocide, the SS showed itself willing and able to define mass death as a technological problem, albeit one whose ends at once grotesquely mirrored and yet differed from those of German armaments factories. In addition to the need to design efficient crematoria, security remained the overarching concern: the IKL was afraid that ordinary German citizens might notice the steady transports of corpses from inside the camp fences and grow suspicious. Nazi municipal officials (who in principle had nothing against brutality towards social outcasts) also felt that the sight of so many dead bodies was unseemly and complained.

Evidence that IKL officers or SS business executives recognized the failing health of inmates as a *production* problem rather than an issue of security appears first within Burböck's SS Labor Action Office in directives that called for the removal of sickly workers from camp labor forces. Burböck repeatedly warned his officers to stop allocating weakened prisoners to SS industrial enterprises.[40] By the summer of 1941, in fact, epidemics and the attrition of inmates was causing a labor shortage within the SS's own companies.[41] Burböck therefore ordered his officers to add a new category to their filing systems and reports: '*Arbeitsunfähig*' or 'Unfit to Work,' and it became one of the few collated in numerical form.[42] He intended this statistical innovation to provide more accurate information about working populations throughout his system so that the Labor Action Führer would, theoretically, be able to distinguish between healthy and incapacitated workers at a glance through their files.

The enactment of this innovation, however, underscores the modern *mismanagement* of the concentration camps, for the statistics facilitated a new method of murder, the Action 14 f 13, in which the IKL culled prisoners systematically for extermination for the first time. The project was originally conceived to purge the mentally ill and physically handicapped, categories that included recidivist criminals because the Nazis considered their social deviance a genetic, that is, medical trait. *Kommandant* staffs had to conduct preliminary selections in preparation for special commissions of doctors who arrived to pass ultimate 'scientific' judgment on those who would be eliminated and those who would be allowed to live.

According to one doctor's testimony, the categories for extermination were broadened over the course of 1941.[43] As another testified, 'In the autumn of 1941 an investigation was conducted on all Jews by the camp physician [at

Buchenwald]. Those that were *unfit for labor* were sorted out [my emphasis].'[44] This was the period directly after Burböck stepped up his complaints about sickly prisoners in the Labor Action. It is also certain that Burböck was in constant contact with the IKL adjutant who was supervising organized selections for extermination at this time. Jews were the largest single group of prisoners selected, but selections included Poles, Czechs, 'Asocials,' and 'Inveterate Criminals.' (In one selection of 293 prisoners, 119 were Jews, the rest fell in other categories.)[45] The most consistent label, that which all selected had in common, however, was 'unfit to work.' At this time, 'special selections' on the basis of these same categories also began at Auschwitz.[46] Starting with the Action 14 f 13, statistical compilations of the 'unfit to work' became a managerial tool for the liquidation of prisoners.

Yet even here, the IKL's style of mismanagement emerges more clearly than rhetoric of well-oiled 'machinery of extermination' or the 'technocrats of death' might imply. For instance, on the 19th and 20th of January, 1942 the doctor Fritz Mennecke selected 214 inmates out of a group of 293 for extermination at Groß-Rosen.[47] The *Kommandant's* staff had first specially selected prisoners upon receiving a quota: Groß-Rosen had to cull 250 prisoners from the camp population. 'The requested number ... was exceeded by 43 in order to have the necessary elbow-room for possible losses...';[48] that is, the camp staff saw the action as a chance to demonstrate initiative and ambitiously provided 293 instead. The Kommandant of Groß-Rosen sent a concluding report in March:

> 214 inmates were selected [i.e. by the doctor from the 293]. From this number 70 were transferred on 3/17/42, and 57 inmates on 3/18/42. Between 1/20 and 3/17/42, 36 selected inmates died. The remainder of 51 inmates consists of 42 Jews who are able to work and 10 other inmates, who have regained their strength owing to a temporary cessation of work (camp closed between 1/17 and 2/17/42) and who will therefore not be transferred.[49]

The tone of the report was one of pride. The Kommandant's staff had enthusiastically culled prisoners for extermination that actually recovered, which demonstrated their weak commitment to productivity as a value in selection for death. Of note, the Kommandant's numbers did not always add up (10 plus 42 does not equal 51!), a further demonstration that the modern control which the Labor Action Office strove to impose was more sham than reality. Camp staff saw the Action 14 f 13 as a convenient excuse to purge unwanted inmates; the desire to excise 'disease' from the body politic overrode any concern for factory management; and doctors helped in the task, even when the possibility existed that the sick might recuperate for meaningful work.

Arthur Liebehenschel, adjutant to the Inspector of Concentration Camps rebuked the mismanagement at Groß-Rosen:

> According to the report ... 42 of the 51 inmates selected for special treatment 14 f 13 became 'fit to work again' which made their transfer for special treatment unnecessary.

This shows that the selection of these inmates is not being effected in compliance with the rules laid down. Only those inmates who correspond to the conditions laid down, and this is the most important thing, who are no longer *fit to work*, are to be brought before the examining commission [emphasis mine].[50]

This protest did not halt or alter the implementation of 14 f 13. Kommandanten put more creativity into seeking administrative loop-holes for killing the 'unfit' than in preserving the 'fit' for production, and they happily applied the new statistical files that Burböck's Labor Action Führer put at their disposal.

By inserting 'unfit to work' on an information sheet, Burböck gave *Kommandanten* and medical personnel a category with which to organize the sick and injured in files that were easily-accessible and easy to compile. Kommandanten, in turn, used Burböck's technique to expand the mode of operations that their calling had long demanded of them: the destruction of prisoners, which could now proceed on a new and mortal scale, in part because of the new tools of information manipulation. Burböck had provided even more than this: he gave them a conceptual bridge. Now *Kommandanten* could believe that they too were actively safeguarding productivity by parsing camp populations into the 'fit' and 'unfit to work'; they simply concentrated most enthusiastically on the elimination of the 'unfit.' This brief digression also casts a new light on the judgments passed upon the Enlightenment, instrumental reason, and the 'iron cage' of bureaucracy which post-war scholars have often blamed for driving banal managers like Burböck to acts of industrial genocide. Far from a 'discourse' of rationality which compelled these perpetrators to ratchet up the scale and scope of extermination, ideological commitment, which placed the cleansing of the body politic foremost, animated their conscious initiative. Modern means of control only accelerated the pace after the SS had already passed substantive judgments upon the bodies of inmates.

After the 14 f 13 program, the SS's system for murder and labor encompassed doctors as well as mid-level managers, as the participation of Dr. Mennecke demonstrates. Mennecke, in fact, strove mightily to meet his own quotas for 'special treatment,' and he wrote extensive letters home expressing his near glee with the entire operation. This was not a civil servant whose moral sensibilities were dulled by weary days totting up figures; rather, he made clear that moral judgments adhered to his statistics, judgments he shared proudly with his wife. Typical of Nazi racial and productivist ideology, he conflated categories of health, criminality, race, and industrial efficiency. '...all [Jews] do not need to be "examined,"' he wrote his wife, 'but ... it is sufficient to take the reasons for their arrest from the files (often very voluminous!) and to transfer it to the reports. Therefore it is merely a theoretical work.' Yet rather than declare a death sentence upon these Jews in racial and criminal terms, the 14 f 13 reports used Burböck's measure of industrial efficiency, 'unfit to work,' and in the process the eugenic presumption that productivity was a function of biology found expression in statistical terms.[51]

THE SS DOCTOR AS INDUSTRIAL MANAGER

Increasingly, the SS deferred to doctors and trained orderlies to supervise its data on the laboring bodies of inmates. The Inspector of Concentration Camps simultaneously channeled its management away from technological systems of production and toward the supervision of prisoners' health. Burböck proved an abject failure, as noted, and the IKL removed him from the SS Labor Action in consequence (before the Ministry of Armaments and War Production came calling); but Burböck's successor used the same techniques and interpreted his job in the same medicalized terms even as he turned the old card files into effective tools.

In March of 1942 Himmler made the executive decision to incorporate the IKL, up until then a free-standing administrative office, into the SS Business Administration Head Office (WVHA) which had previously been responsible only for the SS's industrial enterprises. An experienced industrial manager, Gerhard Maurer, took over the Labor Action 'OC' to be more precise, he reconstituted it after its demise under Burböck. Although Maurer had no advanced education, he had gained experience in accounting and corporate administration during the Weimar period as a factory book-keeper. After the factory closed in the Great Depression, he found work in a bank. By the beginning of 1942, he had also served as a liaison to the IKL and had helped establish links to IG Farben at Auschwitz.[52] He came to his new duties after overseeing the finances of the German Equipment Works, which had employed over 7,000 prisoners before 1942. Likewise his deputy, Karl Sommer, was also an energetic and tenured manager of SS companies, who, according to post-war testimony, possessed a phenomenal memory for numbers and an eye for minute detail. The two had met while Sommer was an employee of the German Earth and Stone Works, the largest SS corporation, and they had already formed a working relationship.[53]

In the Labor Action Office Maurer provided an inspiring and tireless example quite different from his dilettantish predecessor. He quickly established himself as the de facto chief of the newly established Office Group D-Labor Action within the WVHA, and after his wife was killed in a bombing raid on Halle in August of 1944, his subordinates and superior officers alike testified that he worked with redoubled concentration. He nearly eclipsed his nominal superior officer Richard Glücks, who, not unlike Burböck, was an ineffectual manager. Maurer continually absorbed tasks that were ignored and neglected by others until hardly any important document emanated from the Inspector of Concentration Camps without his initials in the letterhead. The chief of the WVHA, Oswald Pohl, came to trust Maurer's judgment almost unconditionally and routinely signed off on his initiatives (as did Glücks).[54]

Yet indicative of SS industrial practice, neither Maurer nor Sommer had ever managed the factory floor. Their experience was confined to the manipulation of statistics, but they lacked the expertise to ensure that their numerical calculations

Ideology Counts: Controlling the Bodies of Concentration Camp Prisoners 189

accurately mirrored the agglomerations of material, machines, and human beings engaged in production. This deficiency had no small consequences as they intervened in the concentration-camp system. Although Maurer and Sommer knew the power of numbers and what to do with them, they had no choice but to rely upon others to generate statistics about the material world. Their entry into the SS Labor Action soon created the basis for a new cooperative relationship with engineers in Speer's Armaments Ministry as well as private industry based on a division of labor: Maurer and Sommer managed prisoner's 'health' while German engineers managed the material world of the factory floor.

Maurer availed himself of the existing role of medical doctors to generate comprehensive statistics on the camp's working populations. He could thus extend control over the material of prisoner's bodies, a more simple and pliable domain than the aggregate technics of the factory floor, for which he lacked the knowledge and experience. Here his efforts proved more successful and began to meet not with the resistance but with the eager cooperation of industry. Burböck had already suggested and partially carried through medical surveillance of camp populations, but only Maurer took effective action to implement it. All post-war testimony confirms that Maurer, not Burböck, finally established standardized card files that differentiated 'fit' and 'unfit' laborers at this time.[55] Moreover, files, such as they existed under Burböck, had varied from camp to camp. Maurer homogenized his system so that information could be transferred throughout a national network of slavery in interchangeable form. He also systematically extended his control over population statistics in a way that Burböck never considered. By introducing standardized forms—a classic technique of modern management—Maurer enabled his office to dictate from Berlin what statistical information subordinates at lower levels had to report, and he thus enforced the uniformity of surveillance, a further step toward the interchangeability of data needed for modern control. As Shoshana Zuboff has noted, these techniques served to discipline both the body of the worker and the eye of management. Statistics from myriad work sites became manageable in polyglot.[56]

Yet contrary to claims that instrumental rationality drove the pace of genocide, Maurer's statistics contained his own ideological reckoning not his alienated conscience (and Maurer cannot, by any stretch of the imagination, be considered an alienated bureaucrat!). Forms, like the one Maurer introduced, are easy to overlook, yet the style of the WVHA was reified in these slips of paper no less than it was extolled in the florid propaganda of Josef Goebbels. They had nine entries, all to be filled with a simple cardinal number; beyond noting entrances and transports from the camps, they exclusively requested information about the condition of prisoners' bodies; and thus the medicalized management of prisoners became rendered in fungible data.[57]

A low-level officer under Maurer reported how this worked in practice:

> Death announcements were ordered alphabetically in a well equipped cellar room ... I performed this activity daily.... Later I collected the incoming population reports, that

190 *Cultures of Control*

came by mail from the individual concentration camps. I collated these together by the following criteria: age of prisoners, entry-date, population, history of transfers, cases of death, special cases of 'special treatment' [systematic extermination].... My activities collating the populations of camps were collected into a general statistical report every month.[58]

This passage might, on the surface, compare well to the similar controls imposed by Andrew Carnegie's steel mills in the 1880s when such methods of modern surveillance were invented. As described by Alfred Chandler,

... each department listed the amount and cost of materials and labor used on each order as it passed through the subunit. Such information permitted ... Carnegie monthly statements and, in time, even daily ones providing data on the costs of ore, limestone, coal, coke, pig iron, spiegel, molds, refractories, repairs, fuel, and labor for each ton of rails produced.[59]

The comparison illustrates, in other words, that Maurer's forms functioned just as any rational tools of administration, but what do they reveal about SS ideals? In one important aspect, the SS's meticulous attention to numbers differed from Carnegie's: instead of measuring the collective inputs and outcomes of material production *in toto*, Maurer measured only his prisoners' health. Maurer did not, for example, demand statistics on the supply of foodstuffs to the camps, nor on the quantity or quality of work performed by fit prisoners. This would have meant seeing the entirety of production as a problem *of the SS*. It was easier and ideologically consistent—as an SS man—to seek the root causes as a medical problem *of the prisoners*. One merely had to drain off the 'poor human material' from the camps like so much waste. Maurer's forms originally carried the heading 'Prisoners Unfit to Work or Prisoners Unavailable for the Labor Action [Maurer's underlining].'[60] Burböck's original term 'Unfit to Work' thus reappeared in refined form with immediate consequences. From now on the main responsibility for maintaining working populations fell within the purview of the camp doctors, quite consistent with the SS Labor Action's medicalized/ eugenic first principles. The IKL's medical officers (*Standortärzte*) now became key organization men in Maurer's administrative scheme.

By 1943, Maurer's ruthless methods began to work. Consider, for example, a letter to Buchenwald at the end of 1942 following an order to transfer 150 unskilled inmates to Auschwitz. The *Kommandant*, in keeping with a long concentration-camp tradition, had ignored any concerns for production and loaded up rail cars with prisoners he found disposable. Under Wilhelm Burböck this decision would have gone unnoticed, and *Kommandanten* had routinely purged their camps of inmates in this way. Under Maurer such local autonomy would no longer do; for now his standardized forms granted him statistical control over the bodies that he ordered about like commodities:

By 12/4/42, out of the total transported from Buchenwald, 18 prisoners have died. Another three are currently lying in the Prisoners Sickbay [*Häftlingskrankenbau*]

suffering from various afflictions. Of the remaining 129, 22 are bodily weak; three have foot injuries, inflammations, and swelling; one has no left arm; one has a deformed wrist; three have frostbite on their fingers. From the entire transport, only 100 are fit for work, two thirds.

'I invite your explanation,' he concluded.[61] From his desk in Berlin, Maurer was now able to monitor the exact details of transactions between one concentration camp in Thüringen and one in Schlesien hundreds of miles apart. He made it clear that when the IKL ordered transports, it expected to get them in exactly the numbers, quality, and condition specified; negligence and willful mismanagement would not go unnoticed nor unpunished.

Before Henry Friedlander's *Origins of Nazi Genocide* and Michael Burleigh's *Death and Deliverance: 'Euthanasia' in Germany 1900–1945* most historical literature on SS doctors concentrated on the small minority of specialists like Josef Mengele who conducted experiments on human subjects. In other words, SS doctors have rarely appeared in their role as everyday workmen of the Holocaust. Instead historians of science and medicine have focused on their horribly barbarous research agendas. Nevertheless, in the larger scheme of the Holocaust, these remain relatively insignificant. If we view these doctors from the vantage of the history of technology, however, their role as industrial managers comes immediately to the forefront. As Maurer's initiatives show, their function as technical managers was much more typical and much more important to the SS (it certainly led to more killing). Doctors applied their specialized training and social position to pass judgment upon the material condition of the human body and rendered those judgments in Maurer's forms. Furthermore, their profession had long been called upon in Germany, even before the Nazi era, to parse the fitness of patients into categories for state insurance policies, veterans benefits, and pension programs. The SS Labor Action simply extended these normal state-administrative tasks to a new and deadly operation, and, as Mennecke's enthusiastic letters to his wife demonstrate, most camp doctors earnestly threw themselves into the effort.

In the eyes of the SS, Maurer's initiative was also a success, for most doctors and medical orderlies genuinely attempted to preserve those who were 'fit to work.' However, preserving the labor capacity of the 'fit to work' went hand in hand with the *destruction* of life. Because slave labor was an open-loop system that managed both the input of fresh workers and their extermination once their bodies had been ground down in Germany's war machine, Maurer also called upon doctors and medical orderlies to cull those 'unfit to work,' which often included isolating, segregating, and killing them (with injections of phenol, for example). Medical personnel disposed of sickly inmates without compunction and often with great alacrity. In fact, purging the sick was justified as a measure which protected working inmates from epidemics, and the Polish historian Tadeusz Paczula has noted that the statistics of the SS doctors often lumped marginal cases in with the sick to make the absolute numbers of healthy workers more reliable. Now with a renewed attention to production, what had begun in the 14 f 13 program never ceased: to preserve the 'fit' the SS was all the more eager to send even those who might have potentially recovered to their deaths.[62]

Nothing illustrates more the new managerial role of SS doctors astride the dividing line between life and extermination than an order issued in December 1942 by Richard Glücks (still officially Inspector of Concentration Camps and Maurer's nominal superior officer). Glücks noted that out of a total of 136,000, 70,000 incoming prisoners had died. This startling mortality rate shocked even the SS and, more important to them, endangered its credibility in the eyes of Germany's armaments managers who relied on the SS to supervise the bodies of prisoners. He therefore wrote to all camp doctors:

> The best doctor in a concentration camp is that doctor who holds the work capacity among inmates at its highest possible level. He does this through surveillance and through replacing [the sick or injured] at individual work stations. Statistics for those able to work should not merely be a figure on a piece of paper; rather they must be regularly controlled by camp doctors ... Toward this end it is necessary that the camp doctors take a personal interest and appear on location at work sites.[63]

The code for Maurer's Labor Action Office (Office D2) in the letterhead of this order clearly marked it as his initiative. He placed responsibility for local labor conditions in the hands of the SS doctors, an unusual move considering production supervision was never part of their training or qualifications, yet he was merely acting consistently with his conception of what constituted the major problem with production.

The order demonstrates an additional facet of Maurer's managerial rationality: the lives of prisoners meant nothing to him, nor for that matter to the SS doctors. The IKL specifically ordered medical officers to remove and replace weak, sick, or injured prisoners. The fate that awaited those 'unfit to work' was common knowledge: as a foreman once snorted to a Jewish laborer at a factory near Auschwitz, 'Whether you stinking Jews work or not, you'll go one way or another into the crematorium and go through the oven.'[64] The SS had already long dealt with the 'unfit to work' though the operation 14 f 13, and since the end of 1941 blueprints for the gas chambers and crematoria at Auschwitz lay on the drafting boards of the WVHA's construction division. Being 'removed' from a work detail meant death, as the steady out-flow of ashes in the cycle of the SS-Labor Action testified. In the grisly world of SS management, Maurer felt no qualms about systematic extermination as a tool to purge the Labor Action of 'poor human material.' In fact, he introduced modern management into the institutions of genocide to accomplish exactly this goal.[65] Historians have noted that armaments work, due to the 'primacy of production' in total war, represented a force for the preservation of prisoners, often their only hope for survival; but this should not be exaggerated: murder and life were managed in tandem.[66]

Camp conditions stabilized during a brief period from the end of 1942 to the beginning of 1944, when the utter collapse of German transport and industry began to destroy what fragile order the IKL could impose. Mortality rates among working prisoners dropped from around 10% a month at the end of 1942 (17.2% in December at Mauthausen) to around 2–3% in 1943.[67] Yet Maurer's modern managerial initiatives can hardly be judged humane on these grounds: the very fact the Office Group D judged a 2–3% mortality rate tolerable speaks for itself, and Maurer had, from the start, fully intended to couple the preservation of working prisoners with the liquidation of the sick and weak. As one shrewd historian of the Nazi Genocide has already pointed out, falling mortality statistics are deceptive, for the population of the camps was increasing geometrically throughout this period.[68] The absolute number of deaths remained constant; only the increase in population brought the 'death rate' down. Furthermore, as is well known, SS doctors appeared daily in their role as industrial managers on the rail platforms at Auschwitz to parse prisoners into two lines: those heading for the gas chambers and those heading for labor details. Most prisoners who arrived in camps like Auschwitz went immediately to their death and never entered the balance sheets of the IKL. The number murdered was larger than the number reserved for the Labor Action by an order of magnitude.[69]

Maurer's office requested that SS doctors act like factory managers, yet his orders entailed much more than metaphor; they were not *like*, they *were* factory managers. Medical personnel assumed the oversight of technological systems and undertook innovations in order to supervise the bodies of prisoners. At a cement factory at Auschwitz, for example, orderlies convinced their plant management to route steam pipes to their medical barracks for newly installed autoclaves in order to sterilize prisoners' clothing. By doing so, orderlies hoped to reduce the danger of epidemic disease from lice and other parasites (which threatened guards and civilian employees of the factory as well as the prisoners). They also took over the daily supervision of the high-pressure autoclaves, becoming, in essence, the technical foremen of an industrial steam laundry.[70]

In another example, orderlies assigned to satellite commandos of Dora-Mittelwerk, which manufactured the V-2 rockets, also installed and operated similar technology. When one reported to his superiors, what he described was far less a medical procedure than an industrialization of sanitation. As the orderly proudly noted, he relied on industrial chemicals and precise time tables:

> The hygienic measures in the camp—cleansing of the toilet installations and the sleeping quarters at regular intervals with chlorine—is carried through on a running basis [*laufend*]. The prisoners' shower bath [with delousing chemicals] remains in steady-running operation [*Betrieb*]; at the time of this report a total of 8,460 prisoners have been bathed and simultaneously controlled for health ...[71]

The commodity processed in this 'operation' (in German the word '*Betrieb*,' or 'operation,' is the same used for 'factory') was 'human material,' the prisoner's body. Quite typical of the statistical vigilance that Maurer had imposed, the medical staff responsible also dutifully reported its results in statistics: out of 4,100 prisoners in this particular commando, 30 were reported dead and 405 in need of medical attention, or 'unfit to work'; careful notice was given to incidents of lice and scabies.

Many have mistaken references to delousing, such as that made above, with the shower stalls built to carry out the gassing of Jews and other camp inmates, something that Holocaust deniers delight in pointing out. In fact, Auschwitz and Mauthausen had both. The systems should not be conflated, but the confusion is, in one sense, not without warrant. For as noted, the task of culling and killing the 'unfit' went hand in hand with daily tallies of those 'fit to work.' The gas chambers and crematoria designed for genocide came to form a seamless web with medicalized factory management in one technological system. Since 1939, Nazi doctors had managed the gassing of human beings during the euthanasia campaign and had invented these techniques in conjunction with the SS as the pace of operations expanded. Henry Friedlander has noted the minute division of labor and other trappings of industrial operations that this entailed.[72] In this context, it is hardly surprising that medical personnel conceived their task as a factory operation (*Betrieb*).

As Friedrich Entress, a doctor assigned to various concentration camps during the war, testified at Nürnberg: 'At the end of 1942 I learned that the construction of four new crematoria with modern gas chambers was begun.' He had seen the plans hung on the walls of the construction bureau at Auschwitz and connected the innovation with mass production:

> The new gas chambers had a unique chute through which the gas was let in and had a modern ventilation system. Adjacent to the gas chambers the cremation ovens were located so that the crematorium could carry through the extermination of the prisoners on an assembly line.[73]

The medical profession had taken part in these innovations at every turn, and in doing so doctors had transformed themselves into technical managers.

No better example can be found than Kurt Gerstein. In August of 1942 he participated in what amounted to efficiency experiments with chemical poisons. 'The entire operation [the Final Solution] must be carried through faster, much faster,' his superior reportedly told him.[74] As SS personnel filled four gas chambers with 750 victims each, every chamber connected to the exhaust pipe of a single diesel engine, Gerstein held a stopwatch and counted off three hours and fourteen minutes until all the occupants were dead. This took so long that the SS decided to switch over its gas chambers to Zyklon-B (prussic acid).

Gerstein is especially interesting because he represents something of a hybrid. He had studied at the Polytechnic University of Marburg (achieving the prestigious Diplom Ingineur in 1931), and before the war he had been a partner in a machine-building firm that made automatic greasing mechanisms for locomotive brakes (De Limon Fluhme & Co.). By the mid 1930s he decided to go back to school and study medicine. In the early forties, the Waffen SS recruited him along with 40 other doctors, and he began to work designing filters and disinfecting apparatus for field troops. By 1942, he had received a promotion to the 'Chief of the Technical Branch of Disinfecting, which also included the branch for strong poison gases for disinfection.'[75] Because of his expertise in both engineering and medicine, the SS called him to the East for its experiments with killing techniques. Thus, it would seem, Nazi terms for the gas chambers as 'disinfecting chambers' and 'sanitary installations' stems less from a need to avoid confronting the actual fact of killing through euphemism than from the actual technological trajectory within which these systems evolved.

The example of Gerstein, the gas chambers, autoclaves, and chemical delousing showers demonstrate that the SS doctors could, when they wished, engage in concerted innovation and technological production. Some, notably Albert Speer and other industrial managers of the Third Reich, ridiculed the SS for its incompetence and technological stupidities, but their criticism entirely misses the point. It overlooks the question of what aspect of slave labor the SS wished to manage competently. Guided by eugenic ideology, the SS invented technological systems and statistical surveillance in order to exert the maximum control over prisoners' bodies; on the other hand, the bulk of its factory managers never conceived industrial output as their central concern. Rather, they set out to 'protect' production from prisoners as their primary task: 'Question: Why are prisoners dangerous?' read a training manual for concentration camp guards under the tutelage of the WVHA:

Answer: Because they can destroy the unity of our nation, lame our power, threaten our victory. They threaten to make it possible for those at home to stab the soldiers at the front in the back, just as they did before [i.e. in 1918].[76]

The SS still believed it was doing service to the war effort by removing prisoners from the *Volksgemeinschaft*. Yet it would be wrong to think that, just because SS officers often made a shambles of armaments factories, technological systems were truly beyond their comprehension; the SS was concerned with the control of technological systems but only to the limited extent that the concentration camps placed genocide and eugenics first and foremost. Managers like Maurer were willing to develop the means to balance war production with killing, but they were never willing to halt the murder, and they devoted the vast bulk of their energy and creativity in order to ensure the ever increasing scale and scope of systems toward that end. It is perhaps possible to construe this as an 'orgy of rationalization' driven by a hegemonic discourse of instrumental reason, but it is far more plausible to take seriously the SS's first philosophy, that is, its judgments of human nature—the gamut of its racial hygiene, loony nationalism, and racial imperialism. Upon the rail platforms of Auschwitz, doctors like Josef Mengele and managers like Maurer converted those moral judgments into managerial policy not vice versa.

NOTES

1 NO-1285, Pohl to Thierack, 16/4/43 (prepared the previous month but filled in with the appropriate statistics). Of note, the prisoners referred to here where of mixed provenance, and not all of them were Jews, who had mostly been killed already by 1943.
2 Michael Thad Allen, 'Engineers and Modern Managers in the SS: The Business Administration Main Office' (Philidelphia: Ph.D. Dissertation, University of Pennsylvania, 1995): 350–424. Richard Overy, *Why the Allies Won* (New York: WW Norton, 1995): 180–207.
3 Quotations from M. Norton Wise, 'Introduction': 5–6 and Ken Alder, 'A Revolution to Measure: The Political Economy of the Metric System in France': 41, both in M. Norton Wise, (ed.), *The Values of Precision* (Princeton: Princeton University Press, 1995). See also Wise, 'Precision: Agent of Unity and Product of Agreement, Part I—Traveling,' in *The Values of Precision*: 92. Andrea Rusnock, 'Quatification, Precision, and Accuracy: Determinations of Population in the ancien Régime,' in *The Values of Precision*: 18. Theodore Porter, 'Precision and Trust: Early Victorian Insurance and the Politics of Calculation,' in *The Values of Precision*: 180–1. See also J. L. Heilbron, 'Introductory Essay' and 'The Measure of Enlightenment,' in Tore Frängsmyr, et al., ed., *The Quantifying Spirit in the 18th Century* (Berkeley: University of California Press, 1990): 1–23, 207–61. For the case of the Enlightenment in Germany, see Reinhart Koselleck, *Preußen zwischen Reform und Revolution. Allgemeines Landrecht, Verwaltung und soziale Bewegung von 1791 bis 1848* (Stuttgart: Ernst Klett Verlag, 1967): esp. 163–217, 663–71. For the case of technological control during this period see Eric Brose, *The Politics of Technological Change in Prussia: Out of the Schadow of Atiquity, 1809–1848* (Princeton: Princeton University Press, 1993).
4 For an interesting comment on Foucault and industrial management: Anson Rabinbach, *The Human Motor: Energy, Fatigue, and the Origins of Modernity* (New York: Basic Books, 1990): 12–18, 236.
5 Götz Aly, *'Endlösung' Vokerverschiebung und der Mord an den europäischen Juden* (Frankfurt am Main: S. Fischer, 1995): 15–17, 100, 133, 236–50. Richard Evans, Charles Maier, and others have argued that the Nazi genocide remains a unique event in all of history precisely because, 'Nazi mass murder was an end in itself,' whereas the collectivization campaigns of the Soviet Union, by contrast, were carried out in the name of 'social reconstruction.' Richard Evans, *In Hitler's Shadow: West German Historians and the Attempt to Escape from the Nazi Past* (New York: Pantheon, 1989): 88. Compare Charles Maier, 'A Holocaust Like All the Others? Problems of Comparative History,' in idem, (ed.), *The Unmasterable Past: History, Holocaust, and German National Identity*

(Cambridge: Harvard University Press, 1988): 82. Statements by Nazi perpetrators suggest, however, that they conceived of their racial demographics as the social reconstruction of Eastern Europe. Compare Götz Aly and Susanne Heim, *Vordenker der Vernichtung. Auschwitz und die deutschen Pläne für eine neue europäische Ordnung* (Frankfurt/M: Fischer, 1993) and Rolf-Dieter Müller, *Hitlers Ostkrieg und die deutsche Siedlungspolitik. Die Zusammenarbeit von Wehrmacht, Wirtschaft und SS* (Frankfurt am Main: Fischer, 1991).

6 Wise, 'Introduction': 4. Alder, 'A Revolution to Measure': 42. Zygmunt Bauman, *Modernity and Ambivalence* (Ithaca: Cornell University Press, 1991) and *Modernity and the Holocaust* (Ithaca: Cornell University Press, 1989). Wolfgang Sofsky, *Die Ordnung des Terrors: Das Konzentrationslager* (Frankfurt am Main: Fischer Verlag, 1993). Detlev Peukert, 'The Genesis of the "Final Solution" from the Spirit of Science,' in Thomas Childers, and Jane Caplan, (ed.), *Reevaluating the Third Reich* (New York: Holmes & Meier, 1993): 234–252.

7 Bauman, *Modernity and the Holocaust*: 98.

8 Wise, 'Introduction': 9.

9 On ideology and bureaucracy, see Shoshana Zuboff, *In the Age of the Smart Machine: The Future of Work and Power* (New York: Basic Books, 1988): esp. 44, 102. In general Clifford Geertz, 'Ideology as a Cultural System,' in David Apter, ed., *Ideology and Discontent* (London: Free Press, 1964): 47–76. Most recent scholarship on the Holocaust has begun to reassess older conceptions of the 'banality of evil' and the moral alienation of 'technocrats' in the genocide. Perpetrators of the Holocaust now appear invariably more dedicated and cognizant of the 'moral' dimension of their work than they did in scholarship of the 1960s, 1970s, and 1980s. See Ulrich Herbert, *Best. Biographische Studien über Radikalismus, Weltanschauung und Vernuft 1903–1989* (Bonn: Dietz, 1996). Hans Safrian, *Die Eichmann Männer* (Wien: Europa Verlag, 1993). Omer Bartov, *Hitler's Army: Soldiers, Nazis, and War in the Third Reich* (Oxford: Oxford University Press, 1992). Michael Thad Allen, 'The Banality of Evil Reconsidered: SS Mid-Level Managers of Extermination through Work,' *Central European History* 30(1997): 253–294. Of course, Daniel Goldhagen, *Hitler's Willing Executioners: Ordinary Germans and the Holocaust* (New York: Alfred A. Knopf, 1996) is also a response to older literature that downplayed the ideological motivations of Nazi killers, however erroneous his own conclusions might be. This trend would seem to be in accord with recent literature on modern bureaucratic milieus and the way in which statistical information contains within it social and moral judgments. See Olivier Zunz, *Making America Corporate 1870–1920*. (Chicago: University of Chicago Press, 1990) and Ronald Kline, 'Ideology and Socieal Surveys: Reinterpreting the Effects of "Laborsaving" Technology on American Farm Women,' *Technology and Culture* 38(1997): 355–85.

10 Zunz, *Making America Corporate*: 49. Ernst Jünger *Der Arbeiter* in *Werke, Band 6* (Stuttgart: Ernst Klett Verlag, 1964 (reprint 1932)) can also be read as a panegyric to the empowerment of collective organization and modern technology.
11 M. Norton Wise, 'Precision: Agent of Unity and Product of Agreement, Part II—The Age of Steam and Telegraphy,' in Wise, M. Norton, (ed.), *The Values of Precision*: 227, compare Porter, 'Precision and Trust': 181.
12 The term 'measurement of men' is analyzed by Michael Adas, *Machines as the Measure of Men: Science, technology, and Ideologies of Western Dominance* (Ithaca: Cornell University Press, 1989) and Stephen Jay Gould, *The Mismeasure of Man* (London: Penguin, 1997).
13 Robert Jay Lifton, *The Nazi Doctors* (New York: Basic Books, 1986). Robert Proctor, *Racial Hygiene: Medicine under the Nazis* (Cambridge: Harvard University Press, 1989). Michael Kater, *Doctors under Hitler* (Chapel Hill: University of North Carolina Press, 1989) and *Das 'Ahnenerbe'. Die Forschugns- und Lehrgemeinschaft in der SS. Organisationsgeschichte von 1935–1945* (Heidelberg: Doctoral Dissertation, Ruprecht-Karl-Universität, 1966). Henry Friedlander, *The Origins of Nazi Genocide: From Euthanasia to the Final Solution* (Chapel Hill: University of North Carolina Press, 1995). Benno Müller-Hill, *Murderous Science: Elimination by Scientific Selection of Jews, Gypsies and Others; Germany 1933–45* (Oxford University Press, 1988). Jonathan Harwood, *Styles of Scientific Thought: The German Genetics Commujnity 1900–1933* (Chicago: University of Chicago Press, 1993). Götz Aly; Peter Chroust, and Christian Pross, *Cleansing the Fatherland: Nazi Medicine and Racial Hygiene* (Baltimore: Johns Hopkins University Press, 1994).
14 Bruno Latour, *We Have Never Been Modern* (Cambridge: Harvard University Press, 1993). Alder's essay, 'A Revolution to Measure' identifies a kind of disenchantment in the abstract statistics generated by Enlightenment thinkers. By substituting, for example, pure geometrical area measurements for more traditional measures of how much land a man and an ox could plow in a day, he argues, geographical spaces were purified of references to the human body and social complexity rooted in local knowledge. On this point, however, he overlooks the much more sophisticated combinations of hybrid statistical controls such as the hourly wage, which as Jürgen Kocka has noted, led to much more powerful management of conglomerations of human and technological environments than traditional measures such that the particular could be accounted for in universalistic equations. Jürgen Kocka, 'Preussischer Staat und Modernisierung im Vormärz: Marxistisch-Leninistische Interpretationen und ihre Probleme,' in Hans Ulrich Wehler, (ed.), *Sozialgeschichte Heute: Festschrift für Hans Rosenberg zum 70. Geburtstag* (Göttingen: Vandenhoeck & Ruprecht, 1974): 211–227 and more generally Jürgen Kocka, *Die Angestellten in der deutschen Geschichte 1850–1980* (Göttingen: Vandenhoeck & Ruprecht,

1981): 12–63 which is a generalized summary of his more in depth study *Unternehmensverwaltung und Angestelltenschaft am Beispiel Siemens 1847–1914: Zum Verhältnis von Kapitalismus und Bürokratie in der deutschen Industrialisierung* (Stuttgart: Ernst Klett Verlag, 1969).

15 Throughout Alfred Chandler, *The Visible Hand: The Managerial Revolution in American Business* (Cambridge: Harvard University Press, 1977) the author notes that the primary innovations in modern management were first developed by engineers but does not comment further on this point.

16 With the exception of construction work, where a much different regime prevailed, SS engineers were rare in the management of slave labor. Allen, 'Engineers and Modern Managers in the SS': 425–75.

17 Raul Hilberg's early and Robert Jay Lifton's more recent judgment that the Nazis used euphemistic language to hide and overcome their moral scruples is rarely born out in the context of how such bureaucratic language was generated; far more often, SS language in particular was rooted in managerial control. Compare Raul Hilberg, *The Destruction of the European Jews* (New York: Holmes & Meier, 1985): 962–7., Lifton, *The Nazi Doctors*: 6, 418–29. Compare Aly, *Endlösung*: 21.

18 Avraham Barkai, *From Boycott to annihilation: The Economic Struggle of German Jews 1933–1943* (Hanover: University Press of New England, 1989): 114.

19 Falk Pingel, 'Die Konzentrationslagerhäftlinge im nationalsozialistischen Arbeitseinsatz,' in Waclaw Dlugoborski, (ed.), *Zweiter Weltkrieg und sozialer Wandel. Achsenmächte und besetzte Länder* (Göttingen: Vandenhoeck & Ruprecht, 1981): 151–63 and *Häftlinge unter SS-Herrschaft. Widerstand, Selbstbehauptung und Vernichtung im Konzentrationslager* (Hamburg: Hoffmann und Campe, 1978): 69–75.

20 Friedlander, *The Origins of Nazi Genocide*: 17.

21 Friedlander, *The Origins of Nazi Genocide*: 81.

22 First quotation after Richard Breitman, *The Architect of Genocide: Himmler and the Final Solution* (New York: Alfred A. Knopf, 1991): 136; second quotation, Himmler to Pohl, 5/3/43, T-175/73.

23 Pingel, *Häftlinge unter SS-Herrschaft*: 119–125. Ulrich Herbert, 'Arbeit und Vernichtung. Ökonomisches Interesse und Primat der "Weltanschauung" im Nationalsozialismus,' in Dan Diner, (ed.), *Ist der Nationalsozialismus Geschichte? Zur Historrisierung und Historikerstreit* (Fischer Verlag: Frankfurt am Main, 1987): 198–236. Hermann Kaienburg, *'Vernichtung durch Arbeit' Der Fall Neuengamme. Die Wirtschaftsbestrebungen der SS und ihre Auswirkungen auf die Existenzbedingungen der KZ-Gefangenen* (Bonn: Verlag JHW Dietz Nachf., 1990): 234–50. Gregor Janssen, *Das Ministerium Speer. Deutschlands Rüstung im Krieg* (Berlin: Verlag Ullstein, 1968): 97–103. The role of the SS, the Armaments Ministry, and private industry in pushing for concentration camp prisoners

in war factories has largely been misinterpreted by historians. Compare Allen, 'Engineers and Modern Managers in the SS': 366–70.

24 Himmler to Glücks, 26/1/42, BDC Hängeordner 643. Here two things must be noted. Himmler's orders in January of 1942 did not yet refer to the armaments push of total war but to labor brigades the SS planned for its own projects for the coming peace, which the SS optimistically believed would come soon. In March the first overtures were made to the SS by officials in Speer's ministry. At that time allocations of labor were redirected to armaments factories. The details of these often misunderstood negotiations can be found in Allen, 'Engineers and Modern Managers in the SS': 350–424.

25 Arthur Liebehenschel to all KLs, 19/1/42, 'Überstellung von Juden,' printed in Harry Stein, *Juden in Buchenwald* (Weimar: Gedenkstätte Buchenwald, 1992): 119. This decree orders the immediate transfer of the 'mit *Fernschreiben* gemeldete Anzahl von arbeitsfähigen Juden dem KGL-Lublin.' It was received in the KL's on the 26th, as the hand-written date on the letter shows. It mentions a Fernschreiben from 8/12/41, however, which I have never found. Himmler's Fernschreiben of 26/1/42 would undoubtedly reinforce this ongoing effort to fill the camps Auschwitz and Lublin with forced laborers. I thank Peter Witte for providing me with this information.

26 Allen, 'Engineers and Modern Managers in the SS': 239–50.

27 NO-2468, Niederschrift, 17/3/42, 'Verlegung von Rüstungsfertigung in Konzentrationslager.'

28 NO-2468, 20/3/42 Letter Schieber to Saur.

29 Karl Heinz Ludwig, 'Vereinsarbeit im Dritten Reich 1933 bis 1945,' in idem, and Wolfgang König, (ed.), *Technik, Ingenieure und Gesellschaft. Geschichte des Vereins Deutscher Ingenieure 1856–1981* (Düsseldorf: VDI Verlag, 1981): 429–454 and 'Widersprüchlichkeiten der technisch-wissenschaftlichen Gemeinschaftsarbeit im Dritten Reich' *Technikgeschichte* 49(1979): 245–54.

30 Janssen, *Das Ministerium Speer*: 69–76 Dietrich Eichholtz, and Joachim Lehmann, *Geschichte der deutschen Kriegswirtschaft. Band 2: 1941–3* (Berlin: Akademie-Verlag, 1985): 65–7, 80–97.

31 Lutz Budraß and Manfred Grieger, 'Die Moral der Effizienz. Die Beschäftigung von KZ-Häftlingen am Beispiel des Volkswagenwerks und der Hanschel Flugzeug-Werke,' *Jahrbuch für Wirtschaftsgeschichte* 2(1993): 89–135.

32 See the monthly reports of Karl Mummenthey, 1940–1941, on the DESt, BAK NS3/1346.

33 NO-1712, Wilhelm Burböck to Dienststellen des HAHB; KL Sachsenhausen, Dachau, Buchenwald, Mauthausen, Neuengamme, und Auschwitz, 27/11/40, 'Neuer Organisationsplan für die Hauptabteilung I/5.' PS-3677, Burböck, 7/11/41.

34 Ibid, PS-3677. Glücks to all Kommandanten, 14/10/41, 'Abstellung von Häftlingskommandos,' BAP, PL5: 42055.
35 Ulrich Herbert, *Fremdarbeiter. Politik und Praxis des 'Ausländer-Einsatzes' in der Kriegswirtschaft des Dritten Reichs* (Bonn: Verlag JHW Dietz Nachf., 1986): 270–71.
36 Op. cit. NO-3668. Wilhelm Burböck to all KL's, 11/12/41, 'Monatsberichte,' BDC SS Personal-Akte Burböck.
37 Glücks to all KL's, 12/2/42, 'Herabminderung der Häftlingszahl für Lagerbetriebe,' BAP, PL5: 42056 also printed in Walter Bartel, et al., ed., *Buchenwald Mahnung und Verpflichtung* (Berlin: Kongress Verlag, 1960): 233.
38 Eugen Kogon, ed., *Nationalsozialistische Massentötungen durch Giftgas: Eine Dokumentation* (Frankfurt/M: Fischer, 1983). Lifton, *The Nazi Doctors* 134–47. Götz Aly, ed., *Aktion T4 1939–1945. Die 'Euthanasie'-Zentrale in der Tiergarten Straße 4* (Berlin: Edition Hentrich, 1987).
39 NO-4353, Theodor Eicke to SS Administration, 21/6/38. See also Klaus Drobisch and Günther Wieland, *System der NS-Konzentrationslager 1933–39* (Berlin: Akademie Verlag, 1993): 435.
40 NO-3665, Burböck to CC's, 23/7/41, 'Strength Report.' Op. cit., NO-1712. Compare with the complaints in Op. cit. Mummenthey, 27/11/40 and 20/12/40.
41 NO-3715, Affidavit of Franz Auer. His estimate is almost certainly too high, as the largest employer of prisoners at this time was the Deutsche Ausrüstungswerke at a little over 7000. Enno Georg, *Die wirtschaftlichen Unternehmunen der SS* (Stuttgart: Deutsche Verlags-Anstalt, 1963): 61. At this time almost all prisoners worked in SS Companies. The chemical concern IG Farben, the only private industry involved with the concentration camps on any scale at this time, was only now beginning to employ prisoners at Auschwitz. Peter Hayes, *Industry and Ideology: IG Farben in the Nazi Era* (Cambridge: Cambridge Univesity Press: 1987): 325–376.
42 Op. cit. NO-3665, Burböck to CC's, 23/7/41.
43 NO-2637, Postwar testimony of Dr. med. Fritz Mennecke.
44 NO-2636, testimony of Fredinand Roemhild.
45 PS-1151, Liebehenschel to KL Kommandants, 10/12/41, 'Former correspondence of 12/11/41.' PS-1234, Kommandant Groß-Rosen to Liebehenschel, 16/12/41, 'Selection of Inmates' and accompanying list of prisoners.
46 Danuta Czech, 'KL Auschwitz as an Extermination Camp,' in Kazimierz Smolen, ed., *Selected Problems from the History of KL Auschwitz* (Oswiecim: Panstwowe Muzeum w Oswiecimiu, 1979): 29–50. Jean-Claude Pressac, *Die Krematorien von Auschwitz. Die Technik des Massenmordes* (München: Piper, 1994): 38–50.
47 Op. cit., PS-1151 NO-907, Menneke's letter to his wife, 25/11/41.
48 Ibid.

49 Ibid.
50 Ibid.
51 NO-907, Dr. Med Fritz Mennecke to his Wife, 25/11/41. Selected letters and other information about Mennecke are also published in translation in Aly, Chroust, and Pross, *Cleansing the Fatherland*: 238–93.
52 NO-557, Gerhard Maurer to Pohl, 15/4/41, 'Monatsbericht für Monat März 1941.' See also Enno Georg, *Die wirtschaftlichen Unternehmungen der SS*: 61. BDC Personal-Akte Maurer and undated, post-war 'Lebenslauf,' BAP PL5: 42063. NO-495, Pohl's Organizational Chart of 3/3/42, BDC Hängeordner: 2635. Order Pohl to Salpeter as Chef Amt III A 31/1/40 and Hohberg to Pohl, 18/9/40; 31/1/40, 'Übernahme der wirtschaftlichen Betriebe der SS in Dachau in die Deutschen Ausrüstungswerke,' BAP PL5: 42055. NI-15149, 1/4/41, 'Bericht über eine Besprechung im KZ Auschwitz 28/3/41.'
53 BDC SS Personal-Akte Karl Sommer, NO-1578, NO-2739, Affidavits of Karl Sommer. NO-1202, Affidavit of Hans Moser.
54 BDC SS Personal-Akte Maurer. NO-1202, Affidavit Hans Moser. NI-1065, Affidavit Karl Sommer and Affidavit Gerhard Maurer in Defense Document Books of Karl Sommer. Paul Zapke, 1/8/52, 'In der Reichsschutzsache Gerhard Maurer,' BAP PL5: 42064.
55 Maurer to all Kommandanten and Arbeitseinsatzführer, 21/11/42, 'Übersicht über Anzahl und Einsatz der Häftlinge im KL,' BAP PL5: 42055. Testimony of Oswald Pohl, Protocol: 1315. NO-1566, Affidavit of Hanns Bobermin. Interrogation of Gerhard Maurer, BAP PL5: 42063.
56 Maurer, 24/6/42, Häftlingseinsatzführertagung am 1/6/42; 22/8/42 Maurer to all Kommandants and Arbeitseinsatzführer, BAP, PL5: 42055. Compare with Bürbock's managerial tools: NO-718, Burböck to all Schutzhaftlagerführer 'E', 28/11/41, 'Assignment of internees' detachments; Regulation IKL dated 14/10/41 Par. E.' Zuboff, *In the Age of the Smart Machine*: 97–124.
57 Ibid, Maurer, 24/6/42.
58 Johannes Tuchel, *Konzentrationslager. Organisationsgeschichte und Funktion der 'Inspektion der Konzentrationslager' 1934–38* (Boppard am Rhein: Harald Boldt Verlag, 1991): 28, see also 286–9. See also Falk Pingel, *Häftlinge unter SS-Herrschaft. Widerstand, Selbstbehauptung und Vernichtung im Konzentrationslager* (Hamburg: Hoffmann und Campe, 1978): 12–33: 150.
59 Alfred Chandler, *The Visible Hand*: 267. Similarly, James Beniger, *The Control Revolution: Technological and Economic Origins of the Information Society* (Cambridge: Harvard University Press, 1986): 224–5 speaks of the data control used by the Western Railroad in the mid-1800's: 'Certainly the most modern aspect of the organizational structure ... was the company's particular attention to regularity in data collection, to formalization of information processing and decision rules, and to standardization of communication with feedback.' See also p. 279.

60 Ibid, Maurer, 24/6/42. Maurer actually used the phrase 'Nicht arbeits- und nicht einsatzfähige Häftlinge,' not 'arbeitsunfähig.'
61 NO-1935, Maurer to Buchenwald, 7/12/42, 'Häftingsüberstellung (Bauhilfsarbeiter) zum KL Auschwitz.' Maurer badgered Kommandanten and Labor Action Officers throughout the war: Maurer to Höss, Kommandant Auschwitz, 4/9/43, 'Abgabe von Juden-Häftlingen,' BAP PL5: 42056; op. cit., Maurer, 21/11/42; and NO-1977, Maurer to Kommandanturen-Arbeitseinsatz, 29/1/45, 'Forderungsnachweise, Zusammenstellung und Übersicht.' Maurer also attempted to used these managerial techniques to impose control upon SS authorities outside the WVHA, but with limited success. See for example: NO-1547, Gerhard Maurer to all KL Kommandanten, 3/1/44, 'Neueinlieferung von Häftlingen; Schutzhäftlinge aus den besetzten Ostgebieten.'
62 Tadeusz Paczula, 'Organization und Verwaltung des ersten Häftlingskrankenbaus in Auschwitz,' in Jochen August, ed., *Dei Auschwitz-Hefte* (Weinheim: Beltz Verlag, 1987): 159–71.
63 NI-10815, Glücks to all KL Lagerärzte, 26/12/42, 'Ärztliche Tätigkeit in den KL.'
64 Kommandoführer to Schutzhaftlagerführer KL Au. III, 20/4/44, 'Meldung wegen Wachvergehen,' Golleschau, Band 3.
65 Maurer's exact order in German for the removal of weak, sick, or injured prisoners reads: 'der [beste Arzt hält] die Arbeitsfähigkeit durch Überwachung und *Austausch* an den einzelnen Arbeitsstellen möglichst hoch.' The Office Group D installed a chief doctor to coordinate all camp medical duties. This was Enno Lolling, a mediocre man who was known to have a history of morphine addiction. Mostly, he indulged his fascination for the medical experiments carried out on human subjects in the camps. (NO-065 and NO-407 Affidavits of Oswald Pohl and NO-444 Affidavit of Dr. Rudolf Brandt.) Otherwise, he was a weak and unimaginative organizer and did almost nothing to monitor the daily medical activities of the camps. Here, as with all operations in the Office Group D, Maurer directed matters, not old hand IKL personnel. BDC SS Personal-Akte Enno Lolling. See *Kaul, Ärzte in Auschwitz*: 76–83, 168, 170. Lifton, *Nazi Doctors*: 384–416. Also 'Diary of Johann Paul Kremer' in *KL Auschwitz as seen by the SS* (Warschau: Interpress Publishers, 1991): 149–215.
66 Herbert, 'Arbeit und Vernichtung': 198–236; 'Rassismus und rationales Kalkül. Zum Stellenwert utilitaristisch verbrämter Legitimationsstrategien in der nationalsozialistischen "Weltanschauung",' in Wolfgang Schneider, (ed.), *Vernichtungspolitik. Eine Debatte über den Zusammenhang von Sozialpolitik und Genozid im nationalsozialistischen Deutschland* (Hamburg: Junius Verlag, 1991): 25–36; and 'Von Auschwitz nach Essen. Die Geschichte des KZ-Außenlagers Humboldtstraße,' in Wolfgang Benz, and Barbara Distel, (ed.), *Sklavenarbeit im KZ. Dachauer Hefte. Studien und Dokumente zur Geschichte der nationalsozialistischen*

Konzentrationslager (Dachau: Verlag Dachauer Hefte, 2(1986)): 13–34. Above all, the chances for survival depended upon what kind of labor the SS selected prisoners for. Prisoners in construction brigades often suffered the most deadly conditions; those assigned to private industry where they were shielded from the SS often enjoyed vastly improved chances of survival.

67 Miroslav Karny, "'Vernichtung durch Arbeit". Sterblichkeit in den NS-Konzentrationslagern,' *Beiträge zur nationalsozialistischen Gesundheits- und Sozialpolitik* 5 (1987): 140–45. Pingel, *Häftlinge unter SS-Herrschaft*: 182–3.

68 Ibid, Miroslav Karny, extrapolates from individual transports which arrived in Auschwitz in August of 1943. Roughly 2/3 of incoming prisoners were immediately gassed. Only 1/3 were withheld for work.

69 At Auschwitz there was a clear separation of prisoners destined for work and those destined for 'Special Treatment' among the entering Schutzhäftlinge at the beginning of 1944: NO-1553, Liebehenschel, 25/3/44, 'Directive for the Information of the Chiefs of the Political Divisions.' See 1469–PS, Oswald Pohl to Heinrich Himmler, 30/9/43, 'Todesfälle in den KL.' Himmler's pride: NO-020, Oswald Pohl to Heinrich Himmler, 5/4/44, 'Konzentrations- und Arbeitslager' and Letter from Himmler on 22/4/44 in reply. Miroslav Karny, 'Vernichtung durch Arbeit': 150–152 points out the variation of estimates between the Reichsministerium für Rüstungs- und Kriegsproduktion, Heinrich Himmler, and officers within the WVHA. Regarding labor at the extermination camps: Adalbert Rückerl, *NS-Vernichtungslager im Spiegel deutscher Strafprozesse: Belzec, Sobibor, Treblinka, Chelmno* (München: Deutscher Taschenbuch Verlag, 1977): 67–8.

70 Michael Thad Allen, 'The Puzzle of Nazi Modernism: Modern Technology and Ideological Consensus in an SS Factory at Auschwitz,' *Technology and Culture* 37(1996): 558–68.

71 Monatsberichte, Sanitätsdienstgrad, Aussenkommando Hans, 22/1/45, M1079/1.

72 Friedlander, *The Origins of Nazi Genocide*: 86–110.

73 NO-2368, Affidavit of Friedrich Entress. Entress's detailed descriptions and most of his chronological account are born out by Pressac, *Die Krematorien von Auschwitz*: 69–96.

74 NI-1553, CIOS Consolidated Advance Field Team VII, interview with Diplom-Ingenieur Kurt Gerstein.

75 Quote: PS-1553, testimony of Kurt Gerstein. NI-1553, CIOS interview with Kurt Gerstein. BDC Personalakte Kurt Gerstein, esp. Lebenslauf.

76 'Lagerordnung für Häftlinge,' with Maurer's memo to Kommandanten 27/7/43 'Bewachung der Häftlinge,' BAP PL5: 42053.

CHAPTER 9

Beasts and Systems: Taming and Stability in the History of Control[1]

David A. Mindell

WILEY POST, EARLY CYBERNETIC HERO

At the controls of his airplane, late on a July night in 1933, Wiley Post fell asleep. High above the Canadian Rockies, the demon of fatigue caught up with the rough-hewn Oklahoma pilot in his Lockheed Vega, *Winnie Mae*, on the final leg of the first solo around-the-world flight. Yet the sleepy Post did not spiral to earth (as Lindbergh had feared he would do when crossing the Atlantic five years before). *Winnie Mae*, without its operator's steady hand, kept flying straight and level. The airplane shown in Figure 17 was equipped with a self-regulating machine: a prototype Sperry automatic pilot.[1]

The autopilot allowed Wiley Post to overcome fatigue, the greatest obstacle facing his record flight. Sometimes the imperfect control system failed and Post flew manually. The two worked together, trading control, playing on each other's strengths and alleviating each other's weaknesses. A day after his nap, Post landed in New York. Public exposure of the successful seven-day flight promoted the Sperry Automatic pilot, and brought automatic control into public view. The *New York Times* declared "the days when human skill and an almost bird-like sense of direction enabled a flier to hold his course for long hours through a starless night or over a fog are over. Commercial flying in the future will be automatic."[2]

By no coincidence did Wiley Post accomplished this feat in such intimacy with a machine. He had already made his mark as a pilot particularly close to his airplane. An early colleague recalled "he didn't just fly an airplane, he put it on.' Another remembered Post as, 'as near to being a mechanical flying machine as any human who held a stick."[3] In Post's words, when flying:

> I tried my best to keep my mind a total blank ... I do not mean that I paid no attention to the business of handling the ship. I mean that I did it automatically, without mental effort, letting my actions be wholly controlled by my subconscious mind.[4]

Post rearranged his instruments to cluster them right in front of his one good eye (the other had been lost in an oilfield accident); he modified his cockpit so he could fly with only one foot on the rudder pedals and one hand on the wheel. Sperry President Preston Bassett later described Post's piloting:

> "As many will recall, Wiley had only one good eye. So, all combined, the setup was that of a man flying around the world with one eye, one arm, one leg, and two instruments. You will see that we are building toward a very good servomechanism."[5]

Basset's characterization of Post's relationship to his machine hints at a principle which underlies the history of control systems. Despite the temptation to equate 'automatic' with 'autonomous,' automatic controls do not set machines free but rather reign them in. Post's autopilot did not replace him; rather it connected him more tightly to the machine, the aggregate forming a single 'human servomechanism' which allowed Post to retain control, even while asleep. Elmer Sperry once referred to the airplane as "a beast of burden obsessed with motion," equating machinery with animals.[6] Like beasts, machinery threatens to run out of control: airplanes spin, ships veer, guns misfire, and systems oscillate. These wild behaviors all enact versions of 'instability,' the precise engineering term for technical chaos. Control systems, like stirrups on a horse, do not create autonomy in the machine but rather harness it, bringing its independence and wildness under the will of human intention.

This paper surveys technologies of control in the United States in several different contexts between the two world wars, arguing that in each engineers used controls to tame rather than to liberate machinery. Sometimes they employed Sperry's 'beast' vision, other times they designed 'systems' to balance out the dangers. Each was characterized by a search for stability, although the meaning of stability varied between, and even within, different settings. Four different cultures of control developed with unique but related assumptions about how to subordinate machinery to human intentions: 1) The Sperry Company, as illustrated by Wiley Post, developed devices to tame unruly machines, such as autopilots for ships and airplanes. 2) The U.S. Navy Bureau of Ordnance and its contractors (the Ford Instrument Company, Arma, and General Electric) developed technologies of 'fire control,' mechanical analog computers which harnessed the awesome power of naval guns to hit distant moving targets. 3) Communications engineers at Bell Laboratories developed feedback amplifiers and feedback theory for long-distance transmission of information, protecting the far-flung network from internal and external demons. 4) Researchers and students in the Electrical Engineering department at MIT worked on problems of instability in electric power systems, building differential analyzers, and electrical simulation machines, from which they developed the 'theory of servomechanism.' The modern theory and practice of control emerged from these four cultures, indeed they laid intellectual foundations for computers and information systems—a vision distinct from the 'autonomous thinking machines' which populate the common understanding of the roots of computing.

This revision of our understanding of control systems, from 'liberating' to 'taming,' connects their history to broader histories of technology as a social force. Is technology out of control? Does technology drive history? Recent scholarship criticizes the myth of autonomous progress and argues instead for a vision of technology based on human choice and decision.[7] Control systems, because they create 'automatic' machinery, are particularly vulnerable to this myth of autonomy. Automata resemble artificial life, modern robots mechanical humans. From Galatea to Frankenstein, autonomous technology has a metaphoric appeal and mythic significance, but it coexists with another venture: the search for techniques to bring technology, or technologies, under control.

BEASTS: SPERRY GYROSCOPE

In the early twentieth century, increases in size, performance, and complexity pressed new machines to the edge of stability. Ships and airplanes became too big and too fast for people to pilot with traditional methods. Sperry Gyroscope's control systems reined in the machines, adding precision to their power. The company rarely designed the machines themselves, but rather added to developments in other areas. These reigns, Sperry's regulators, brought human operators into new combinations with new machinery, bringing technological power into the range of human reaction and endurance.

Sperry's biographer, Thomas P. Hughes, has shown that Elmer Sperry's work consistently improved existing machinery through the application of feedback mechanisms—making smoother arcs, steadier power, straighter courses, and more stable flying.[8] Elmer Sperry started several companies, but Sperry Gyroscope, founded in 1910, was the last and the largest (it became the Sperry Corporation after Elmer Sperry's death in 1930). *Elmer Sperry's greatest contribution may have been the very notion of a company that specialized in control systems as a discrete technology.

Elmer Sperry had a simple vision, but one with distinct advantages: seeing machines as 'beasts,' led to simple, tightly-coupled controllers which were easy to manufacture and use, critical features during wartime mobilization. Sperry company engineers in the twenties and thirties worked out a set of compromises on complexity, cost, and automation. In time, the company came to see the control system as making the machine the extension of a person, literally grafted onto the senses, the body, or the mind, "technological appendages."[9] By 1940, they could write coherently about "the inability of the unaided man to operate his weapons":

> His airplanes have become so big and fly so far that he must have automatic pilots instead of flying by hand. The machine gun turrets must be moved by hydraulic controls. The targets of his antiaircraft guns now move so fast in three dimensions that he

can no longer calculate his problems and aim his gun. It must all be done automatically else he would never make a hit ... There has come into being a whole new field of scientific accessories to extend the functions and the skill of the operator far beyond his own strength, endurance, and abilities.[10]

This vision, which equates automation with "the extension of skills" (rather than their replacement), culminated three decades of control engineering at Sperry. During that time, Sperry engineers worked out their ideas and expressed themselves by building machinery with different degrees of automation, different roles for the human operator, and different types of systems. They developed skills not only in negative feedback mechanisms but also in electrical instrumentation, data transmission, analog computing, and power drives. What eventually emerged as Sperry's 'philosophy' of automation during the second world war was the result of engineering research, commercial imperatives, production compromises, and military demands.

Sperry Gyroscope tried several failed product lines for every one that stayed in production; they had great difficulty developing complex, high-performance control systems and deploying them in the field. The company's main attempt at selling large scale control systems, its naval Fire Control System, failed, as eventually did its antiaircraft fire control and other systems projects (as the next section demonstrates, other companies such as the Ford Instrument Company and General Electric, succeeded). In fact, the company's history with automatic machinery is as remarkable for its difficulties as for its successes. When World War II came, the company's value lay in its production capacity and engineering vision as much as its research department. The major pre-war product line, antiaircraft fire control, was discontinued at the height of the wartime boom because of manufacturing problems.[11]

While Sperry Gyroscope did not succeed in building systems, the beast vision had remarkable staying power. Just before World War II, Sperry engineers began work on a number of new products which assured the company's success during the war: simple and easily manufactured fire controls for aircraft. Early World War II bombers did not use central gun directors to coordinate their guns. Machine gunners defending B-17 flying fortresses (themselves using Sperry instruments and autopilots) from attacking fighters each worked individually, with no coordination or centralized control (except through voice intercom). The Sperry Corporation produced these individual controls, hydraulic turrets for machine gun defenses of B-24 and B-17 bombers. These devices allowed a gunner to rapidly and smoothly swing around himself and his machine guns to end off attacking airplanes. Sperry Gyroscope built on strengths in aviation instruments and its corporate tradition of reference and measuring devices (going back to Elmer Sperry's original gyrocompass). The company built instruments of perception; gyroscopic sensors coupled to visual indicators called "lead computing sights" imposed scales on the gunners' vision and indicated where to aim. Vickers, a Sperry subsidiary, built instruments of articulation, the electro-hydraulic power controls which moved the turrets.[12]

These machines, especially the famous 'Ball Turret' shown in Figure 19, comprised a popular image of mechanized air combat during World War II. Their production occupied much of Sperry's wartime resources. These simple but effective machines placed heavy emphasis on the human operator, aiding his mind and his body at critical points but leaving command in his hands. At Sperry, at least, the beast vision survived—gunners tamed the unruly guns with their own hands. During the war Sperry made not the most advanced or intricate products, but rather those that effected simple, tight assemblages of mechanical and human functions and which could be produced in large numbers.

Sperry trumpeted its vision of automation as the extension, rather than replacement, of the human operator's capabilities, brought forth by its experience between the wars:

> There has come into being a whole new field of scientific accessories to extend the functions and the skill of the operator far beyond his own strength, endurance, and abilities.

Mass conscription in wartime would have little affect without increases in production; Sperry argued its products brought the wartime mobilization of manpower together with the mobilization of industry.

> Over a billion dollars of this material [control systems] must be produced by us within the next two years. But this billion dollars' worth of technical equipment will fill the

Figure 17 Wiley Post and his Lock-heed Vega, *Winnie Mae*, in which he made his around-the-world solo flight with a Sperry Autopilot. Sperry Company Sales Pamphlet, 'Round the World with the Sperry Pilot.'

210 *Cultures of Control*

Figure 18 The box above the control stick. Sperry Company Sales Pamphlet, 'Round the World with the Sperry Pilot.'

vital gap between the one hundred billion dollars' worth of weapons and the thousands of men who must operate them. Without this equipment, neither men nor weapons would be effective.[13]

Sperry's control systems united the beasts procured by the services with the men who would ride them into battle.

SYSTEMS: NAVAL FIRE CONTROL

Where Sperry built control systems which tightly coupled operators to their machines and guns, another tradition of control engineering before World War II, that of naval fire control, employed large, multi-faceted systems to control gunfire and its associated machinery. These systems tamed not only machines

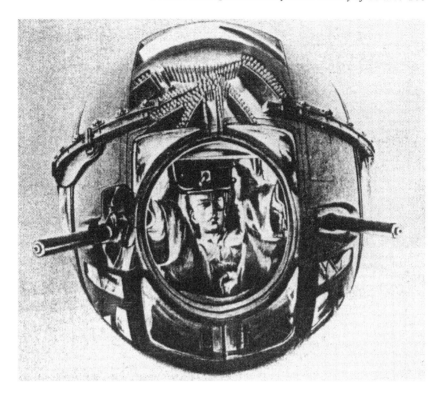

Figure 19 Sperry ball turret for the B-17 bomber, 1941. Note eyepiece for lead-computing sight. Roswell Ward, 'Aircraft Turrets: Description of Product Development and History,' February 16, 1944.

but the human operators as well, integrating information, interconnecting people, and replicating social and political relationships. As Sperry's parallel success demonstrates, this evolution, from beasts to systems, was neither linear, causal, nor complete; but naval fire control did develop systemic controls unmatched in their precision.

In the first decades of the twentieth century, the size and speed of turbine-powered warships, combined with advances in gun and powder technology, created a 'revolution in naval gunnery.' At the turn of the century, typical naval engagements took place at ranges between 2,000 and 4,000 yards. World War I battleship main batteries could shoot 20,000 yards, which increased to 34,000 yards by World War II and to 40,000 with wartime advances.[14] Firing shells and military utility, however, are not the same thing, and the revolution precipitated an attending crisis. Shooting to great distances exceeded the ability to hit anything that far away. As ranges grew, accuracy in aiming became critical; errors of fractions of a degree, difficult to eliminate from guns mounted on a moving, pitching platform, caused shells to miss their targets altogether. Naval guns were

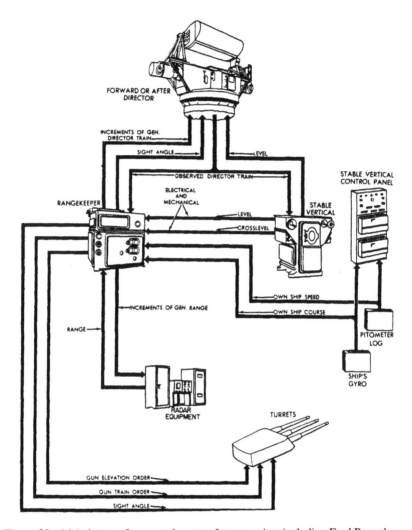

Figure 20 Main battery fire control system from a cruiser including Ford Rangekeeper Mark 8, Arma Stable Vertical Mark 6, and General Electric Mark 34 Director, circa 1940. Note optical rangefinder integrated into director. This diagram leaves out all human operators, of which perhaps fifty would be required to run such a system. (from 'Main Battery Fire Control System CA 68, 72, and 122 Class Ships,' OP 1387 U.S. Navy Bureau of Ordnance, 1948, 82).

useless if they could not be governed; they had power but not precision. Stability of the firing platform, and of the mathematical solution to the firing equations, became paramount concerns.

Fire control systems were the primary solution to these stability problems. They incorporated a host of factors, including the range and bearing of the

Figure 21 M-9 Gun director designed by Bell Laboratories during World War II under an NDRC fire control contract. The device employs negative feedback amplifiers for electrical computations, and could be connected into a radar set for automatic operation. Property of AT&T Archives. Reprinted with permission of AT&T.

target, the pitch and roll of the firing vessel, wind speed, air temperature, and ballistics. Then calculated the proper angle and elevation for the guns and transmitted that data to the gun turrets, along with orders to fire. Until about 1930, the leading edge of this technology concerned 'surface fire,' aiming the main guns on destroyers, cruisers and battleships to hit distant targets, usually other ships. In 1916, Elmer Sperry's first employee, a talented mechanical designer named Hannibal Ford, left the company to form his own firm, the Ford Instrument Company. Soon thereafter Ford produced the 'Ford Rangekeeper,' a gunnery computer which would be the core of shipboard fire control systems for generations. Working exclusively for the secretive and powerful U.S. Navy Bureau of Ordnance, the Ford Instrument Company became, in the words of a British observer, "the secret Fire Control Design Section of the U.S. Navy."[15]

On the eve of World War II, the typical fire control system included a set of diverse machines, as shown in Figure 20. Instruments such as rangefinders, telescopes, and eventually radar observed the target and sent data into a centralized 'plotting room,' protected by armor and buried deep in the hull. There a 'rangekeeper' or 'computer' built by the Ford Instrument Company calculated the course and speed of the target, correcting it with inputs from the ship's gyro and speed log. A gyroscopically-controlled 'stable element' built by the Arma Engineering Company corrected the solution for the pitch, roll, and yaw of the firing ship, and ordered the guns to fire when the ship rolled to a specified point (the mechanical replacement of 'continuous aim firing' introduced by Admiral

Figure 22 SAGE computerized air defense system. Operators use 'light guns' to designate targets on a computer-driven radar screen. (Claude Baum, *The System Builders: The Story of SDC* (Santa Monica, California: System Development Corporation, 1981).

Sims at the turn of the century).[16] General Electric, an experienced system builder, made the data transmitters that tied these elements together, electric motors to automatically turn the turrets, and switchboards to program the system for different configurations. This basic system, comprised of elements from these three companies, remained largely unchanged throughout World War II and after. The diagram shows only machinery, but that obscures a critical point: human operators attended the machinery at every point, serving as system integrators' the intellectual glue that held the system together. At the same time, the system disciplined their own behavior into a prescribed series of inputs and outputs.

Fire control systems shifted the location of what John Keegan has called "the face of battle," the point where a complex system of fighting actually meets the enemy.[17] Down in the plotting rooms, officers fought by operating machines and supervising data flows. Naval gunnery was mediated by a "system of information," and by the 1930s no realm of warfare had become as mechanized, precise, and remote.[18] This displacement could not proceed in isolation; it necessarily accompanied parallel shits in infrastructure. As technology changed, the critical industries shifted from gun manufacturers to instrument manufacturers to electrical and electronics companies.

Despite its precision, fire control systems between the World Wars remained both cumbersome and 'open loop:' human operators still closed the primary feedback loop, observing targets (even with radar), plotting shell splashes, and making corrections.

A new threat, however, pushed these methods to their limits. Aircraft were making the battleship itself obsolete, and increasingly the ship's resources went toward defending itself. To successfully fight aircraft, fire control needed to be quicker and less centralized, if less precise. Yet when traditional fire control methods were applied to the antiaircraft problem, the powerful guns could not follow attacking airplanes and maintain their stability. For example, the Mark 37 "antiaircraft gunfire control system," introduced in 1939, had "fully automatic rate control," which automated the feedback loop for estimating the target's course, previously a manual activity. The tracking and convergence feature of the Mark 37, however, had a stability problem in its servo loop which caused the turret motors to oscillate. The numerous, connected feedback loops interacted and fed back on each other in ways not well understood, and "when the first complete director-to-gun system was tested, operation was entirely unsatisfactory."[19] Against fast-moving torpedo planes and dive-bombers, naval fire control had reached the limits of stability.

STABILITY IN TELEPHONE TRANSMISSION: BELL LABS

Despite the tight human-machine couplings of Sperry's products, and the networks of devices which supported fire control, in neither did engineers use an explicit concept, much less a theory, of 'feedback' or 'control.' These ideas emerged from two remaining threads: Bell Laboratories and MIT. Each developed conceptual tools for understanding stability and taming the wildness of systems.

A famous story describes how in 1927, one morning a young Bell Laboratories engineer named Harold Black rode the ferry into Manhattan to work and conceptualized the feedback amplifier, which quickly became a fundamental element of modern electronics:

> I suddenly realized that if I fed the amplifier output back to the input, in reverse phase, and kept the device from oscillating (singing, as we called it then), I would have exactly what I wanted: a means of canceling out the distortion in the output.[20]

'Singing,' the bugaboo of high-gain amplifiers, was the latent instability of the telephone network. If the output of an amplifier accidentally leaks into the input, it breaks into oscillation or 'sings,' which renders it useless for amplifying voice signals. Nevertheless, Black proposed to invert the output of the amplifier and then feed it back into the input on purpose. The feedback amplifier promised to renew a signal while introducing minimal distortion, an ideal solution for repeater amplifiers for AT & T's national long-distance network. Black's idea at first received little support; few could believe this feedback would produce anything but an oscillating mess.

Part of what Black saw as 'resistance' from his peers resulted from differing notions of 'stability.' Black considered *stability of transmission* the key point of the amplifier: the feedback loop reduced distortion, which meant it also increased the amplifier's long-term stability. Repeater amplifiers inhabited a far-flung geographical network with daily and seasonal variations in temperature and humidity. These variations would alter the performance of vacuum-tubes—a significant problem with the carefully-matched components, finely-tuned to squeeze every last bit of distance or clarity out of the long wires. Black believed his amplifier solved this stability problem, making the gain of the amplifier depend on a small number of reliable, passive components, not on the notoriously quirky and sensitive vacuum tubes. In his initial patent application, Black explicitly rejected 'singing' as the kind of stability problem which concerned him:

> Another difficulty in amplifier operation is instability, *not used here as meaning the singing tendency*, but rather signifying *constancy of operation* as an amplifier with changes in battery voltages, temperature, apparatus changes including changes in tubes, aging, and kindred causes ... Applicant has discovered that the stability of operation of an amplifier can be greatly improved by the use of negative feedback.[21] [emphasis added]

He acknowledged the other meaning of stability, but assigns it unequivocal second billing:

> Applicant uses negative feedback for a purpose quite different from that of the *prior art* which was to prevent self-oscillation or 'singing.' To make this clearer, applicant's invention is not concerned, except in a very secondary way ... with the singing tendency of a circuit. Its primary response has no relation to the phenomena of self-oscillation.[22]

Black's retrospective construction of his famous ferry ride smoothed over this critical difference At the time, however, other, more mathematically-oriented engineers at Bell Labs could not ignore it. They found the tendency to break into self-oscillation the fundamental problem with Black's idea.[23] Feedback of the type Black was proposing was something they explicitly tried to avoid in their designs. The simplest form of the unwanted phenomenon occurred in telephone handsets, new in the 1920s: if you laid them down on a table, sound from the speaker would travel through the table into the microphone, where it would be amplified through the speaker, and repeat the cycle indefinitely, causing the phone to 'howl' (the same phenomena as the whistling feedback in a public address system).[24]

Beasts and Systems: Taming and Stability in the History of Control 217

Two other engineers, more sophisticated theorists than Black, took up his idea and addressed this latter stability problem. They showed that feedback amplifiers could work extremely well, but only under certain constrained conditions which would prevent them from singing. Harry Nyquist created the "Nyquist stability criterion," a simple graphical method for analyzing the stability of a feedback amplifier (or, as it turned out, any feedback system).[25] Hendrik Bode invented another graphical method which provided guidelines for actually designing amplifiers for both maximum response and maximum stability, which became known as the 'Bode plot.' Bode began a landmark 1940 paper on the topic of feedback with a caveat:

> The engineer who embarks upon the design of a feedback amplifier must be a creature of mixed emotions. One the one hand, he can rejoice in the improvements in the characteristics of the structure which feedback promises to secure him. On the other hand, he knows that unless he can finally adjust the phase and attenuation characteristics around the feedback loop so the amplifier will not spontaneously burst into uncontrollable singing, none of these advantages can be actually realized.[26]

Bode and Nyquist's work showed that in electronics, mathematical analysis could tame the demon of instability, and that negative feedback amplifiers themselves could stabilize transmission in the telephone network. This seems exactly the type of insight the Navy needed for the oscillating servos on its gun turret mounts, but it would be a decade these techniques were applied to other types of feedback systems.

THE STABILITY PROBLEM IN ELECTRIC POWER: MIT

One other type of electrical network traversed the countryside during these years: electric power systems.[27] In the 1920s, regional systems were connecting into interregional and national grids, the proposed 'superpower' systems. These networks had a number of generators (at hydroelectric, steam, and coal stations), each with feedback devices regulating voltage and frequency. Generators drove transmission lines connected to a series of loads, including factories, streetcar systems, and residential areas. These complex networks, however, tended to become unstable in response to changes in operating conditions, but the dynamics of the process remained mysterious.[28]

How did the numerous devices hung on a power network interact? How did the networks respond to external disturbances, short-lived events or 'transients,' such as lightning strikes, sudden applications of load, and short circuits? When a factory started up, for example, or a section of the grid tripped off, a transient in the form of a traveling wave moved through the network. How would this transient propagate? Would it exceed the power limits at certain points? How would the system react to the transient? Ideally, the network would damp the transient and it would die away after a short time. If however, the transient induced oscillations which caused it to grow with each cycle, the system became unstable: progressive oscillations caused it to shake itself apart or to fail by exceeding its power limits.

Beginning in 1923, Vannevar Bush, a young electrical engineering professor at MIT, began to investigate these transient phenomena.[29] Because of the obvious industrial importance of the work, and because of Bush's interest in the topic, network stability became a major topic at MIT during the twenties. By 1930, one in five students in electrical engineering studied transients or stability. A number of men who would later become leaders in electrical engineering looked at power system stability as students during this period. Harold Edgerton, for example, first employed stroboscopic photography to examine the "pulling into step" phenomenon of large generators reacting to transient network events.[30] Similarly Frederick Terman, who would later build Stanford's electrical engineering program and become "the father of Silicon Valley," wrote a 1924 doctoral thesis, "The Characteristics and Stability of Transmission Systems." Terman addressed the proposed 'superpower' systems and argued the problem of stability intimately related to the behavior of governors, regulators, and feedback mechanisms. Not only did individual devices affect transient behavior, Terman wrote, but the characteristics of the network itself had much in common with feedback controls.[31]

In 1925, Bush and colleague R. D. Booth published a paper in the transactions of the AIEE entitled 'Power System Transients' which put Terman's work into a larger engineering context.[32] As their central problem, Bush and Booth asked the following question:

> What is the degree of stability of such a network [a system of power stations connected by transmission lines and operating close to its power limits] when subjected to disturbances of the types likely to be encountered in practice?[33]

To address this problem, Bush and Booth proposed a "point by point" method of calculation, whereby one starts with the steady-state of the system and then calculates how it changes for incremental parameters of time during a transient. One can then piece together how a system behaves for a certain time interval. For this laborious computation Bush first conceived his famous research program in calculating machines. During the 1930s, the program produced the Product Integraph several generations of Differential Analyzers, and the Cinema Integraph.[34]

While working on these machines, Harold Hazen, a student of Bush, developed a servomechanism to couple the output of one calculation stage to the input of another. From this work, in 1934 Hazen wrote a paper, "The Theory of Servomechanisms," which formulated a taxonomy and a general mathematical description of feedback systems. He argued that the same principles applied to the speed control of steam turbines and water wheels, the stabilization of ships by gyroscopes, the operation of gyrocompass repeaters, the automatic stabilization and guiding of aircraft, and "in fact the automatic recording or control of almost any measurable or measurable and controllable physical quantity."[35] Hazen not only proposed this general theory of servomechanisms, but suggested it could be expanded to a general theory of systems. He added "entire closed-cycle control systems are dynamically similar to servo-mechanisms and their operation is investigated by the same methods," thus breaking down the distinction between servos and larger control systems. Numerous diverse kinds of machinery could be considered analytically identical.

One of Hazen's key insights into servos was that they fundamentally worked like amplifiers. Where the helmsman on a ship uses a small input of his muscles to control a huge vessel, and where a fire control officer uses delicate instruments to move heavy guns, Hazen's servos took low-power inputs and amplified them into high-power outputs. This point might suddenly have connected servomechanisms to feedback amplifiers at Bell Labs, but Hazen did not make the final conceptual leap. He proposed the rudiments of a comprehensive theory, but one not yet complete.

On another key point, Hazen departed from classic principles of governors and regulators (which Sperry and the Navy both shared). In those settings, as well as with simple machine controls, one would set the speed of an engine or the course of a ship, and the regulator or servo would 'hunt' back and forth while setting on that speed or course. The feedback mechanism maintained *consistency* in the face of external "disturbances," such as changes in load, wind, or target speed. One cared less about the machine's behavior while the setpoint was changing than about its ultimate stability. Sperry's automatic pilots kept airplanes straight and level, but they did not provide high maneuverability. Rangekeepers in naval fire control had several minutes to converge on their solutions. In contrast, Hazen saw feedback problems as similar to rapid, transient phenomena, so his analysis emphasized high-speed dynamic response, rather than 'steady stage' solutions. The systems still had to be stable, but now he cared about how they behaved while the setpoints were changing. Hazen's theory allowed engineers to make 'high-performance' servomechanisms which were also stable, which spoke to exactly the problem the Bureau of Ordnance was facing with its antiaircraft gun directors. Hazen made the conceptual leap from *regulation* to *control*.

CONTROLLING TECHNOLOGY

By the late 1930s, the four threads of control were beginning to drift together. Sperry Gyroscope funded research at MIT in aircraft controls. MIT taught a special course in control engineering for naval fire control officers. Bell Laboratories began integrating radars with fire control systems for the navy. Students at MIT began conversations with Harry Nyquist about similarities between feedback amplifiers and servomechanisms. Still, in 1940, the four cultures of control remained largely separate realms, separated by military secrecy, industrial concerns with patents and proprietary development, and plain narrow-mindedness. 'Control' remained tied to specific contexts—Sperry's beasts, fire control's systems, Bell Lab's amplifiers, and MIT's servomechanisms.

In 1940, however, a concerted, government effort began to merge these threads. That year, Vannevar Bush founded the National Defense Research Committee (NDRC) to enlist the nation's scientific talent in research and development for war. A special committee devoted to fire control began guiding the direction of development. They studied and incorporated prior work from Sperry, the navy, Bell Labs, and MIT. Through a research program of eighty research contracts—commissioning 'fundamental' studies as well as development—the NDRC brought the four threads together and combined their specialties: close coupling of human operators to machinery, distributed systems of sensors and computers, application of electronic feedback amplifiers, and use of theoretical tools to establish stable, high-speed response. This combination produced new techniques for stabilizing servos, automated antiaircraft systems (such as the M-9 gun director, developed at Bell Labs, Figure 21), advances in computing devices and information theory, and the seeds of Norbert Wiener's cybernetics. In this world, the interruption of noise replaced the threat of instability as the order-opposing demon.

World War II cemented the idea that government in the United States could direct the development of technologies to further national goals, what some have called 'command technology.'[36] This trend, of course, accelerated after the war, particularly with weapons, space, electronics, and computing. By fairly direct roots (often the same engineers), the NDRC's work in fire control spawned the large command, control, and information systems which characterized the era of nuclear standoff (Figure 22). Also, during the Cold War, the world these systems helped create, and their technological politics, contributed to the sense of alienation and powerlessness which gave rise to critique of technology seen as an abstract force in need of taming.

Kubrick's *Dr. Strangelove*, for instance, comically frames the struggle between democracy and communism as a struggle between human-directed and automatic systems for firing missiles (the so-called "doomsday machine"). Cultural fears of technology out of control responded to the pervasiveness in our culture of technologies of control. Descendants of these systems remain in place today, for example, as internetworking and as the tightly coupled human/computer interfaces which characterize personal computing. They continue to raise, in different forms, related questions about the control of technology.

Is technology a beast, portending wildness or a system, responding to rational direction? Do control systems increase the human ability to direct technology? Or does the promise of control fuel an obsessive drive for technological power? As we enter a period defined more by distributed information than by military command and control, these historical questions frame an anxious paradox.

Sitting at a personal computer, we experience our most powerful and intimate relationship with a machine: the thrill of control, extending our powers. In that very moment, however, we sense an abstract and impersonal force: the specter of technology, threatening instability. Wiley Post might have mulled similar ironies as he drifted off to sleep, entrusting his life to *Winnie Mae* and a Sperry automatic pilot.

NOTES

1 *New York Times,* July 23, 1933, 1. Wiley Post, "Destination—New York," *Sperryscope* 7 (no. 2, October, 1933). Post's Sperry Autopilot remains on display at the Smithsonian Air and Space Museum. Events dramatically demonstrated the importance of the Sperry Autopilot in keeping Post alert. The day after Post landed, two Italian aviators were critically injured landing in New York after a transatlantic flight. The cause of the crash: pilot fatigue. *New York Times,* July 24, 1933.
2 *New York Times,* July 24, 1933, 2.
3 J. H. Conger, quoted in Stanley R. Mohler and Bobby H. Johnson, *Smithsonian Annals of Flight Number Eight: Wiley Post, His Winnie Mae, and the World's First Pressure Suit* (Smithsonian Institution Press, 1978), 5. Will Harris in *Oklahoma City Times,* August 16, 1935, 1, quoted in Mohler and Johnson, 116.
4 Wiley Post and Harold Gatty, *Around the World in Eight Days: The Flight of the Winnie Mae,* (Hamilton, Ltd., 1932), 26.
5 Preston R. Bassett, "Servomechanisms: Controlling Vehicles in the Air," address given before the New York Section of the American Rocket Society, *Sperryscope* 13 (no. 1, second quarter, 1953), 22.
6 Quoted in Hughes, *Elmer Sperry,* 173. Sperry's words have a biblical ring: "Of all vehicles on earth, under the earth and above the earth, the airplane is that particular beast of burden which is obsessed with motions, side pressure, skidding, acceleration pressures, and strong centrifugal moments ... all in endless variety and endless combination."
7 Merritt Roe Smith and Leo Marx, eds., *Does Technology Drive History? The Dilemma of Technological Determinism* (MIT Press, 1994). Langdon Winner, *Autonomous Technology: Technics-out-of-Control as a Theme in Political Thought* (MIT Press, 1977).
8 Thomas P. Hughes, *Elmer Sperry: Inventor and Engineer* (The Johns Hopkins University Press 1971), 284–85.
* To avoid confusion, I will use the terms "Sperry," or "Sperry Gyroscope" to refer to the company, and "Elmer Sperry" to refer to the man, except where syntax makes "Sperry" clearly the man.
9 Hughes, *Elmer Sperry,* 173.

10 "Introduction," to Sperry Company History, n.d., probably 1942, Sperry Gyroscope Papers, Box 40. Similarly, Sperry Corporation President Thomas Morgan wrote in the 1943 Annual Report, "The primary value of Sperry's military products is that they extend the physical and mental powers of the men in the Armed Forces enabling them to hit the enemy before and more often than the enemy can hit them." This annual report refers not only to Sperry Gyroscope but to the other companies, including Ford Instrument, under the Sperry Corporation.

11 David Mindell, "Antiaircraft Fire Control and the Development of Integrated Systems at Sperry, *IEEE Control Systems*, April, 1995.

12 "Aircraft Fire Control," Sperry Gyroscope Company report, Sperry Gyroscope Company papers, Hagley Museum and Library, Box 22. "Aircraft Computing Sights," Sperry Gyroscope Company, Vinson report edited by Roswell Ward, February 8, 1944. Report, Sperry Gyroscope papers, Box 40. Roswell Ward, "Aircraft Turrets: Description of Product Development and History," February 16, 1944. Sperry Gyroscope papers Box 40.

13 "Introduction," to Sperry Company History, n.d., probably 1942. Sperry Gyroscope Papers, Box 40.

14 Estimates of gun ranges vary depending on whether one measures the distances at which fleets conducted battle practice, the distance of historical engagements, or the distance theoretically possible under ideal conditions. Elting Morison reports that Admiral Sims trained gunners to fire at 1600 yards, to prepare for battle conditions at 6,000 yards in *Admiral Sims and the Modern American Navy* (Houghton Mifflin, 1942), 142. For other estimates, see *Administrative History of the U.S. Navy in World War II*, Volume 79, *Fire Control* (United States Navy, 1946), 2–3. Rodrigo Garcia Y Robertson, "Failure of the Heavy Gun at Sea, 1898–1922," *Technology and Culture* 28 (no. 3, 1987) 539–557. For a detailed assessment of accuracy at long ranges, see W. J. Jurens, "The Evolution of Battleship Gunnery in the U.S. Navy, 1920–1945," *Warship International* (no. 3, 1991), 240–71.

15 Ford Instrument Company, "Report on Organziation and War Activities of the Ford Instrument Company, Inc.," June, 1919. National Archives RG 74 E-25 Box 2740, Subject File 36276/110. "Secret fire control design section:" Captain H.J.S Brownrigg, RN, "'Ford,' Fire Control System: Interviews with representatives of Ford Instrument Coy. of New York," IQ/DNO (January–June, 1919) Naval Library, Ministry of Defence, London. Courtesy Jon Tetsuro Sumida.

16 Elting Morison, "Gunfire at Sea," in *Men, Machines, and Modern Times* (Press, 1966).

17 John Keegan, *The Face of Battle: A Study of Agincourt, Waterloo, and the Somme* (Penguin Books, 1976), 32.

18 The term "system of information," was used as early as 1905 by an official U.S. Navy Body to describe fire control. Fire Control Board quoted in

Norman Friedman, *US Naval Weapons*, (Conway Maritime Press, London, 1983), 28. Friedman's book also contains an excellent survey of fire control technology.
19 Rowland and Boyd, *US Navy Bureau of Ordnance*, 377–8. United States Navy, *Administrative History of the U.S. Navy in World War II*, 146–7.
20 Harold S. Black, "Inventing the Negative Feedback Amplifier," *IEEE Spectrum* 14 (December, 1977), 54–60. Harold S. Black, "Stabilized Feedback Amplifiers," *Bell System Technical Journal* 13 (1934).
21 Harold S. Black, "Wave Translation System," Patent no. 2,102,671, page 2.
22 Black, "Wave Translation System," 2.
23 Harry Nyquist, Discussion of H.S. Black, "Stabilized Feed-Back Amplifiers," *Elec. Eng.* (September, 1934), 1311–12.
24 Harvey Fletcher, "The Theory of the Operation of the Howling Telephone with Experimental Confirmation," *Bell System Technical Journal* (no. 1, January, 1926), 27–49. Fletcher's paper does not employ the terms "stability" or "feedback," in its analysis, although it does analyze electro-acoustic circuits which greatly resemble canonical feedback systems.
25 Harry Nyquist, "Regeneration Theory," Bell System Technical Journal 11 (1932) 126–47. Also see Bennett, *A History of Control Engineering 1930–1955*, 82–84.
26 H. W. Bode, "Relations Between Attenuation and Phase in Feedback Amplifier Design," *Bell System Technical Journal* 19 (July, 1940), 421–454. For other discussions of this paper, see Bennett, *A History of Control* Engineering, 84–86.
27 For the history of the interconnection of electrical power networks, see Thomas P. Hughes, *Networks of Power: Electrification in Western Society, 1880–1930* (Johns Hopkins University Press, 1983).
28 For a contemporary review of the subject, see C. L. Fortescue, "Transmission Stability: Analytical Discussion of Some Factors Entering into the Problem," *Trans. A.I.E.E.* 26 (February, 1927), 984–994 and discussion 994–1003.
29 Karl L. Wildes and Nilo A. Lindgren, *A Century of Electrical Engineering and Computer Science at MIT 1882–1982* (MIT Press, 1985), Chapter 4. Vannevar Bush, *Operational Circuit Analysis* (John Wiley and Sons, Inc., 1929).
30 Harold Edgerton, "Abrupt Change in Load on a Synchronous Machine" (S.M. Thesis, MIT 1927), and "Benefits of Angular-Controlled Field Switching on the Pulling-into-Step Ability of Salient-Pole Synchronous Motors" (Sc.D. thesis, MIT, 1931). Wildes and Lindgren, *A Century of Electrical Engineering*, 145–7.
31 Frederick Terman, "The Characteristics and Stability of Transmission Systems" (Sc.D. diss., MIT, 1924), 1.
32 Vannevar Bush, "Power System Transients," *AIEE Trans.* 44 (1925), 229–30.

33 Ibid., 229.
34 Harold Hazen, O. R. Schurig, and M. F. Gardner, "The M.I.T. Network Analyzer, Design and Application to Power System Problems," *AIEE Trans.* 49 (July 1930), 872–875. Vannevar Bush, H. R. Stewart, and F. D. Gage, "A Continuous Integraph," *Jour. Frank. Inst.* 211 (January 1927), 63–84. Vannevar Bush and Harold Hazen, "Integraph Solution of Differential Equations," *Jour. Frank. Inst.* 211 (December, 1927), 586–88. Vannevar Bush, "A New Machine for Solving Differential Equations," *Jour. Frank. Inst.* 212 (no. 4 1931), 447–488. Vannevar Bush, "A New Type of Differential Analyzer," *Jour. Frank. Inst.* 240 (no. 4, October, 1945), 255–326. Gordon S. Brown, "The Cinema Integraph: A Machine for Evaluating a Parametric Product Integral," (Sc. D. thesis, MIT, 1938). Larry Owens, "Vannevar Bush and the Differential Analyzer: The Text and Context of an Early Computer," *Technology and Culture* 27 (no. 1 1986), 87.
35 Harold Hazen, "Theory of Servo-Mechanisms," *Jour. Frank. Inst.* 218 (September 1934), 279–331. Harold Hazen, "Design and Test of a high-performance Servo-Mechanism," *Jour. Frank.Inst.* 218 (November 1934), 543–580.
36 William McNeill, *The Pursuit of Power: Technology, Armed Force and Society Since A.D. 1000* (Chicago University Press, 1982), Walter McDougall, *The Heavens and the Earth* (Basic Books, 1985).

CHAPTER 10

Liquifying Information: Controlling the Flood in the Cold War and Beyond

Mark D. Bowles

Many observers argue that since 1945 information has become central to our cultural identity. They proclaim our era as the 'information age,'[1] periodizing it as the 'mode of information,'[2] in which our 'information society,'[3] surrounded by the technologies of our 'information culture,'[4] breeds new workers in an 'information economy,'[5] who daily experience the effects of the 'information revolution,'[6] all while living in an 'information environment,'[7] which induces 'information anxiety,'[8] and 'information fatigue.'[9] These words—age, society, culture, economy, revolution, environment, anxiety, and fatigue—all subsumed by the privileged term information, indicates, at the very least, the widespread perception that a fundamental cultural shift has occurred, powerfully affecting how we work, live, and experience our lives. But, the enthusiastic emphasis placed on the production of information has led to a crucial problem. A culture which so highly values information is susceptible to the problem of information overload. Since World War II, the fear of too much information has become one of our most significant yet historically neglected intellectual concerns.

At the very dawn of this information age, coexisting with the beginnings of the Cold War, a growing number of intellectuals[10] interpreted information as an external force that needed to be managed due to the perception of imminent overload. In 1945 Vannevar Bush, the analog computing innovator and statesman of science, initiated much of the debate and argued that overpublication created the potential for science to become 'bogged down in its own product, inhibited like a colony of bacteria by its own exudations.'[11] Describing the situation as a 'morass' of publications, he said that in physics alone there were thousands of journals and lamented that because of this exponential growth in literature, no one could keep up with them all. In contrast, nearly 50 years later, historian Steven Lubar wrote, 'I swim in an ocean of information.'[12] With computers, televisions, telephones, newspapers, CDs, and videos, there is an 'astonishing electronic information infrastructure surrounding me—surrounding us all.' The distinction between the statements of Bush and Lubar are significant. While Bush was struggling in his boggy marsh, Lubar portrayed himself as swimming comfortably through his information waterscape. Though each intellectual suggested that he was surrounded by information, his perception of control was very different.

The problem represented by information overload centers on the need for *control*. To understand my use of this contested term, I borrow JoAnne Yates's definition from her *Control Through Communication*. She argued, from her managerial perspective, that control was the 'mechanism through which the operations of an organization [were] coordinated to achieve desired results.'[13] Transforming this definition from a managerial to a (post-1945) information context, we can extract the following: the preferred mechanism used for organizing information was the computer; the organizations at play were the various professional groups concerned with the problem of information overload, most often dominated by scientists and engineers; finally, the desired result these professionals sought was the ability to use the new computer systems to manage as well as search for specific information.

Information is yet another ambiguous term. Since World War II, its meaning has gone through a tremendous transformation. In 1994 historian and self-proclaimed neo-Luddite Theodore Rozak wrote, 'Information has had a remarkable rags-to-riches career in the public vocabulary over the past forty years,' now achieving the 'exalted status of a godword.'[14] Prior to World War II, information was of marginal importance, referring only to simple facts such as someone's name, date of birth, or telephone number. Rozak argued that since the war, a 'widespread public cult' has grown, fetishizing the importance of information. An examination of the two editions (1933 and 1989) of the *Oxford English Dictionary* reveals these changing meanings. While both editions listed 7 general definitions of the word, the 1989 entry doubled in size and added an 8th section listing attributive and combinations of the word. This included some twenty-two new combinations with the word information: content, desk, explosion, flow, gap, office, service, storage, system, transfer, work, carrying, gathering, giving, seeking, bureau, officer, processing, retrieval, revolution, science, and technology. Clearly, information no longer was just a 'fact,' but had grown into what Rozak called a fetish or thing.

It was Claude Shannon who was one of the first to reinterpret information in this way. Shannon postulated a theory to mathematically describe an information source and determine the number of bits of information per second that this source gave off.[15] Warren Weaver wrote that for Shannon's mathematical theory, 'the amount of information is defined...to be measured by the logarithm of the number of available choices.'[16] Thus, Shannon cast information in quantifiable terms that could be physically measured. Today, the predominant way to conceptualize information remains as a physical object. In 1991, Michael K. Buckland argued that information systems (computers, libraries, museums, etc.) used what

Figure 23 A flood of information. Artist's conception. Watercolor by Gyorgy Szalay (1998). Collection of the author. Photograph: author.

he called 'information-as-thing.'[17] These types of 'things' included the physical bits manipulated by computers, books stored at libraries, and artifacts displayed at museums. The bits, books, and artifacts all represented information as a physical entity.

The professionals who shaped the debate over information overload since 1945 also defined information as a physical object and a thing. This allowed them to quantify, count, and measure its tendency to increase and become out of control. For them, information represented all forms of written knowledge, and its management consisted of new computer systems to inscript, preserve, store, retrieve, disseminate, and use the information. Thus, I use the phrase 'information out of control' to mean the ways in which various professional groups adopted computing technology to better manage an exponential growth of quantifiable information.

In this essay, I will focus primarily on the *perceptions* of the loss of control over information and the meanings ascribed to it since 1945.[18] I will begin by describing what intellectuals called the 'information crisis' in the generation after World War II and explore how they perceived too much information as a dangerous Cold War threat. In the second section, I will analyze these perceptions by studying how various groups used language, specifically metaphor, in relation to the information problem. Information as a physical entity was 'liquefied' as scientists, engineers, and other intellectuals employed drowning and flooding metaphors in the early part of the Cold War to describe their fear of an information 'deluge.' They believed that the exponential growth of specialist publications were out of control with the result being a potential to falter as a nation before the rising 'tide' of communism.

In the final section, I will again use the water metaphor as a tool to understand our current notions of information overload. This analysis reveals the diverging ways in which an intellectual elite and the general public conceptualize their relationship with information. Today intellectuals rarely use the flooding rhetoric to indicate a loss of control. Instead, they express confidence in their abilities to manage information through new water metaphors such as 'navigating' through 'pipelines' of information and manipulating it through 'fire-hoses' of data, metaphorically placing themselves in control of their information production. In contrast, the general public (those who are now encountering massive amounts of information for the first time from computer, print, and broadcast media) are the ones who express being 'swamped' by information and 'drowned' in data. Thus, by examining the metaphorical liquefaction of information and the associated transition in language usage we can better understand one of the central concerns associated with the information age.

THE INFORMATION CRISIS IN THE COLD WAR

After World War II, intellectuals interpreted the problem of information overload as a Cold War threat. Information, as a new commodity, was viewed as yet another benchmark with which to compare the United States to the Soviet Union. With rumors of a vast scientific information service existing behind the Iron Curtain, the information race quickly became as important as the space race. However, this race was not about which nation generated or produced the most information. The contest was framed as a race for which superpower would learn to control the information that they already had, thereby assuring the continued advancement of their political ideologies.

It was Vannevar Bush who initiated the concern over information excess. One of this century's outstanding technological utopians and engineering heroes, Bush designed and built the most popular and powerful computing machines prior to the war.[19] In 1945 he wrote a prophetic article called 'As We May Think' for the *Atlantic Monthly* which predicted a future where a desktop computing machine called the 'memex' managed everyone's daily information needs. Bush believed scientists needed such a machine because of the 'growing mountain of research' that confronted them during their daily work, bogging them down from advancing their ideas.[20] He said that the methods scientists and other intellectuals used for communication were archaic.²¹ Specifically, he argued that the 'printed page, the library, the spoken word, the lecture...[were] no longer adequate.'²² He attacked libraries in particular because he felt they could not cope with the task of making available all the increasing amounts of relevant literature. Bush said that the problem was not that scientists published too much, but instead they were unable to effectively manage and access the records that existed.²³

A condensed version of 'As We May Think' also appeared in *Life* magazine, emphasizing Bush's belief that while specialization was one way to deal with the problem, a 'bridge between disciplines' became necessary.²⁴ He lamented that the most advanced technology used for managing information was developed during the days of the 'square-rigged ships.' He also argued that his memex would free the future investigator so that he was no longer 'anchored.' Bush's argument (nautical metaphors included) generated the momentum towards what became the widespread perception of the *information crisis* for the generation after World War II.

The fear of too much information quickly spread throughout the United States' intellectual communities. Scientists used statistical methods in an attempt to gauge the total store of information and determine exactly its tendency to increase exponentially. One observer calculated that the total characters of text residing in the Library of Congress was 200 trillion, while another increased this figure to 380 trillion.²⁵ J. C. R. Licklider in his influential *Libraries of the Future* investigated the rate at which information was increasing. He argued that 'the size of the store is doubling every 15 to 20 years, which makes the current rate of growth about 2 million bits per second.'²⁶ These scholars perceived this information explosion as a national threat, and as one observer wrote, the United States was in the midst of an 'information crisis.'²⁷

While Bush was the first to identify the problem, it was MIT mathematician Norbert Wiener who presented a study of how information could become out of control. At the end of World War II, Wiener began working on a theory of messages. While the field of electrical engineering had a theory concerning the transmission of messages, Wiener was more interested in studying 'messages as a means of controlling machinery and society.'²⁸ Calling this new theory 'cybernetics,' he attacked the related problems of both the control and communication of information. As one observer argued, one of the problems associated with the 'Cybernetic Revolution' itself was the 'information explosion.'²⁹

For Wiener, information was the content mechanisms or individuals exchanged between themselves and their world. Information was 'feedback,' the essential component required to manage either a mechanical or biological system. Wiener believed that modern life placed greater demands than any previous era on the process of exchanging information. He warned that 'our press, our museums, our scientific laboratories, our universities, [and] our libraries' had to keep up with these new information needs, else they would 'fail in their purpose.'

Rather ironically, while Wiener argued the importance of feedback, he claimed that excessive feedback presented a great danger. For example, he said, there was a 'well-known tendency of libraries to become clogged by their own volume.'[30] Like Vannevar Bush, Wiener believed that overpublication could create a situation in which bad science would drown out good science. Wiener stated that too many inconsequential articles were being published, disguised as important scientific advances. He said that because so many scholars published solely for the 'intellectual prestige of becoming a priest of communication' the value of the information itself dropped 'like a plummet.'[31] He feared that 'If a new Einstein theory were to come into being as a Government report in one of our super-laboratories there would be a really great chance that nobody would have the patience to go through the trash...to discover it.'[32] As Wiener concluded, 'To live effectively is to live with adequate information.'[33]

Wiener further defined his concern by introducing a political component to his argument. He expressed a warning to any country that might fail to manage its information. The nation with the greatest security, would be the one 'whose informational and scientific situation is adequate to meet the demands that may be put on it.'[34] Again Wiener used the word 'adequate' to describe information, illustrating his concern that too much could be just as dangerous as not enough. Wiener's political interpretation of information overload became one of the dominant concerns during the Cold War.

As the Cold War intensified, the problems of information overload were consistently interpreted within a political framework. One of the main reasons for this interpretation was that scientists and engineers, for the first time, became actively aware of their political influence. As G. Pascal Zachary wrote in his biography of Vannevar Bush, 'the lesson of World War II...[was that] the outcome of war was now decided, as much as anything, by a nation's scientific and engineering wizards.'[35] Historian David Noble argued, 'the scientists emerged from the war with a larger-than-life image (and self-image) as genuine national heroes.'[36] But, could these new United States' heroes keep up with their counterparts in Soviet Union? What were the consequences to the nation if they could not?

Many in government, science, and industry believed that Soviet supremacy in science and technology represented the first falling domino in the demise of democracy. While Sputnik was one very obvious example of Soviet technical capability, an analogous, less obvious, but more foreboding advance also worried a segment of the U.S. scientific community—VINITI the Soviet Information Service. VINITI or the 'Soviet All-Union Institution of Scientific and Technical Information' was a massive abstracting and translation service that also housed an advanced punched-card computing machine for searching scientific literature. This Soviet institute, established in 1952, was rumored to employ over 25,000 people to abstract, translate, automate, and disseminate the world's scientific information to communist countries. They reviewed over 12,000 foreign journals each year and in 1959 they abstracted over 700,000 articles in a variety of different languages.[37] Soviet scientists and intellectuals used this information to further advance their own significant capabilities, as exemplified by their successful launch of Sputnik. A comparable U.S. academic information service employed only 45 people working in relative academic obscurity.

The Soviet Union appeared to be addressing Norbert Wiener's warning for 'adequate information' far better than the United States. While cybernetics was initially rejected in the Soviet Union, Slava Gerovitch argued it was used as a 'vehicle of reform in the post-Stalinist system of science....'[38] Wiener discovered this himself during a trip he took to Moscow in 1960 to attend the First Congress of the International Federation on Automatic Control, when the editors of the leading Soviet philosophy journal told him that many Soviet scholars were exploring the application of cybernetics to various aspects of human knowledge.[39] Earlier that same year a Russian newspaper, *Ekonomicheskaia Gazeta*, published an article on the importance of cybernetics for the Soviet state. The author, A. I. Berg, wrote that the science of cybernetics could enable a socialist society to sustain its economy while releasing 'millions of working people for genuine creative activity.'[40] But, cybernetics had not only the potential to free economic laborers, it also had the potential, through the creation of cybernetic machines to enable its scientists and engineers to become more productive. Soviet scientists worked to construct a type of information machine that, according to Berg, would control the 'avalanche-like growth of every type of scientific information.'[41] This would enable scholars to keep up with their reading material, allowing them to pursue the highest levels of intellectual activity and to capture the lead in world science.

A. I. Mikhailov, the director of VINITI, crystallized this growing perception of a Soviet information problem in a 1962 article entitled the 'Problems of Mechanization and Automation of Information Work.'[42] He argued that there was an 'avalanche of literary production' and that the great armies of scientists and engineers were creating a 'snowballing' amount of information. Mikhailov believed that this output made it increasingly difficult for these workers to master the vast intellectual treasure within their own fields. He also argued that the amount of information was increasing exponentially and doubling every 12 to 15 years. The result of this 'mounting tide' and overwhelming quantitative growth of scientific and technical literature was a greater need than ever before to develop new information services.

The 1957 Symposium on Systems for Information Retrieval held in Cleveland, Ohio, demonstrated the perceived threat that the Soviet Union's information capabilities posed. Over 900 people attended this 3-day symposium, and sponsorship came from 25 professional, academic, industrial, and governmental organizations including the American Chemical Society, Case Institute of Technology, Bell Telephone Company, and the U.S. Department of Commerce. A day before the symposium began, the advisory board met and read a letter by R. A. Fairthorne, Senior Scientific Officer from the British Royal Aircraft Establishment. Fairthorne wrote that the immediate problem was 'the challenge of the vast USSR activities in information work...and how to keep ahead of it.'[43] The advisory board agreed with Fairthorne's assessment of the central problem and concluded that the Soviet 'threat...justifies the mobilization of all possible efforts' to solve the information control crisis.[44]

As the Cold War intensified, the race to manage scientific and technical information remained one of the leading national problems. In 1959 Allen Kent, an eminent information specialist, wrote that 'There is active competition between the USSR and the United States in making effective use of the world's published scientific knowledge' and that there was a 'chance that we may be beaten out by the Russians by using research tools developed here, but exploited more forcefully in Russia.'[45] He believed at this time that had the United States set up an equivalent information center to VINITI, the U.S. could have beaten the Soviets in launching the first orbiting satellite.[46]

The apparent answer to this Soviet threat and the preferred plan to bring information back under control was through new computer systems housed in specialized information centers. According to Alvin M. Weinberg, director of the Oak Ridge National Laboratory, by 1963 there were about 400 specialized information centers throughout the United States.[47] The best known of the academic information centers and sometimes called the 'World's most advanced information retrieval system' was the Center for Documentation and Communication Research (CDCR), located at Western Reserve University in Cleveland, Ohio.[48] A MIT librarian wrote that the CDCR had a 'supremacy' and a 'uniqueness' in areas of machine literature searching that was 'unrivaled' by any other information center.[49] These devices included new punch-card computing machines which became the heart of their information service for scientific and industrial researchers. In 1960 the *Wall Street Journal* reported that the CDCR was able to transfer 150,000 abstracts into a machine readable form, and their computers could search 100,000 articles per hour.[50] Throughout the 1950s and 1960s, many saw the CDCR as the 'forerunner of a vast centralized national information facility that [would] increase the efficiency of the research worker and safeguard American scientific progress' and 'challenge the Soviet Union in the science race.'[51]

In 1961 *Business Week* reported on how the CDCR was directly serving the nation in the Cold War. A metallurgist who was 'stalled by a missile production problem' sent a request to the CDCR asking, 'What has been written on impact-forming of metals at high speeds?' The CDCR searched 50,000 abstracts with their computers and provided an answer to the metallurgists' question in under half an hour.[52] As the *Washington World* reported in 1962, the CDCR was 'quietly revolutionizing' the science of storing and retrieving data by managing the 'deluge of information' while playing a vital role in protecting the United States from the Soviet threat.[53]

Despite the intensity of the perception of the information crisis, the Cold War was neither lost nor won on the basis of which nation learned to manage its information. However, the information crisis was extremely important in pro-

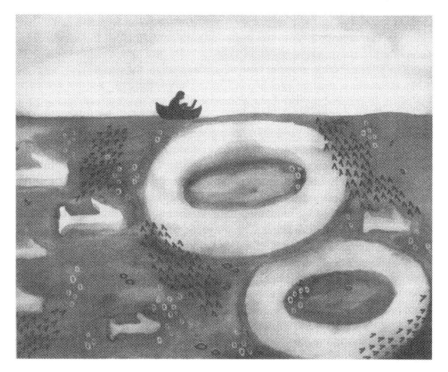

Figure 24 Floating on the information sea. Artist's conception. Watercolor by Gyorgy Szalay (1998). Collection of the author. Photograph: author.

viding a justification for establishing the computer as the main tool for information processing. The information crisis also initiated the widespread use of a water metaphor to describe the problem of too much information. This metaphor was exemplified by the journalist at *Washington World* writing of the deluge of information, the director of VINITI expressing concern over the snowballing amount of information (an appropriate water metaphor for the Siberian climate), and by Vannevar Bush warning of the scientist who was anchored in the morass of information. By historically analyzing the use of this water metaphor, it can serve as a cultural indicator to gauge the confidence that groups or individuals perceived that they possessed to control their information.

WATER AS CONTROL METAPHOR IN THE COLD WAR

The metaphor has been called one of the great mysteries of human speech leading to a number of questions concerning its function in language. Why is human speech littered with examples of metaphor? What meanings do metaphors have? Is there a relationship between metaphor and thought? Scholars

234 *Cultures of Control*

are now beginning to focus on the role metaphor plays in everyday life. They argue that that metaphor is central to the English language, that the meanings of metaphor are more about thought than words, and that by understanding metaphor we can better interpret how we conceptualize the world.[54]

My argument is that the problems of information overload, and the information crisis in particular, have been conceptualized and understood through the use of metaphor. While many different metaphors were used to illustrate this problem (from jungles to mountains to explosions), the water metaphor became the central method of describing the impact of too much information. An examination of the various water metaphors in the generation after World War II illustrates the similar textual imagery used to describe the inability to control information. These examples include: a 'morass,' a 'cascade of new knowledge,' a 'flood of information,' a 'mounting flood of scientific data threatens to swamp the researcher,' a 'flood of reports,' a 'flood of words,' a 'flood of literature,' a 'flood of knowledge,' a 'flood of technical knowledge,' a 'flood of technical information,' a 'deluge of scholarly enquiry,' 'drowned in a flood of our scientific knowledge,' an 'outpouring of printed materials,' 'engulfed in a tidal wave of paper,' 'engulfed in a sea of information,' a 'growing ocean of known fact,' and 'knee deep in paper and facing a rising tide.'

These statements, all occurring from 1949 to 1963, are important not only for the similar metaphorical imagery of water and a failure to contain it but also because they represented the thoughts of a group of highly influential experts. The above phrases came from the Director of the Office of Scientific Research and Development, a computer specialist at the Center for Documentation and Communication Research, the director of the Oak Ridge National Laboratory, an Eisenhower White House press release, a government advisory committee to President Kennedy, a United States Circuit Judge, an eminent historian of science, an army colonel who was head of the military Scientific Information Section, an engineer at the Stanford Research Institute, a journalist at the *Wall Street Journal*, an English professor at the University of Colorado, a past president of the American Association for the Advancement of Science, a historian at the University of Minnesota, the director of the MIT libraries, an IBM ad campaign, a well-known science fiction author, and the library school dean at Western Reserve University.[55] During these years, perception of an information crisis spread to every professional discipline and was intensely debated by leading experts in their respective fields. This common perception was in a very important way communicated through the 'metaphor-theme' of information as an overpowering source of water.[56]

From 1945 to the early 1960s, this perception dominated the debates concerning the future of scholarly inquiry as well as the future of the nation itself. But, in 1963 there arose a reaction against these concerns of an information flood beginning only after World War II. The most carefully-argued critic was Derek J. de Solla Price who first published *Little Science, Big Science* in this year. Treating scientific activity as a measurable entity, he supported what many of those who feared an information flood were arguing—that scientific literature was increasing exponentially. But, after a detailed analysis, Price concluded that this growth rate has been a constant attribute of science since the 17th century and that 'Scientists have always felt themselves to be awash in a sea of scientific literature.'[57]

This fear of too much information has been a consistent feature not only of science, but all forms of intellectual inquiry. For example, in 1751 Voltaire published *The Age of Louis XIV,* after spending nearly 20 years of research and writing. He lamented that other 18th century authors did not take such care and time in crafting their own books for publication and wrote, 'it has become so easy to write mediocre stuff that we have been flooded out with books.'[58] In 1820 Washington Irving published his widely popular collection of essays entitled *The Sketch Book.* In one essay, he described how the rate of publication had dramatically increased in the 19th century. Irving claimed that by 1820 everyone, it seemed, was a writer and able to 'pour' themselves into print. He believed that the consequences of this were alarming, resulting in a 'stream of literature [that] has swollen into a torrent—augmented into a river—expanded into a sea.' Soon the 'modern genius' would be 'drowned in the deluge.'[59] These examples from Voltaire and Irving illustrate both the continuity of this concern, the consistent use of water metaphors to explain the problem, and the belief that each generation confronted exponential information growth.

However, in the generation after World War II intellectuals did rhetorically link their concerns of information overload to a unique problem—expressions of Communist expansion during the Cold War. This linkage had an earlier history. In *Male Fantasies*, Klaus Theweleit examined the water metaphors used by the German Freikorps—a private and volunteer army existing in the aftermath of World War I. After studying the correspondence, novels, and autobiographies of these men he found a recurring metaphorical usage to describe Bolshevism. This rhetoric included terms such as the red flood, Bolshevistic flood, the dam had broken, and the raging Polish torrent. By the end of World War II, this became one of the central ways to describe the spread of communism. In 1946 Clark Clifford wrote that the Soviet Union was able to 'flow into the political vacuum of the Balkans.'[60] The New York *Herald Tribune* quoted a congressman as saying, 'If that [communist] bridgehead is ever established Great Britain and the United States will be Islands in a great red ocean.'[61] In 1949, U.S. Supreme Court testimony in Terminiello v. Chicago articulated the fear of being 'drowned in this tidal wave of Communism.'[62] That same year, when *Time* magazine honored Harry Truman as its Man of the Year, they wrote of 'the ebb and flow of Communism's tide.'[63] As a young U.S. congressman, John F. Kennedy spoke of the need to prevent the 'onrushing tide of Communism from engulfing all of Asia.'[64] In 1952 Miklós Horthy, the governor of Hungary, wrote in his memoirs of being 'overwhelmed by the Communist flood.'[65] And in 1955, British Prime Minister Anthony Eden wrote, 'In stemming the tide of Communism...Britain has a central and crucial part to play.'[66] Thus, this metaphorical rhetoric shows the similarity between how intellectuals and politicians conceptualized the information crisis and communist expansion.

Why was a water metaphor used for describing both information overload and the threat of communism? Suggesting an answer to the general question of why metaphors are used, Sam Glucksberg and Boaz Keysar said that metaphors are *not* merely ornamental, they are necessary. They are 'necessary for casting abstract concepts in terms of the apprehendable.'[67] The idea of too much information as well as the concept of spreading communism were abstract concepts that required something familiar in order to comprehend it. Water and the plethora of familiar imagery associated with it provided a range of experiences with which to explain and describe one's relationship to information as well as communism. Our experience with large bodies of water is one of awesome, untamed power and impenetrable mystery. As the last geographical frontier on our planet, the sea remains uncharted and uncontrolled. The strength of tides, the threat of rising rivers, and the possibility of floods always keeps the possibility of drowning and destruction close at hand. The water metaphor was also used because it enabled one to understand the potential role computers and containment could play in helping to manage this flood. Both information and communism were believed to be forces that the U.S. had to impose control in order to preserve democracy. Thus, the metaphor helped powerful social groups to sell and persuade others of their answers—new technology and military containment—to the most pressing problems of this era.

WATER AS A CONTROL METAPHOR IN THE LATE 20TH CENTURY

By the late 20th century, water remained a compelling metaphor for expressing the relationship between information and the problems of overload. However, there were a great number of new words added to the water metaphor lexicon illustrating a changing experience with information. It was Norbert Wiener who anticipated this changing experience when he coined 'cybernetics' in 1947. While he used the term to name his new science of information, the word itself came from the Greek *kubernetes* meaning helmsman.[68] Conjuring up a nautical image of navigating a vessel through water, Wiener's metaphor, unlike any other of his or any previous time, signified the potential mastery over water/ information through the semblance of a one steering a ship. The helmsman manuvered his ship (representing a packet of controlled information) through the uncontrolled waterscape of information noise. The 'control' inherent in Wiener's own metaphor, anticipated the dominant way intellectuals would speak about information in the 1990s.

New water metaphors now reflect the heroic mastery and the thrill required to calm the information forces. For example, the name for randomly exploring the world-wide-web is 'surfing.' A 'pipeline' is the term used to describe the conduit for information on the internet. 'The WELL' is one of the most influential on-line communities.[69] The Electronic Frontier Foundation calls the internet a 'global watering hole,' leaving the impression that the user can come, take a satisfying drink of water, and then leave when content.[70] Historian Edward Tenner likened information to a 'fountain,' while cyberspace author Donald B. Whittle referred to information as a 'fire-hose.'[71] Astronomer Clifford Stoll wrote in *Silicon Snake Oil* of the 'stream of data' that the internet delivered on demand with the modem acting as a 'faucet, letting information flow into the computer.'[72] The head of an internet audio company referred to the new Internet 2 as a 'digital spigot.'[73] One of the most popular self-help guides for the internet was called *Navigating the Internet* which began by asking the reader to 'Imagine yourself as a navigator out in the ocean' needing the proper 'charts' to find your way.[74] *Netscape Navigator* remains one of the leading internet browsers in the 1990s. And, of course, if anyone copied this or any other software package illegally they would be referred to as a 'pirate.' Thus, all of these terms—surfing, pipelines, wells, watering holes, fountains, fire-hoses, faucets, spigots, navigators, and pirates—all signify control. Their linkage to information indicates a level of confidence that the 'flooding' crisis of information during the Cold War was long over.

The most eloquent representation of the similarities between the powers of the ocean and the lure of mastering computer hardware and software, is Tracy Kidder's 1981 Pulitzer Prize winning novel *The Soul of a New Machine*.[75] This was the true story about Data General Corporation's efforts to produce a new computer. Kidder entitled the prologue 'A Good Man in a Storm,' and introduced the protagonist, Tom West, a brilliant computer programmer, on board a small sailing ship in the ocean. As West led a team of programmers past their endurable limits to create a machine inscribed with their own souls, Kidder's portrayal of a man who was always confidently 'weathering a storm' became the perfect analogy to the forces he confronted while building the new machine. By giving the first description of West at sea, Kidder reinforced this heroic image by describing him as the 'ghost of an old-fashioned virtuous seaman,' who rarely slept over the four-day voyage, and was able to manipulate the sails so expertly that even the captain worried that his ship had never gone so fast. Thus, the dual image of West, mastering the forces of the sea as well as the process of building an information machine, was essential for understanding the parallels between the ancient mariner and the contemporary programmer.

This shift in the use of the water metaphor is indicative of how we view the problem of too much information as the 20th century comes to a close. As our information systems become more pervasive and complex in our society, many envision themselves as emerging from the 'flood of data' and conquering the force itself as illustrated by the new nautical metaphors. We now have a host of powerful terms, which did not exist in the generation after World War II, depicting humanity regaining command over its own creation, imposing control for the first time in our information age.

However, not everyone shares this perception of information control. The general public have not had the opportunity to experience the so-called information age until recently. As librarian L. B. Heilprin wrote in 1961, the information 'crisis has not yet reached the attention of the public.'[76] During the Cold War, the public was insulated from the information crisis because their lives did not center around the production of information as did, for example, the scientists and engineers who continually had to maintain a detailed familiarity with the publications in their specialty. As Allen Kent illustrated in 1958, although Sputnik raised tidal waves of 'hysterical alarm' throughout the U.S., the Soviet information service VINITI and the information crisis in general cast not even a 'ripple.'[77]

In the 1990s, while the technological elite might feel accustomed to our information culture, it is the general public that now characterizes itself as drowning. They see themselves as being flooded by information delivered by internet, television, and print media. For example, Dennis M. Wint, president of the Science Center (which he describes as 'a bridge between the scientist and the community') writes, in our society today 'we are swamped by information [and] have no control over it.'[78] Also representing the wider community, New Zealand poet Keith Newman wrote a lyrical poem called 'Information Overload' describing the 'Technology tidal waves rushing to our shores…Drowning in data, thirsty for truth.'[79] A further example is Paul Gilster's frontispiece in *Digital Literacy* which has a small personal computer surrounded by flowing water with the caption reading 'Somewhere within the flood of on-line information there is a creative universe just for you.'[80]

The theme of control, or lack of it, was also an important component of science fiction writer William Gibson's matrix which he described as a massive internet-like computer communication network. In his award-winning book *Neuromancer*, he defined the matrix as a 'sea of information' with its lethal data or 'black ice' constantly threatening a simulation overload.[81] In 1994, *Wired* executive editor Kevin Kelly (called a 'popular scientist' by the *London Spectator*) wrote a book entitled *Out of Control* describing how information has 'flooded our environment with messages, bits, and bytes.' He said that this 'unmanaged data tide is at toxic levels' and concluded that like the 'raw explosion of steam' our information production is more than we can control.[82] In 1996, British psychologist David Lewis diagnosed a malady he called Information Fatigue Syndrome which was the psychological and physical effects of the inability to control information. He said that, 'it's sensible to stand under a faucet, not Niagra Falls' for ones information needs.[83]

These examples (from a museum curator, a poet, a journalist, a science fiction author, a popular scientist, and a psychologist) represents the popular view of information in the 1990s, and demonstrates a common experience of information out of control. Thus, as our cultural identity continues to center around the production of information, our perceived interaction with the information waterscape, from drowning to surfing, will serve to indicate our level of control as we experience the information age.

ACKNOWLEDGEMENTS

The author would like to thank Miriam Levin, The Charles Babbage Institute, and the CWRU works-in-progress group: Molly Berger, Cathy Kelly, David Hochfelder, and Patrick Ryan.

NOTES

1 In President Clinton's 1998 State of the Union Address he argued that 'Quietly, but with gathering force, the ground has shifted beneath our feet as we have moved into an *Information Age*, a global economy, a truly new world.' In total he used this term three times throughout the speech. William Jefferson Clinton, State of the Union Address, 1998, as found at, http://www.whitehouse.gov/WH/SOTU98/address.html.
2 Mark Poster, *The Mode of Information: Poststructuralism and Social Context* (The University of Chicago Press, 1990), 5–14.
3 James R. Beniger, *The Control Revolution: Technological and Economic Origins of the Information Society* (Harvard University Press, 1986), 21.
4 Steven Lubar, *InfoCulture: The Smithsonian Book of Information Age Inventions* (Houghton Mifflin Company, 1993), 4.
5 Daniel Bell, *The Coming of a Post-Industrial Society* (New York: Basic Books, 1976), ix-xxii.
6 This term is frequently used to describe our seemingly discontinuous jump into an information society. One example of its usage is the 'Center for Information-Revolution Analysis (CIRA)' at the RAND Graduate School in Santa Monica, California. CIRA's 'mission' is to 'understand the social, economic, political, and personal' results of the information revolution. As found at, http://www.rand.org/centers/cira.
7 Gordon Pask and Susan Curran argued that 'The rapid proliferation of computation, communication, and control devices is coming to form what we call the "information environment."' Pask and Curran, *Micro Man: Computers and the Evolution of Consciousness* (New York: Macmillan, 1982), 2–3.
8 Richard Saul Wurman, *Information Anxiety* (New York: Doubleday, 1989).
9 Psychologist, David Lewis, diagnosed a specific, clinical problem resulting from too much information and called it 'information fatigue syndrome.' 'New Independent Research Reveals Cost of the Information Revolution,' Reuters Press Release, 14 October 1996, http://www.fireworks.org/firefly/clients/reuters/press/141096.htm.
10 While the OED defines 'intellectual' as a 'person possessing or supposed to posses superior powers of intellect' I make no qualitative judgement in labeling a group or an individual an intellectual. Instead, I use this term for anyone who centers a significant portion of their work around the

production and consumption of written words, thus susceptible to the problems of too much information. This includes scientists, engineers, humanists, executives, lawyers, physicians, and governmental leaders.
11 Vannevar Bush, 'For Man to Know,' *The Atlantic Monthly* (August 1955): 32.
12 Lubar, *InfoCulture*, 2–4.
13 JoAnne Yates, *Control Through Communication: The Rise of System in American Management* (The Johns Hopkins University Press, 1989), xvi.
14 Theodore Rozak, *The Cult of Information: A Neo-Luddite Treatise on High Tech, Artificial Intelligence, and the True Art of Thinking* (University of California Press, 1994), xiii–xiv, 3.
15 Claude Shannon, *The Mathematical Theory of Communication* (The University of Illinois Press, 1949), 10.
16 Warren Weaver, 'Recent Contributions to the Mathematical Theory of Communication,' as found in, Shannon, *The Mathematical Theory of Communication*, 100.
17 Michael Buckland, 'Information as Thing,' *Journal of the American Society for Information Science* 42, 5 (1991): 351–360.
18 These meanings were sometimes distorted by the self-serving interests of computer systems developers as well as Cold War anxieties. A full treatment of the meanings and perceptions of information overload from World War II to the present can be found in the author's forthcoming dissertation from Case Western Reserve University.
19 Bush built his differential analyzer in 1933 as a machine to solve differential equations. For a cultural study examining the reception of the differential analyzers in Britain and America, see Mark D. Bowles, 'U.S. Technological Enthusiasm and British Technological Skepticism in the Age of the Analog Brain,' *Annals of the History of Computing* 18, 4 (October–December 1996): 5–15.
20 Vannevar Bush, 'As We May Think,' *Atlantic Monthly* 176 (July 1945): 101–108.
21 Bush, 'For Man to Know,' *The Atlantic Monthly* (August 1955): 32.
22 Bush, 'Science and Medicine,' *Medical Annals of the District of Columbia* 22, 1 (January 1953): 5.
23 Bush, 'As We May Think,' 101–108.
24 Bush, 'As We May Think: A Top U.S. Scientist Foresees a Possible Future World in Which Man-Made Machines Will Start to Think,' *Life* (10 September 1945): 112–124.
25 C. P. Bourne, 'The World's Technical Journal Literature: An Estimate of Volume, Origin, Language, Field, Indexing, and Abstracting' (Stanford Research Institute, 1961), Case Western Reserve University (CWRU) Archives, Box 45VC4. J. W. Senders, 'Information Storage Requirements for the Contents of the World's Libraries,' *Science* 141 (1963): 1067–1068.
26 J. C. R. Licklider, *Libraries of the Future* (The MIT Press, 1965), 14–15.

27 Joseph N. Bell, 'Crisis! How Can We Store Human Knowledge,' *Popular Mechanics* (November 1962): 104.
28 Norbert Wiener, *The Human Use of Human Beings: Cybernetics and Society* (Doubleday Anchor Books, 1954), 15.
29 J. Rose, *The Cybernetic Revolution* (Harper & Row Publishers, Inc., 1974), vii.
30 Norbert Wiener, *Cybernetics: or Control and Communication in the Animal and the Machine* (New York: The Technology Press, 1948), 184.
31 Wiener, *The Human Use of Human Beings*, 134.
32 Wiener, 'Science, Monkeys, and Mozart,' *The Saturday Review* 37 (20 November 1957): 15–16, 46–48.
33 Wiener, *Cybernetics*, 15.
34 Wiener, *The Human Use of Human Beings*, 121–122.
35 G. Pascal Zachary, *Endless Frontier: Vannevar Bush, Engineer of the American Century* (The Free Press, 1997), 1.
36 David F. Noble, *Forces of Production: A Social History of Industrial Automation* (Oxford University Press, 1984), 10.
37 D. B. Baker and M. Hoseh, 'Soviet Science Information Services,' *Chemical and Engineering News* 38, 2 (11 January 1960): 70. Francis Bello, 'How to Cope with Information,' *Fortune* (September 1960): 189.
38 Slava Gerovitch, 'Striving for "Optimal Control": Soviet Cybernetics as a "Science of Government,"' in this volume.
39 'Norbert Wiener Visits the Editorial Offices of Our Journal,' *Voprosy Filosofii* 9 (1960), as found in, 'Norbert Wiener's Discussion with Soviet Philosophers,' *Soviet Review* 1, 4 (November 1960): 3.
40 A. I. Berg, 'Cybernetics and Life,' *Ekonomicheskaia Gazeta* (12 June 1960), as found in, Berg, 'Cybernetics and Society,' *Soviet Review* 1, 1 (August 1960): 43–55.
41 Ibid., 51.
42 A. I. Mikhailov, 'Problems of Mechanization and Automation of Information Work,' *Revue Internationale de la Documentation* 29, 2 (1962): 49, 51.
43 R. A. Fairthorne to Advisory Board Meeting, 29 March 1957, CWRU Archives, Box 45DB.
44 Notes from the advisory board meeting, 14 April 1957, CWRU Archives, Box 45DB.
45 Allen Kent to Arnold E. Keller, editor of *Management and Business Automation,* 24 April 1959, CWRU Archives, Box 45ED.
46 As Kent recalled later, 'while this sounds crazy now, this was the sort of thing that one said in Washington when one sought grants to pay for the developments in this field.' Allen Kent, interview with author, 20 March 1998.
47 Alvin M. Weinberg, 'Science, Government, and Information,' presentation at the American Management Association, March 1963, CWRU Archives, Box 45VC4.

48 'US Organizations, Particularly in the Chemical Field, Use Advanced Information Retrieval on Expanding Scale,' *Chemical and Engineering News* (24 July 1961), 96, as found in, CWRU Archives, Box 45DC50. 'Western Reserve Up-Dates Information Retrieval Center,' *Business Automation* (July 1961): 40, as found in, CWRU Archives, Box 45DC50. Russell W. Kane, 'WRU Plans World Center Here for Document Study,' *The Plain Dealer* (17 January 1958): 1, as found in, CWRU Archives, Box 45DC50.
49 Robert E. Booth to Jesse Shera, 2 March 1956, CWRU Archives, Box 45VC3.
50 'Scientists Use New Computers to Index Technical Reports,' *Wall Street Journal* (7 April 1960), as found in, CWRU Archives, Box 45DC50. 'Technical Literature Research Speeded by Electronic Library,' *Wire Recorder* 5, 1 (March 1960), as found in, CWRU Archives, Box 45DC50.
51 '"Search Selector" Project Seen Leading to Central Data Setup,' *Electronic News* (29 August 1960): 25, as found at, CWRU Archives, Box 45DC50. Bud Weidenthal, 'Says Billions Wasted in Research Duplication,' *Kent Press* (9 March 1961): 16–C, as found in, CWRU Archives, Box 45DC50. Hubert Humphrey said, the CDCR should receive 'top priority in our industrial mobilization—and scientific program.' 'WRU to bid for $350,000 to back document center,' *Plain Dealer* (29 May 1959), as found in, CWRU Archives, Box 45DC50.
52 'Machines Take Toil Out of Searching,' *Business Week* (8 July 1961), as found in, CWRU Archives, Box 45DC50.
53 'Computers Trying to Catch Up, Digest Growing Store of World's Knowledge,' *Washington World* (2 October 1962).
54 Michael J. Reddy, 'The Conduit Metaphor: A Case of Frame Conflict in Our Language About Language,' *Metaphor and Thought* 2d ed. Andrew Ortony (Cambridge University Press, 1993), 164–201. George Lakoff, 'The Contemporary Theory of Metaphor,' *Metaphor and Thought*, 203–204.
55 These individuals were: Vannevar Bush, Allen Kent, Alvin M. Weinberg, anonymous White House press release, Kennedy Science Advisory Committee, Derek J. de Solla Price, John R. Brown, Fred J. Walker Jr., Charles P. Bourne, anonymous *Wall Street Journal* journalist, Louis Sawin, Chauncey D. Leake, Philip D. Jordan, Vernon D. Tate, IBM, Isaac Asimov, and Jesee Shera.
56 Max Black said that a 'metaphor-theme' is 'available for repeated use, adaptation, and modification by a variety of speakers or thinkers on any number of specific occasions.' Max Black, 'More About Metaphor,' *Metaphor and Thought*, 24.
57 Derek J. De Solla Price, *Little Science, Big Science* (Columbia University Press, 1963), 15.
58 Voltaire, *Age of Louis XIV*, trans. Martyn P. Pollack, (New York: Everyman's Library, 1961), 371.

59 Washington Irving, *The Sketch Book*, ed. George Philip Krapp (Chicago: Scott, Foresman and Company, 1906), 185.
60 Clark M. Clifford, 'Report to President Truman,' September 1946, as found in, *Major Problems in American History Since 1945: Documents and Essays*, ed. Robert Griffith (D. C. Heath and Company, 1992), 106.
61 Everett M. Dirksen (Republican of Illinois). 'Dirksen Backs Foreign aid to Stave Off Reds,' *New York Herald Tribune* (19 Nov 1947).
62 Terminiello v. Chicago, No. 272, Supreme Court of the United States, decided 16 May 1949, as found at, http://www.bc.edu/bc_org/avp/cas/comm/free_speech/terminiello.html.
63 'Harry S. Truman: Fighter in a Fighting Year,' *Time* 51, 1 (3 January 1949), 9.
64 John F. Kennedy speech as found in, Howard Zinn, *A People's History of the United States* (Harper & Row Publishers, 1980).
65 Miklós Horthy, *Memoirs*, as found at, http://www.msstate.edu/Archives/History/hungary/horthy/20.htm.
66 Anthony Eden, 'A Personal Statement by the Prime Minister,' 1955, 'United For Peace and Progress: The Conservative and Unionist Party's Policy,' as found at, //160.5.16.48/depts/po/table/time/con55.htm.
67 Sam Glucksberg and Boaz Keysar, 'How Metaphors Work,' *Metaphor and Thought*, 420.
68 Frederick Adams, 'Cybernetics,' *The Cambridge Dictionary of Philosophy*, ed. Robert Audi (Cambridge University Press, 1995), 173.
69 Katie Hafner, 'The Epic Saga of the Well,' *Wired* (May 1997): 99.
70 Adam Smith, 'EFF's Guide to the Internet, v. 3.0' 6 February 1995. The Electronic Frontier Foundation describes itself as 'a non-profit civil liberties organization working in the public interest to protect privacy, free expression, and access to public resources and information on-line as well as to promote responsibility in new media.' The Electronic Frontier Foundation web page, http://www.eff.org.
71 Edward Tenner, 'The Impending Information Implosion,' *Harvard Magazine* (November-December 1991): 33. David B. Whittle, *Cyberspace: The Human Dimension* (W. H. Freeman and Company, 1997), 387.
72 Clifford Stoll, *Silicon Snake Oil: Second Thoughts on the Information Highway* (Doubleday, 1995), 35, 50, 56.
73 Chuck Melvin and Zach Schiller, 'The Internet's Pulse Quickens,' *The Plain Dealer* (15 February 1998).
74 Richard J. Smith and Mark Gibbs, *Navigating the Internet* (Sams Publishing, 1994), xxv.
75 Tracy Kidder, *The Soul of a New Machine* (Avon Books, 1981).
76 L. B. Heilprin, 'On the Information Problem Ahead,' *American Documentation* (1961): 6, 14.
77 Allen Kent, 'Soviet Documentation—A Trip Report,' Confidential Seventh Special Report of the CDCR, 31 December 1958, CWRU Archives, Box 45A2.

78 Dennis M. Wint, 'Dealing With Information Overload,' *St-Louis Post-Dispatch* Section C, Page 2 (17 December 1992).
79 Keith Newman, 'Information Overload,' http://www.wordworx.co.nz/overload/htm, 1996.
80 Paul Gilster, *Digital Literacy* (John Wiley & Sons, Inc., 1997).
81 William Gibson, *Neuromancer* (Ace Books, 1984), 43, 239.
82 Kevin Kelly, *Out of Control: The New Biology of Machines, Social Systems, and the Economic World* (Addison-Wesley Publishing Company, 1994), 125–128.
83 Bridget Murray, 'Data Smog: Newest Culprit in Brain Drain,' *American Psychological Association Monitor* (March 1998): 42.

BIBLIOGRAPHY

Anonymous. '"Search Selector" Project Seen Leading to Central Data Setup,' *Electronic News* (29 August 1960).
Anonymous. 'Computers Trying to Catch Up, Digest Growing Store of World's Knowledge.' *Washington World* (2 October 1962).
Anonymous. 'Dirksen Backs Foreign Aid to Stave Off Reds.' *New York Herald Tribune* (19 Nov 1947).
Anonymous. 'Harry S. Truman: Fighter in a Fighting Year.' *Time* 51, 1 (3 January 1949).
Anonymous. 'Machines Take Toil Out of Searching,' *Business Week* (8 July 1961).
Anonymous. 'Norbert Wiener's Discussion with Soviet Philosophers.' *Soviet Review* 1, 4 (November 1960): 3–12.
Anonymous. 'Scientists Use New Computers to Index Technical Reports,' *Wall Street Journal* (7 April 1960).
Anonymous. 'Technical Literature Research Speeded by Electronic Library,' *Wire Recorder* 5, 1 (March 1960).
Anonymous. 'US Organizations, Particularly in the Chemical Field, Use Advanced Information Retrieval on Expanding Scale,' *Chemical and Engineering News* (24 July 1961), 96.
Anonymous. 'Western Reserve Up-dates Information Retrieval Center,' *Business Automation* (July 1961).
Anonymous. 'WRU to bid for $350,000 to back document center,' *Plain Dealer* (29 May 1959).
Audi, Robert ed. *The Cambridge Dictionary of Philosophy*. Cambridge University Press, 1995.
Baker, D. B. and M. Hoseh. 'Soviet Science Information Services.' *Chemical and Engineering News* 38, 2 (11 January 1960): 70–75.
Bell, Daniel. *The Coming of a Post-Industrial Society*. Basic Books, 1976.
Bell, Joseph N. 'Crisis! How Can We Store Human Knowledge.' *Popular Mechanics* (November 1962).
Bello, Francis. 'How to Cope with Information.' *Fortune* (September 1960).
Beniger, James R. *The Control Revolution: Technological and Economic Origins of the Information Society*. Harvard University Press, 1986.

Berg, A. I. 'Cybernetics and Society,' *Soviet Review* 1, 1 (August 1960).
Bowles, Mark D. 'U.S. Technological Enthusiasm and British Technological Skepticism in the Age of the Analog Brain,' *Annals of the History of Computing* 18, 4 (October-December 1996): 5–15.
Boyle, R. R. 'The Nature of Metaphor,' *Modern Schoolman* 31 (1954).
Buckland, Michael K. 'Information as Thing.' *Journal of the American Society for Information Science* 42, 5 (1991).
Bush, Vannevar. 'As We May Think.' *Atlantic Monthly* 176 (July 1945).
Bush, Vannevar. 'As We May Think: A Top U.S. Scientist Foresees a Possible Future World in Which Man-Made Machines Will Start to Think.' *Life* (10 September 1945).
Bush, Vannevar. 'For Man to Know.' *The Atlantic Monthly* (August 1955).
Bush, Vannevar. 'Science and Medicine,' *Medical Annals of the District of Columbia* 22, 1 (January 1953).
Gibson, William. *Neuromancer.* Ace Books, 1984.
Gilster, Paul. *Digital Literacy.* John Wiley & Sons, Inc., 1997.
Griffith, Robert ed. *Major Problems in American History Since 1945: Documents and Essays.* D. C. Heath and Company, 1992.
Hafner, Katie. 'The Epic Saga of the Well.' *Wired* (May 1997).
Heilprin, L. B. 'On the Information Problem Ahead.' *American Documentation* (1961).
Kane, Russell W. 'WRU Plans World Center Here for Document Study.' *The Plain Dealer* (17 January 1958).
Kelly, Kevin. *Out of Control: The New Biology of Machines, Social Systems, and the Economic World.* Addison-Wesley Publishing Company, 1994.
Kidder, Tracy. *The Soul of a New Machine.* Avon Books, 1981.
Licklider, J. C. R. *Libraries of the Future.* The MIT Press, 1965.
Lubar, Steven. *InfoCulture: The Smithsonian Book of Information Age Inventions.* Houghton Mifflin Company, 1993.
Mikhailov, A. I. 'Problems of Mechanization and Automation of Information Work.' *Revue Internationale de la Documentation* 29, 2 (1962).
Murray, Bridget. 'Data Smog: Newest Culprit in Brain Drain.' *American Psychological Association Monitor* (March 1998).
Noble, David F. *Forces of Production: A Social History of Industrial Automation.* Oxford University Press, 1984.
Ortony, Andrew, ed. *Metaphor and Thought,* 2nd ed. Cambridge University Press, 1993.
Pask, Gordon and Susan Curran. *Micro Man: Computers and the Evolution of Consciousness.* Macmillan, 1982.
Poster, Mark. *The Mode of Information: Poststructuralism and Social Context.* The University of Chicago Press, 1990.
Price, Derek J. De Solla. *Little Science, Big Science.* Columbia University Press, 1963.
Rezun, Miron. *Science, Technology and Ecopolitics in the USSR.* Praeger, 1996.
Rose, J. *The Cybernetic Revolution.* Harper & Row Publishers, Inc., 1974.
Rozak, Theodore. *The Cult of Information: A Neo-Luddite Treatise on High Tech, Artificial Intelligence, and the True Art of Thinking.* University of California Press, 1994.

Senders, J. W. 'Information Storage Requirements for the Contents of the World's Libraries.' *Science* 141 (1963): 1067–1068.

Shannon, Claude. *The Mathematical Theory of Communication.* The University of Illinois Press, 1949.

Smith, Dinitia. 'Librarians' Challenge: Offering the Internet to All.' *New York Times* (6 July 1996).

Smith, Richard J. and Mark Gibbs. *Navigating the Internet.* Sams Publishing, 1994.

Stoll, Clifford. *Silicon Snake Oil: Second Thoughts on the Information Highway.* Doubleday, 1995.

Tenner, Edward. 'The Impending Information Implosion.' *Harvard Magazine* (November-December 1991).

Weidenthal, Bud. 'Says Billons Wasted in Research Duplication,' *Kent Press* (9 March 1961).

Whittle, David B. *Cyberspace: The Human Dimension.* W. H. Freeman and Company, 1997.

Wiener, Norbert. *Cybernetics: or Control and Communication in the Animal and the Machine.* New York: The Technology Press, 1948.

Wiener, Norbert. 'Science, Monkeys, and Mozart.' *The Saturday Review* 37 (20 November 1957).

Wiener, Norbert. *The Human Use of Human Beings: Cybernetics and Society.* Doubleday Anchor Books, 1954.

Wint, Dennis M. 'Dealing With Information Overload,' *St-Louis Post-Dispatch* (17 December 1992).

Yates, JoAnne. *Control Through Communication: The Rise of System in American Management.* The Johns Hopkins University Press, 1989.

Zachary, G. Pascal. *Endless Frontier: Vannevar Bush, Engineer of the American Century.* The Free Press, 1997.

Zinn, Howard. *A People's History of the United States.* Harper & Row Publishers, 1980.

CHAPTER 11
Striving for 'Optimal Control': Soviet Cybernetics as a 'Science of Government'

Slava Gerovitch

Historian Paul Josephson has recently written: 'It is unimaginable that a culture so fascinated with the potential of science to build communism, a culture whose achievements in the 1950s included the hydrogen bomb, nuclear power, tokamaks, and Sputnik, could dismiss the promise of cybernetics.'[1] As with many other unimaginable things, this was precisely what happened in the Soviet Union. In the early 1950s, cybernetics was labeled a 'reactionary imperialist utopia' and a 'pseudo-science.' A few years later, however, this attitude was completely reversed and cybernetics was hailed as a 'progressive' science 'in the service of communism.'[2] This puzzled historians even more: David Holloway has found it 'paradoxical' that it is not in the West, but in the Soviet Union, where cybernetics was originally condemned as a "pseudo-science," that this "general ideology" has been most widely adopted.'[3]

The early rejection and later embrace of cybernetics in the Soviet Union would look less enigmatic if one realized that cybernetics was not an immutable entity; it was reshaped and reinterpreted in different social and political contexts. The diversity of Soviet culture, the convolutions of Soviet history, and the cultural metamorphoses of cybernetic concepts together created the phenomenon of Soviet cybernetics.

This article examines the evolution of cybernetics in the Soviet Union from the object of unbridled criticism during the anti-cybernetics campaign in the early 1950s, to a vehicle of reform in the post-Stalinist system of science in the late 1950s and early 1960s, to a convenient tool of bureaucracy in the late 1960s. I focus on Soviet cyberneticians' attempts to extend the cybernetic concept of control from automated military command to the planning and management of the national economy. I also discuss the diverse roots of the opposition to cyberneticians' radical projects. This paper questions the common assumption that Soviet cybernetic research was directly commissioned by and enjoyed a uniform support of the political authorities, who wished to achieve a greater control over Soviet society. It emphasizes the active role Soviet scientists and engineers played in shaping the agenda of Soviet cybernetic research.

'Optimal planning and control' was a motto of Soviet cybernetics. Cyberneticians saw the main problem of Soviet economy in the inefficiency of mechanisms of data-collection, information-processing, and control, and offered a solution based on mathematical modeling and computer-aided decision-making. They hoped that cybernetics would provide an opportunity to reform Soviet planning and management practices without cardinal political change; they believed that computers could produce a politically neutral, 'optimal' solution. One memoirist aptly named those cyberneticians 'considents,' or half-consenting dissidents.[4] Soviet cyberneticians were looking for a technological solution to an inherently political problem; by its own nature, however, their project was doomed to play a political role.

NORBERT WIENER: A CYBERNETIC CRITIQUE OF CAPITALISM AND COMMUNISM

In his *Cybernetics, or Control and Communication in the Animal and the Machine* (1948), the American mathematician Norbert Wiener put together concepts from physiology (homeostasis), psychology (behavior, goal, learning), control engineering (control, feedback), and communication engineering (information, signal, noise, channel) and created a new language to be equally applicable to living organisms, inanimate machines, and human society.[5] For Wiener, humans and machines were two kinds of *communicative organisms*, or *control systems*, which, *operating* in a certain *environment*, pursued their *goals* to *increase order* and achieve *better organization* by exchanging *information* with other systems via *feedback*.

The origins of this 'cyberspeak' can be traced to Wiener's wartime work on antiaircraft gun control. Wiener participated in efforts to design an antiaircraft predictor, which included a human operator as part of the machine.[6] He formulated a major problem of this project as follows: 'The speed of the airplane has made it necessary to compute the elements of the trajectory of the antiaircraft missile by machine, and to give the predicting machine itself communication functions which had previously been assigned to human beings.'[7] Trying to 'incorporate mathematically' fighters into their weapons, Wiener and his collaborators first suggested that the enemy pilot behaved like a servomechanism, then stated the same about the Allied antiaircraft gunner, and finally extended this idea to the whole of human nature.

Two years after *Cybernetics*, Wiener published *The Human Use of Human Beings: Cybernetics and Society*, in which he developed a cybernetic critique of pervasive controls over social communication under McCarthyism in America and under Stalinism in Russia. Wiener hoped that the cybernetic vision of society as a self-regulating device would make people realize that 'the control of the means of communication was the most effective anti-homeostatic factor driving society out of equilibrium. He condemned 'a society like ours, avowedly based on buying and selling,'[8] contended that 'information and entropy are not conserved, and are equally unsuited to being commodities,'[9] and argued against military secrecy and an unfair patent law.

Wiener did not believe in the ability of the 'invisible hand' of free market to establish an economic and social equilibrium, or homeostasis in cybernetic terms. 'There is no homeostasis whatever,' he wrote. 'We are involved in the business cycles of boom and failure, in the successions of dictatorship and revolution, in the wars which everyone loses.'[10]

Wiener proclaimed the advent of the 'second industrial revolution,' which would bring about 'the automatic factory and the assembly line without human agents.'[11] In his view, this carried 'great possibilities for good and for evil.'[12] On the evil side, it was 'bound to devalue the human brain.'[13] '[T]he skilled scientist and the skilled administrator may survive' this revolution, wrote Wiener, but 'the average human being of mediocre attainments or less has nothing to sell that it is worth anyone's money to buy.'[14]

In December 1948, *Le Monde* published a review of *Cybernetics* by the Dominican friar Père Dubarle, who contemplated the possibility of the creation of an all-powerful *'machine à gouverner,'* which would make the State 'the only supreme co-ordinator of all partial decisions.' 'In comparison with this,' wrote Dubarle, 'Hobbes' *Leviathan* was nothing but a pleasant joke. We are running the risk nowadays of a great World State, where deliberate and conscious primitive injustice may be the only possible condition for the statistical happiness of the masses: a world worse than hell for every clear mind.'[15] Wiener responded by saying that the real danger of *machines à gouverner* was 'that such machines, though helpless by themselves, may be used by a human being or a block of human beings to increase their control over the rest of the human race.'[16]

Wiener employed his cybernetic language to criticize both capitalism and communism: 'A sort of *machine à gouverner* is thus now essentially in operation on both sides of the world conflict, although it does not consist in either case of a single machine which makes policy, but rather of a mechanistic technique which is adapted to the exigencies of a machine-like group of men devoted to the formulation of policy. Père Dubarle has called the attention of the scientist to the growing military and political mechanization of the world as a great superhuman apparatus working on cybernetic principles.'[17] Wiener did not make clear, however, the difference between those cybernetic principles that could provide for social equilibrium, and those that led to the 'mechanization of the world.'

SOVIET COLD WAR-ERA PROPAGANDA AND A CRITIQUE OF CYBERNETICS

When Wiener's *Cybernetics* came out in 1948, it caused stormy debates on both sides of the Atlantic. Responses ranged from the praise to 'one of the most influential books of the twentieth century' (*The New York Times*) to the warning against 'la seduction d'idées confuses qu'on ... propose sous couleur de généralités scientifiques' (*La Pensée*). The Soviet press, however, met cybernetics with uniform hostility.

The anti-cybernetics campaign in the Soviet press was but one in a series of politically charged debates about relativity theory and quantum mechanics, the history of Western philosophy, the theory of resonance in structural chemistry, genetics, linguistics, political economy, physiology, and mathematical logic, which were shaking Soviet science throughout the late 1940s and early 1950s.[18] The major events of the anti-cybernetics campaign spanned over the period from 1950 to 1955, when a number of sharply critical articles appeared in Soviet scholarly journals and popular periodicals, such as *The Literary Gazette*, *Nature*, *Science and Life*, *Problems of Philosophy*, and *Moscow University Herald*. Main anti-cybernetics arguments went as follows:

(1) *An anti-mechanist argument: humans are not machines.* Soviet critics believed that cybernetics extrapolated technical models of self-regulation beyond their area of legitimate application. They charged that cyberneticians committed a serious philosophical mistake by reducing the laws governing biological, psychological, and social processes to physical and mechanical laws. Such reductionism constituted a mechanist deviation from dialectical materialism, the official Soviet philosophy of science.

(2) *An anti-idealist argument: the critique of the concept of information as a non material entity.* Paradoxically, cybernetics proved vulnerable to accusations in both mechanism (the reductionist representation of humans as machines) and idealism (the construction of reality based on pure mathematical formulas). The latter, like the former, was considered a philosophical deviation from dialectical materialism, only in the opposite direction.

(3) *An anti-technocratic argument: society is not a machine.* Soviet critics condemned the extrapolation of cybernetic reasoning into social science and labeled cybernetics a 'technocratic theory.'[19] They charged that cyberneticians 'make a fetish of technology',[20] 'claim that society needs them to step in, "scientifically" explain, and "fix" the "malfunctioning" of society,'[21] and pretend that acute social problems 'can be solved with exact mathematical formulas.'[22]

(4) *The critique of the concept of the 'second industrial revolution.'* Soviet critics wrote that 'contemporary cyberneticians go out of their way to lower man, to show that man can be completely – and should be – replaced by machine.'[23] In the critics' view, cybernetics was directed against the Marxist theory of class struggle and served the 'purpose of convincing proletarians that any struggle for altering the social order is meaningless and useless. Every human part [of the social mechanism] must fulfill a certain function and not even dream of changing its status; otherwise, it will be thrown away and replaced with a spare part.'[24]

(5) *An anti-militarist argument: cybernetics in the service of imperialism.* Under the cover of cybernetics, the critics argued, Western military establishment enrolled scientists and engineers to build electronic, remotely controlled, automated weapons. 'Whom Does Cybernetics Serve?', inquired the title of one

the most well-known anti-cybernetics publications. The answer was clear: '[Cybernetics] brings all its admitted, practical achievements in the construction of computers, together with profoundly reactionary theories, to the altar of war.'[25]

My findings in the recently made available Party archives indicate that the anti-cybernetics campaign was a by-product of a large-scale propaganda campaign aimed at 'criticizing and destroying' contemporary 'bourgeois' philosophy and sociology. In the context of the Cold War, cybernetics was attacked not as a scientific theory or a trend in engineering, but as a philosphical doctrine. Party authorities issued no order specifically against cybernetics, but the campaign acquired an impetus from the officially endorsed anti-American crusade. In June 1949, the Secretariat of the Party Central Committee adopted a 'Plan for the Intensification of Anti-American Propaganda,' which prescribed several topics for an anti-American campaign, including the topic, 'Science in the Service of American Monopolies.'[26] Cybernetics came in handy—with its American origins, philosophical claims, and military connections. *The Literary Gazette* soon published the first direct attack on cybernetics, then other newspapers and magazines joined in; they too had to fulfill their quota for articles critical of American military ideology. The more anti-cybernetics articles were published, the stronger grew the public impression that this campaign reflected an officially sanctioned negative attitude toward cybernetics, and more and more critics were happy to join the chorus. The 'snowball effect' rather than a thorough plan seems better to describe the mechanism of this campaign, which may be true for some of the other ideological campaigns in Soviet science and technology.

A CRITIQUE OF STALINISM AND THE LEGITIMIZATION OF CYBERNETICS

After Stalin's death in March 1953 and the advent of political 'thaw,' Soviet scientists began to speak openly against ideological controls in science, pointing to the cases of relativity theory, genetics, and cybernetics as examples of unacceptable interference by 'philosophers' (meaning Party ideologues) in the affairs of the scientific community. The 'rehabilitation' of cybernetics became a matter of reaffirming the intellectual autonomy of the scientific community.

In August 1955, three authors—Sergei Sobolev, Aleksei Liapunov, and Anatolii Kitov—published an article in the leading philosophical journal *Problems of Philosophy*, in which, for the first time in open Soviet press, they took a favorable view of cybernetics and effectively legitimized this field. All three played a prominent role in Soviet military computing. Sobolev supervised mathematical research for the Soviet nuclear weapons program; Liapunov was a pioneer of Soviet computer programming and the author of some of the first programs with military applications; and Kitov was an organizer of the first computing centers under the Ministry of Defense. The three authors presented cybernetics as a unity of the theory of information, the theory of computers, and the theory of automated control systems. They argued that the 'introduction of electronic calculating machines into control systems makes it possible to implement so-called optimal control' and pointed to the 'great economic and military significance' of cybernetics.[27]

The change of attitude toward cybernetics in the Soviet Union was marked by the shift from the discussion of particular control mechanisms in organisms and machines to the elaboration of a general theory of control. As an indicator of this transition, the translation of Wiener's key term, 'control,' underwent a notable change during the 1950s. When referring to Wiener's *Cybernetics, or Control and Communication in the Animal and the Machine*, Soviet critics in the early 1950s translated the word 'control' in the title literally as *kontrol'*.[28] In contrast, Soviet partisans of cybernetics in the 1958 Russian translation of Wiener's book chose a different Russian word—*upravlenie*—to render the term 'control.'[29] Both *kontrol'* and *upravlenie* in Russian mean 'control,' but nuances differ. The word *kontrol'* has such meanings as 'checking' and 'examining,' while *upravlenie* refers to 'management,' 'administration,' 'direction,' and 'government.'[30] Russian speak of *kontrol'* when the controller is bound by a preset goal and only monitors the controlled process but does not direct it (e.g., quality control). *Upravlenie* is a much more ambitious undertaking; it applies to administrative decision-making and often involves setting new policies (e.g., managing a factory or governing a state).

Mathematician Gelii Povarov—the editor of the Russian translation of *Cybernetics*—took special care to trace the genealogy of cybernetics to the French physicist André-Marie Ampère, who in 1834 introduced the word *cybernétique* to denote the science of government in his classification of sciences.[31] Ampère listed *cybernétique* among the political sciences, and some of the Soviet proponents of cybernetics took it on as a very political science indeed.

When Engineer Vice Admiral Academician Aksel Berg was reporting on the prospects of cybernetic research before the Presidium of the Soviet Academy of Sciences on April 10, 1959, he referred to Ampère in the very first sentence. Berg stated the main task of cybernetics to be the elaboration of methods of controlling production in the entire national economy for optimal performance. He mentioned Wiener only at the end of his talk, stressing that the latter 'reused' the term 'cybernetics.'[32] The Presidium decided to establish the Council on Cybernetics with Berg as chairman and Aleksei Liapunov as his deputy.

Liapunov—arguably the most prominent Soviet advocate of cybernetics, widely reputed to be the 'father of Soviet cybernetics'—involved Berg in the cybernetic movement soon after Berg's retirement (because of ill health) from the post of deputy Minister of Defense in charge of radioelectronics in 1957. To establish the Council on Cybernetics was Liapunov's idea, and he asked Berg to chair it, hoping to capitalize on Berg's influence to the advantage of the new field.

Berg proved a very skillful spokesman for cybernetics. He hired several specialists whose task was to reconcile cybernetics with dialectical materialism. This was duly accomplished. Philosophers loyal to cybernetics managed to include the concept of information in the canonical list of categories of dialectical materialism, asserted that information was 'inextricably connected with matter,' and spoke of the 'dialectical rotation of information and noise.'[33] Under the Council's aegis, Berg organized several sections—mathematical, economic, biological, physiological, linguistic, philosophical, psychological, etc.—each engaged in propagating cybernetics in the corresponding field.

Berg argued that only cybernetics could analyze control processes in each of these fields and provide methodological guidelines for optimal control. As one member of his Council put it, 'cybernetics is not merely a science of control, but of *optimal* control, or optimization of control.'[34] Berg stressed that cybernetic ideas had great political significance: 'However unusual this may seem to certain conservatives, who do not desire to comprehend elementary truths, we will build communism on the basis of most broad use of electronic machines, he argued.' These machines, aptly called "cybernetic machines," ... will provide the solution of the problem of *permanent optimal planning and control*.'[35]

Berg mobilized his Cybernetics Council in publishing a volume appropriately entitled, *Cybernetics Must Serve Communism*, in time for the opening of the Twenty-Second Congress of the Communist Party in 1961. His efforts paid off: a new Party program adopted at this Congress mentioned cybernetics among the sciences called to play a crucial role in the creation of the material and technical basis of communism. While the draft program stated mildly that it was 'necessary to organize the wide use of cybernetics,' the final version adopted at the Congress asserted positively that cybernetics 'will be widely applied' in production, research, planning, and management.[36] Popular press began to call computers 'machines of communism.'[37]. Cyberneticians chose cyberspeak as their language of negotiation with the authorities, hoping to limit the discussion to technical terms; instead, cyberspeak itself became politicized.

MILITARY CYBERNETICS: AN ENGINEERING APPROACH TO COMMAND

In July 1953, at the height of the anti-cybernetics campaign, director of the newly established Institute of Scientific Information Dmitrii Panov submitted to the Party Central Committee a secret report entitled, 'On small-size electronic computing devices and their application for control purposes.' He wrote: 'From materials published in American journals, it is clear that the USA is conducting extensive work on designing various electronic control devices.'[38] Panov went on to cite examples of control devices used in American aircraft and anti-aircraft gunnery and flight control; he pointed out that the greater efficiency of the F-86 aircraft over the Soviet MIG-15 demonstrated during the Korean war might be due to the F-86's on-board automated control system. 'Besides gun control and flight control,' continued Panov, 'electronic computing devices are being applied more widely for automated control of complex industrial installations (for example, in chemical industry).'[39]

In the fall of 1953, soon after Berg's appointment as deputy Minister of Defense, he asked then his subordinate Anatolii Kitov to prepare a report on computers and cybernetics.[40] Soon Berg put Kitov in charge of organizing the first military computing centers. As soon as the Soviet Union started serial production of digital computers, they were put into operation in several military installations. Computer Center No. 1 of the Ministry of Defense was created on August 3, 1954; shortly thereafter the Navy Computer Center and the Air Force Computer Center were established. Computer Center No. 1 provided computing support for the nuclear weapons, ballistic missile, and space programs. Closely following Western developments in cybernetics and computing, Soviet military specialists soon realized that 'universal' digital computers and the algorithms of 'optimal' control could be used to control not only automated weapons, but also military units, and the Center began to elaborate the principles of automated troop control.[41]

Soviet proponents of automation of military decision-making viewed military units as cybernetic systems, and termed their approach 'military cybernetics.' Historian David Holloway has noted that 'inasmuch as [the cybernetic approach] legitimates the transfer of concepts and techniques from engineering to troop control, military cybernetics may be seen as the engineer's approach to command.'[42] Using techniques of operations research, statistical analysis, and control theory, Soviet specialists tried to formalize military problems, define the criteria of effectiveness of control, and develop methods of 'optimizing' military decisions. 'Optimization' in this context, however, as Holloway remarked, meant little more than 'improvement.'[43] One Soviet specialist in 'military cybernetics' later described the excitement of those days as follows: 'Under the slogan, "Automata can do everything!", attempts were made to exclude a human being from troop control systems entirely. The idea that automata could indeed replace humans in troop control was supported by the fact that by that time control of certain types of weapons was successfully automated... All practical attempts to fully eliminate humans from troop control systems, however, resulted in failure.'[44]

In 1961, Computer Center No. 1 was renamed the Central Scientific Research Institute No. 27 and devoted its activity to more modest tasks of introducing means of automation into separate military activities, instead of trying to build a wholly automated chain of command. Besides technical difficulties, the automation of military control faced a political problem linked to the controversy over the nature of military command. Soviet technical officers—proponents of the cybernetic approach—insisted on quantitative methods of decision-making, which traditional commanders opposed. Specialists in military tactics feared that 'military cyberneticians' might take their place and fiercely resisted cybernetic innovations. 'A machine can neither replace the commander, nor, more importantly, eliminate military tactics,' argued one critic. 'Outside cybernetics' field of vision, there appear such cardinal questions about controlling military forces as the relationship between objective and subjective factors, the distribution of functions among control organs, the role of ideological and psychological factors, etc.'[45]

In the early 1970s, however, the Soviet military leadership concluded that the USSR was lagging behind the US in the area of troop control. The Institute No. 27 was subordinated directly to the General Staff and charged with the task of elaborating an 'ideology of complex automation of armed forces control' and the creation of an automated control system for the entire Armed Forces of the USSR.[46]

The idea of trusting control to 'control machines' found particularly strong support in the Soviet space program. Echoing the slogan, 'Control everything controllable and make controllable everything else,' which Soviet authors picked up from the German control engineer Hermann Schmidt in early 1941,[47] the Soviet cosmonaut Konstantin Feoktistov espoused the opinion that 'every operation that can be automated on board a spaceship should be automated.'[48] The Soviet belief in the power of automation to overcome human errors exceeded even that of Americans. Former longtime executive director of the American Institute of Aeronautics and Astronautics James Harford has argued that 'the Russians preferred to depend on automated systems, with the cosmonaut in a passive role. The Americans, however, gave the astronaut a more controlling role, allowing him to override automated systems.'[49]

Cybernetics not only provided technical means for military control; most importantly, it provided the ideology of controllability of complex tasks. 'There are no unknowable phenomena, only unknown ones,' argued Berg. Likewise, there are no uncontrollable processes, only those in which the complexity of the task is not yet matched by the methods and means for its solution. Cybernetics broadens the range of controllable processes; this is its essence and its major merit.'[50]

FROM CONTROL OF WEAPONRY TO CONTROL OF THE ECONOMY

The initiative in applying computers for economic information processing and management came from Soviet computer engineers and the military; the economic establishment was rather skeptical and less than supportive. As a result, developments in the field of automated economic management were heavily influenced by the Soviet conception of automated military command. The same people often worked in both fields and applied the same concept—the concept of cybernetic control—to both management and command. In late 1959, Kitov began working on a report later submitted to the Central Committee, in which he put forward an ambitious proposal to create a nationwide unified automated control system for both the military and the civilian sectors of the economy on the basis of the network of computing centers developed under the Ministry of Defense. Kitov argued that 'the national economy as a whole may be regarded as a complex cybernetic system, which incorporates an enormous number of various interconnected control loops with various levels of subordination.'[52] He believed that the application of cybernetic methods, computer modeling in particular, would place economics on a solid scientific foundation: 'Computer modeling makes it possible to forecast economic processes and to conduct mathematical experiments in economics. Thereby economics turns into an exact experimental science.'[53]

In 1960, Berg, Kitov, and Liapunov published an article in the leading Party journal *Communist*, in which they argued that 'in contrast to capitalist countries, where different companies create separate automated control systems for themselves, [in the USSR,] in the conditions of socialism, it is quite possible to organize a unified complex automated system to control the national economy. It is obvious that the effect from such automation would be much greater than from the automation of control at individual enterprises.'[54] They outlined the contours of the proposed 'Unified State Territorial Network of Information Computer Centers' for automatic statistics collection, planning, distribution of resources, banking, and transportation control.

Liapunov and Kitov envisioned that a unified automated management system for the national economy would make it possible to 'fully implement the main economic advantages of communism: the centralized control and the planned economy. This would secure full harmony and close correspondence between the political and economic structures of the communist state and the technical means for controlling the national economy.'[55]

Not only individuals, but also whole institutions made a transition from working on the automation of control in the military sector to solving problems of civilian economy. In the late 1950s, the Institute of Electronic Control Machines designed a specialized computer M-4 for automatic radar stations. Later the Institute was transferred from the Academy of Sciences to the State Economic Council under the State Planning Committee and shifted the focus of its research to economic problems. In the early 1960s, the Institute built another specialized computer–M-5–for economic applications and elaborated a proposal for a far-reaching price reform based on computer calculations of 'optimal' prices.[56] The experience acquired at solving military problems, however, was not lost. One of the Institute's economists V. D. Belkin, for example, compared controlling the national economy to launching a missile. Before the advent of computers, he reasoned, calculating a missile trajectory took too much time, which precluded the construction of guided missiles. Now computers could calculate missiles' trajectories faster than the missiles moved, and this made it possible to guide missiles in real time. Similarly, Belkin argued, the introduction of computers in economics and the ability to calculate and correct economic plans in a timely fashion would offer new opportunities,[57] envisioning a kind of 'guidance system' for the national economy.

OPTIMAL PLANNING: A VEHICLE OF ECONOMIC REFORM OR AN OBSTACLE TO IT?

Disturbed by the economic difficulties the Soviet Union was experiencing in the early 1960s, Soviet leadership allowed an open discussion of the possible directions of economic reform. Liberal economists suggested to introduce profit as a criterion for evaluating economic performance, decentralize economic management, and stimulate market mechanisms in the Soviet economy. Cyberneticians, on the other hand, saw the solution in automation, optimal planning, and centralization of information-processing.[58] Both sides appealed to Western experience for supporting evidence. In July 1962, the Academy of Sciences dispatched a group of experts to visit the USA and England. Upon their return, the delegates reported that cybernetics, or 'science of control,' was already widely employed in the West for optimal planning and industrial management. They concluded: 'The delegation believes that now is the time to utilize in a serious way the achievements of "science of control" in the planned economy of our country in order to put solid scientific principles in the foundation of economic management and production control with the help of operations research methods.'[59]

At the November 1962 Central Committee Plenum, Nikita Khrushchev acknowledged that profit could play a significant role in the evaluation of the work of an individual enterprise; at the same time he called on his Party comrades to borrow widely Western 'rational' managerial techniques; in a planned economy, he argued, these techniques would be even easier to implement than under capitalism.[60]

During the same month of November 1962, then deputy chairman of the USSR Council of Ministers Aleksei Kosygin met with director of the Kiev Institute of Cybernetics mathematician Viktor Glushkov to discuss the prospects for creating a statewide automated economic management system. The latter suggested to build a unified state network of computer centers to automate economic management on a national scale. Soon the Council of Ministers established a commission with Glushkov as chairman to prepare a detailed proposal.[61]

Work in this field greatly intensified after the 1963 decree of the Party Central Committee and the Council of Ministers, 'On Improving the Supervision of Work on the Introduction of Computer Technology and Automated Management Systems into the National Economy.' Each of the major participating agencies created a special institution to work on problems of automated management: the Central Economic Mathematical Institute of the Academy of Sciences; the Main Computer Center of the State Planning Committee; and the Scientific Research Institute for Design of Computer Centers and Economic Information Systems of the Central Statistical Administration. To coordinate all work in this area, the Council of Ministers State Committee for the Coordination of Scientific Research set up the Chief Administration for the Introduction of Computer Technology into the National Economy.

The Central Economic Mathematical Institute was established with the active support of Berg's Council on Cybernetics and became the hotbed of 'economic cybernetics.' In 1964, the Institute's Nikolai Fedorenko together with Glushkov published a joint proposal for a 'unified system of optimal planning and management' on the basis of a three-tier unified state network of computer centers, which would have linked tens of thousands of nodal computing centers collecting 'primary information,' 30–50 basic computing centers in major cities, and the head center controlling the entire network and serving the government. The authors spoke of the potential 'optimal decision-making on the national scale' through 'processing the entire body of primary economic information as a whole.'[62]

The idea of optimal planning met stern opposition from traditional economists. Specialists from the State Planning Committee tried to preserve their right to manipulate economic plans; optimal planning would have transferred this authority from them to mathematical economists. The opponents of economic optimization argued that it was difficult to define the concept of economic 'optimum' and to settle on a single criterion for optimization; they proposed to start with local planning, and only later move up to a nationwide system, while cyberneticians argued that local optimization without reforming national economic policy was impossible; the critics also contended that economic processes were far too complex for linear models and questioned the utility of linear programming for optimal planning.[63]

In the eyes of Party and government bureaucrats, the aspiration of Soviet cybernetics to become a 'science of government' presented a thinly veiled political challenge to the existing administrative structure and to their authority within this structure. A mathematical analysis of information flows threatened to expose serious flaws within the system of Soviet industrial management and could become a powerful tool in the hands of scientists and engineers who challenged the authority of professional managers. Liapunov, for example, argued that cybernetic analysis often showed that 'certain data required for expedient control are, in fact, neglected, while much information, which is collected with great effort, has no function to fulfill, for this information does not effect decision-making in industrial management.... As a result, many agencies and information channels duplicate one another and make no real impact on production. Detailed mathematical modeling of production control would help to expose those links in control system that are not needed and would possibly help find a more rational system of control in general.'[64] This project could hardly gain support from the managers who feared to become superfluous control links.

Reformer economists, on the other hand, looked at Glushkov's proposal as too centralized and suppressing local initiative. Glushkov admitted that his project for a nationwide computer network would cost more that the space program and the atomic project taken together.[65] Reformer economists argued that Glushkov's project would not only conserve the obsolete forms of centralized economic management, but also suck up the resources needed for economic reform: '[The construction of] the pyramids of Egypt was one of the factors that turned that fertile ancient country into a desert. If an economically meaningless decision is vigorously put into effect, this would ruin the economy. According to the blueprint of a unified state network of computer centres, these centers would spread over the country as a kind of pyramids, designed by talented mathematicians and able engineers with the participation of unqualified economists.'[66] Modernization of management, argued reformers, was a political, not technical problem.

Eventually traditional managers and economists realized that there were many ways to skin the cybernetic cat without necessarily losing their grip on power. Instead of building a nationwide territorial management system, they proposed to develop automated management systems along branch ministry lines; this approach was announced as the official policy in the decree of the Central Committee and the Council of Ministers of March 1966.[67] Instead of sharing information/power with an interbranch agency for automated management, each ministry created an automated management system to serve its internal needs. These systems often proved incompatible with one another and with that of the State Planning Committee. Computer technology now served to strengthen centralized control within each ministry. The growing power of ministries quickly reduced the autonomy of individual enterprises to a minimum, and, in conjunction with the political reaction to the 1968 events in Czechoslovakia, effectively buried economic reforms.

Soviet cyberneticians never succeeded in their attempts to establish control over the national economy in the sense of *upravlenie*; their mission proved limited to creating information management systems of the *kontrol'* type. Mathematics and computing were now 'on tap, not on top' of economic decision-making; the Central Economic Mathematical Institute became just another information-supplying agency for the State Planning Committee. The idea to reform the government with the help of a nationwide automated management system was abandoned.

Having their political options limited, Soviet cyberneticians tried to work out a mathematical solution to the political problem of reforming the Soviet economy. This mathematical solution, however, proved inadequate precisely because it attempted to bracket politics and the complexities that politics brought with it. Glushkov tried to win the support of Soviet leadership for the implementation of a nationwide automated control system; he did not realize that such a reform from above would only reinforce the existing managerial structures rather than remedy the problem. Instead of upsetting the existing power structures, cybernetics was enrolled to serve them.

CONCLUSION

The fate of Wiener's cybernetics was deeply ironic. He developed a cybernetic critique of the manipulation of social communications for political purposes; the same cybernetic framework, however, served to develop pragmatic theories of political control. After Hiroshima, he became an outspoken critic of the military-industrial complex; cybernetic theory, however, was widely used for military purposes. He insisted that information could not be a commodity; nevertheless, cybernetic techniques of measuring and processing information facilitated the marketing of this product. He spoke against military secrecy, while information theory greatly improved cryptography. In Wiener's view, cybernetic knowledge was to liberate rather than further enslave an individual. Cybernetic analysis of control mechanisms, however, helped create new controls in addition to the old ones.

The cybernetic doctrine has proved amenable to various interpretations serving different, sometimes opposite, political causes. While Wiener thought that his cybernetic analysis would be devastating for communist society, Soviet interpreters quietly omitted the most damning passages from his books and made the cybernetic doctrine a foundation of their vision of communism.

While Wiener hoped that his cybernetic analysis would expose the flaws of both capitalism and communism; cybernetics was widely used on both sides of the Atlantic to supply decision-makers with technological solutions to political and organizational problems. The idea of rational automated control had equally powerful appeal to all cyberneticians, whether they were representatives of the 'bourgeois' middle class or members of the Communist Party.

Rather than being inspired by professional Soviet ideologues, the ideology-laden discourse around cybernetics was produced by Soviet cyberneticians themselves, who tried to impose cybernetic schemes on the semi-reluctant Soviet leadership. Cybernetics was not the only science mentioned in the new Party program, but cyberneticians capitalized on this fact more successfully than many other professional groups. Eventually, however, the attention of the Party and government authorities was attracted to the concepts of automated control and rational management not so much by the ideological argument about the 'full harmony' between cybernetic control and the political and economic structures of Soviet society, but rather by the reports about recent developments in that direction in the West. The American air defense system SAGE, for example, served as a direct inspiration for the Soviet project of a unified statewide information transmission system for civilian purposes.[68] The economic and military projects of Soviet cyberneticians could hardly be called unique products of Soviet ideology. Projects for large-scale optimal planning and military command and control have also been developed in the West with a completely different set of ideological assumptions.[69]

American and Soviet cybernetic systems shared both basic design ideas and inherent deficiencies. One Western observer has noted that 'the wonder is not that the [Soviet economic information] system works badly but that it works at all' and added: 'To be fair, I must admit that during several weeks' work with large U.S. military logistics systems, I was filled with the same wonder—albeit of a lesser order of magnitude.'[70]

NOTES

1. Paul R. Josephson, *New Atlantis Revisited: Akademgorodok, the Siberian City of Science* (Princeton University Press, 1997), p. 123.
2. On the history of Soviet cybernetics, see B. V. Biriukov, ed., *Kibernetika: proshloe dlia budushchego. Etiudy po istorii otechestvennoi kibernetiki* (Nauka, 1989); Richard D. Gillespie, 'The Politics of Cybernetics in the Soviet Union,' in Albert H. Teich, ed., *Scientists and Public Affairs* (MIT Press, 1974), pp. 239–98; Loren R. Graham, *Science, Philosophy, and Human Behavior in the Soviet Union* (Columbia University Press, 1987), ch. 8; David Holloway, 'Innovation in Science—the Case of Cybernetics in the Soviet Union,' *Science Studies* 4 (1974): 299–337; Lee Kerschner, 'Cybernetics: Key to the Future?' *Problems of Communism* 14 (November–December 1965): 56–66; D. A. Pospelov and Ia. I. Fet, *Ocherki istorii informatiki v Rossii* (OIGGM SO RAN, 1998).
3. David Holloway, *Technology, Management and the Soviet Military Establishment*, Adelphi paper no. 76 (Institute for Strategic Studies, 1971), p. 19.
4. I. B. Novik, "Normal'naia Izhenauka,' *Voprosy istorii estestvoznaniia, tekhniki*, no. 4 (1990): 14.
5. Norbert Wiener, *Cybernetics, or Control and Communication in the Animal and the Machine*, 2nd ed. (MIT Press, 1965).
6. On the history of cybernetics in the US and its links to military research, see Paul Edwards, *The Closed World: Computers and the Politics of Discourse in Cold War America* (MIT Press, 1996); Peter Galison, 'The Ontology of the Enemy: Norbert Wiener and the Cybernetic Vision,' *Critical Inquiry* 21 (Autumn 1994): 228–66; Steve J. Heims, *Constructing a Social Science for Postwar America: The Cybernetics Group, 1946–1953* (MIT Press, 1993); Idem, *John von Neumann and Norbert Wiener: From Mathematics to the Technologies of Life and Death* (MIT Press, 1980); David A. Mindell, "Datum for Its Own Annihilation": Feedback, Control, and Computing, 1916–1945,' (Ph.D. dissertation, MIT, 1996).
7. Norbert Wiener, *The Human Use of Human Beings: Cybernetics and Society* [1950, 1954] (Avon Books, 1967), pp. 201–202.
8. Wiener, *Cybernetics*, p. 161.
9. Wiener, *The Human Use*, p. 159.
10. Wiener, *Cybernetics*, p. 159.
11. Ibid., p. 27.
12. Ibid., p. 28.

13 Ibid., p. 27.
14 Ibid., p. 28.
15 Quoted in Wiener, *The Human Use*, pp. 245–47.
16 Wiener, *The Human Use*, pp. 247–48.
17 Ibid., p. 249.
18 On the history of these campaigns, see Graham, *Science, Philosophy, and Human Behavior*.
19 Materialist [pseudonym], 'Whom Does Cybernetics Serve?' [1953], trans. Alexander D. Paul, *Soviet Cybernetics Review* 4:2 (1974): 37.
20 Materialist, 'Whom Does Cybernetics Serve?', p. 37.
21 T. K. Gladkov, 'Kibernetika—psevdonauka o mashinakhe, zhivotnykh, cheloveke i obshchestve,' *Vestnik Moskovskogo universiteta*, no. 1 (1955): 64.
22 Iu. Klemanov, "Kibernetika" mozga,' *Meditsinskii rabotnik* (25 July 1952): 4.
23 Materialist, 'Whom Does Cybernetics Serve?', p. 37.
24 Gladkov, 'Kibernetika,' p. 65.
25 Materialist, 'Whom Does Cybernetics Serve?', p. 44.
26 'Plan meropriiatii po usileniiu antiamerikanskoi propagandy na blizhaishee vremia'; Russian Center for Preservation and Study of Documents of Recent History [Rossiiskii tsentr khraneniia i izucheniia dokumentov noveishei istorii (RTsKhIDNI)], f. 17, op. 132, d. 224, ll. 48–52.
27 S. L. Sobolev, A. I. Kitov, and A. A. Liapunov, 'Osnovnye cherty kibernetiki,' *Voprosy filosofii*, no. 4 (1955): 146. The same issue of *Problems of Philosophy* featured an article by philosopher Ernest Kol'man, who also contributed to exonerating cybernetics from ideological sins.
28 See, for example, Materialist [pseudonym], 'Komu sluzhit kibernetika', *Voprosy filosofii*, no. 4 (1955): 217; B. E. Bykhovskii, 'Kibernetika—dmerikdnskaia Izhenauka,' *Priroda*, no. 7 (1952): 127.
29 See Norbert Viner [Wiener], *Kibernetika, ili Upravlenie i sviaz' v zhivotnom i mashine*, trans. I. V. Solov'ev and G. N. Povarov, 2nd ed. (Sovetskoe radio, 1968).
30 O. S. Akhmanova, ed., *Russian-English Dictionary*, 15th ed. (Russkii iazyk, 1991), pp. 250, 668.
31 See Viner, *Kibernetika*, pp. 56–57 (editor's footnote). Povarov later published the book, *Ampère and Cybernetics*, in which he further elaborated this vision of cybernetics as a political science; see G. N. Povarov, *Amper i kibernetika* (Sovetskoe radio, 1977).
32 A. I. Berg, 'Osnovnye voprosy kibernetiki,' [1959] in Idem, *Izbrannye trudy*, vol. II (Energiia, 1964), pp. 34–38.
33 I. B. Novik, *Kibernetika: filosofskie i sotsiologicheskie problemy* (Gospolitizdat, 1963), pp. 58, 80.
34 Ibid., p. 33.
35 A. I. Berg, 'Kibernetika i nauchno-tekhnicheskii progress,' [1962] in Idem, *Izbrannye trudy*, vol. II, p. 152 (emphasis original).

36 Gillespie, 'The Politics of Cybernetics,' p. 272.
37 V. D. Pekelis, 'Chelovek, kibernetika i bog,' *Nauka i religiia*, no. 2 (1960): 27.
38 D. Iu. Panov, 'O malogabaritnykh elektronnykh vychislitel'nykh ustroistvakh i ikh primenenii dlia tselei upravleniia,' July 13, 1953, Center for Preservation of Contemporary Documentation [Tsentr khraneniia sovremennoi dokumentatsii (TsKhSD)], f.5, op. 17, d. 412, l. 78.
39 Ibid., l. 82.
40 A. I. Kitov, 'Rol' akademika A. I. Berga v razvitii vychislitel'noi tekhniki i avtomatizirovannykh sistem upravleniia,' in V. I. Siforov, ed., *Put' v bol'shuiu nauku: akademik Aksel' Berg* (Nauka, 1988), p. 131.
41 '27 TsNII—stareishaia nauchnaia organizatsiia Ministerstva oborony,' *Chelovek i komp'iuter*, no. 21–22 (1996): 4.
42 Holloway, *Technology*, p. 2.
43 Ibid., p. 19.
44 A. B. Pupko, *Sistema: chelovek i voennaia tekhnika (filosofsko-sotsiologicheskii ocherk)* (Voenizdat, 1976), p. 174.
45 G. Telyatnikov, 'Cybernetics and the Theory of Troop Control,' trans. John Schneider, *Soviet Cybernetics: Recent News Items*, no. 13 (1968): 28.
46 '27 TsNII', p. 4.
47 See S. Sobolev *et al.*, 'Lzhenauchnye raboty Instituta avtomatiki i telemekhaniki Akademii nauk SSSR,' *Bolshevik*, no. 9 (1941): 90.
48 V. D. Pekelis, *Cybernetic Medley*, trans. Oleg Sapunov (Mir, 1986), p. 287.
49 James Harford, *Korolev: How One Man Masterminded the Soviet Drive to Beat America to the Moon* (John Wiley & Sons, 1997), p. 163.
50 Berg, 'Kibernetika i nauchno-tekhnicheskii progress,' p. 155.
51 B. N. Malinovskii, *Akademik V. Glushkov: Stranitsy zhizni i tvorchestva* (Naukova dumka, 1993), pp. 83–84.
52 A. I. Kitov, 'Kibernetika i upravlenie narodnym khoziaistvom,' in A. I. Berg, ed., *Kibernetiku—na sluzhbu kommunizmu*, vol. 1 (Gosenergoizdat, 1961), p. 207.
53 Ibidem.
54 A. Berg *et al.*, 'Radioelektroniku—na sluzhbu upravleniiu narodnym khoziaistvom,' *Kommunist*, no. 9 (1960): 23.
55 A. A. Liapunov and A. I. Kitov, 'Kibernetika v tekhnike i ekonomike,' *Voprosy filosofii*, no. 9 (1961): 88.
56 This proposal was shelved. The Institute's director Isaak Bruk clashed with the State Planning Committee leadership and was forced to retire; see B. N. Malinovskii, *Istoriia vychislitel'noi tekhniki v litsakh* (Kit, 1995), pp. 192–93.
57 V. D. Belkin, 'Kibernetika i ekonomika,' in Berg, ed., *Kibernetiku—na sluzhbu kommunizmu*, vol. 1, p. 191.
58 On this debate, see Mark R. Beissinger, *Scientific Management, Socialist Discipline, and Soviet Power* (Harvard University Press, 1988); Martin

Cave, *Computers and Economic Planning: The Soviet Experience* (Cambridge University Press, 1980); John P. Hardt *et al.*, eds. *Mathematics and Computers in Soviet Economic Planning* (Yale University Press, 1967); Richard W. Judy, 'The Economists,' in H. Gordon Skilling and Franklyn Griffiths, eds., *Interest Groups in Soviet Politics* (Princeton University Press, 1971), pp. 209–51.

59 'Otchet delegatsii Akademii nauk SSSR o poezdke v Angliiu i SShA,' 1962, Russian Academy of Sciences Archive (Arkhiv Rossiiskoi Akademii nauk [ARAN]), f. 395, op. 17, d. 39, ll. 58–59.

60 N. S. Khrushchev, *Razvitie ekonomiki SSSR i partiinoe rukovodstvo narodnym khoziaistvom: doklad na Plenume TsK KPSS 19 noiabria 1962 goda* (Politizdat, 1962).

61 Malinovskii, *Akademik*, p. 95.

62 V. Glushkov and N. Fedorenko, 'Problemy vnedreniia vychislitel'noi tekhniki v narodnoe khoziaistvo,' *Voprosy ekonomiki*, no. 7 (1964): 87–92.

63 See 'Ekonomisty i matematiki za "Kruglym stolom,"' *Voprosy ekonomiki*, no. 9 (1964): 63–110.

64 A. A. Liapunov, 'O roli matematiki v sovremennoi chelovecheskoi kul'ture,' [1968] in Idem, *Problemy teoreticheskoi i prikladnoi kibernetiki* (Nauka, 1980), p. 305.

65 Malinovskii, *Akademik*, p. 102.

66 G. Kh. Popov, *Problemy teorii upravleniia* (Ekonomika, 1970), p. 160.

67 *Resheniia partii i pravitel'stva po khoziaistvennym voprosam*, vol. 4 (Politzdat, 1968), pp. 21–27.

68 A. Kharkevich, 'Informatsiia i tekhnika,' *Kommunist*, no. 17 (1962): 93–102.

69 See Edwards, *The Closed World*.

70 Richard W. Judy, 'Information, Control, and Soviet Economic Management,' in Hardt *et al.*, eds., *Mathematics*, p. 32.

Index

Absolute monarchy, 72
Abstracting, 230
Abstracts, 232
Action 14F13, 184–187
Advertising, 147
Age of Improvement, 57
Agriculture, 51, 69
Aircraft, 205–244
Air defense systems, 214
Airing, 91
Air quality, 90–92
Alarm systems, 172
Alder, Kenneth, 53, 71
Ambition, 85
America, 44
American industry, 157
American legal system, 42
Americans, 1
American system, 42
Amiens Chamber of Commerce, 75
Ampère, A. M., 24
Amplifiers, 217
Anaemia, 91
Antagonism, 54
Anti-aircraft gunfire control system, 215
Anti-cybernetics, 250
Antidust League, 116
Anti-idealists, 250
Anti-mechanists, 250
Anti-militarists, 250–251
Anti-technocrats, 250
Architectural structures, 43–44
Architecture, 'techno-nostalgic', 135–150
Arcosanti, 135

Armorican Automobile Club, 116
Arts, 58–59
Associationism, 54
Asylums, 56
Auner par pouces (cloth trade), 71
Automobile associations, 123, 125
Automobile Club of France, 116
Automobile clubs, 114
Automobile magazines, 125, 131
Automobile, *see* Automobile *aspects and also* Motor cars
Automobile tourism, 114
Automobility, 113–114
Aviation industry, 205–224

Babbage, Charles, 24–25
Bacilla fright, 90
Backstage, in domestic situations, 94–95
Bacon, Francis, 13–14
Bacteria, 88–90
Bacteria fright, 102
Bad air concept, 90–92
Ball Turret, 209, 211
Banks, 54–55
Barnes, Joe, 141–142
Basque Country, 127
Battleships, 215
Bauman, Zygmunt, 178–179
Beard, Charles A., 3–4
Beavais spinners, 76
  see also Spinning
Bedrooms, 95–96
Behavioral psychology, 29
Belgian Touring Club, 117

Belgium, 129–130
Belkin, V. D., 256–257
Bellamy, Edward, 137
Bell Laboratories, 158, 165, 170, 215–217, 219–220
Benthamite 'model prison', 56
Berg, Aksel, 252–254, 256–257
Bernstein, R., 19
Biarritz, 124
Bicycles, 117
Bigelow, Julian, 6
Biology, 43, 59, 100
Biometricians, 161
Black, Harold, 215–216
Blood, 59
Bode, Hendrik, 217
Bode plots, 217
Booth, R. D., 218
Bossenga, Gail, 71
Bourdieu, Pierre, 60
Bourgeois (aristocratic) drivers, 131
Bowls, 168
Braudel, Fernand, 47
Bridge (card game), 27
Brown, Denise Scott, 142–143
Buckland, Michael, K., 226–227
Burböck, Wilhelm, 182–187
Bush, Vannevar, 218, 220, 225, 228, 230

Calvinistic faith, 84
Canal construction, 54
Canals, 55
Capitalism, 248–249
Capitalist economy, 46
Car, see under Automobile and also Motor car aspects
Carbon microphone, 158–160
Caritat, Marie Jean Antoine Nicolas, 50
Carlyle, T., 99
Carnegie, Andrew, 190
Cattle, 117, 124
Celebration (Disney town), 138–146
Center for Documentation and Communication Research (CDCR), 232
Certeau, Michel de, 60–61
Chance, 153–174
Change, 75

Childbirth, 93
Christians, 3
Cinema Integraph, 218
Circulation -
  blood, 59
  economic theory of, 44–45, 48, 51–55, 59
  ideology of, 52–55
Circumstances, 49
City, cultural response to, 58, 60
City households, 104
City of Towers, 135
Civil engineering, 2, 43, 54
Civilizing mission, 95
Cleaning, 81–105
Cleaning, craze of, 103
Cleaning mania, 102
Cleaning zeal, 102
Cleanliness, 129
Cloth industry (France, 1780s), 69–80
Code de la route, 131
Cognition, 173
Cognitive devices, in manufacturing control, 153–176
Cognitive mapping, 61
Cold War, 220, 225–239, 249–251
  information crisis and, 228–233
Columbia, Maryland, 145, 149
Comfort, of the home, 102
Command, 253–254
Command technology, 220
Communications, 6, 15, 19, 28, 140, 145
Communicative action, 13–14
Communism, 220, 235–236, 248–249
'Competent female', 100–103
Computers, 5, 30
Concentration camp prisoners, 177–195
Concretization, 163–164
Condillac, 18
Condorcet, Marquis de, 28–29, 47–48, 50–54, 69
Connective structures, 43–44
Consciousness, technology, 103–105
Constructed memory, 135–149
Consumers, 77
Consumption, 91
Consumption junction, 105
Consumption society, 105

Continuous aim firing, 213–214
Control, 1
  and the aircraft industry, 205–224
  of concentration camps, 177–204
  historical definitions of, 21–31
  of machines, 4–5
  manufacturing, 153–176
  and moral competence, 101–102
  motor cars, 126–128, 130–132
  nature of, 41–68
  'optimal', 247–264
  of others, and domesticity, 100–101
  of technological systems, 5
  water as a metaphor for, 233–238
  weaponry to economy, 255–257
Control charts, 155, 160–164
  as cognitive tools, 167–168
Control contexts, 13–39
  historical, 14–21
Control systems, 7
Controlling technology, 219–221
Cooking, 128–129
Corsica, 127
Corvée (road repair), 53
Cotton, 76
Cotton thread, 75–76
Country, and cultural response, 58
Countryside, 59, 126–127
Crédit Mobilier, 55
Cultural control, 126–128, 130–132
Cultural geographers, 43–44, 60
Culture, 16, 135–149
Cybernetics, 28, 229, 231, 236
  military, 253–255
  soviet, 247–266
Cybernétique, 24
Cycling associations, 117

Daily cleaning ritual, 103
Daily dusting ritual, 90
Darwin, Charles, 24
Daividoff, Leonore, 84
Dead corners, and cleaning, 89
Definitions, 21
De Jaucourt, Marquis, 18
De la Mettrie, Julien Offray, 18
Delousing, 193–194
Demi hollande (cloth), 77

Democracy, 220
De Saint-Simon, Comte Claude, Henri, 54
De Saunier, Baudry, 115
Descartes, René, 13–14
De Solla Price, Derek J., 234–235
Determinism, 156
Deterministic philosophy, 157
Development, 42
Developmental biologists, 6, 7
Devices, 153
Diderot, Denis, 18, 20, 28–29
Differential analyzers, 218
Diffusion, of scientific information, in space, 51
Dirt, 83
Dirty world concept, 84
Discipline, 86
Disease, 90
Dishwashing, 92
Disney's New Town concept, 135–149
Dogs, 120, 124
Domestic appliances, 104
Domestic chemistry, 100
Domestic culture, 82
Domesticity, 81–111
Dream City, 137
Driving licences, 125, 130
Dry sweeping, 89–90
Duany, Andres, 138
Dunod (publisher), 115
Du Pont de Nemours, Pierre-S., 48
Dust, 88–90
Dust neurosis, 90

Economic advantage, 164
Economic progress, 48–52
Economy, 255–257
Eden, Anthony, 235–236
Edgerton, Harold, 218
Edison, T., 3
Edu-tainment, 137
Eisner, Michael, 137
Electrical generators, 15
Electricity, 4, 5, 25–26, 104
Electric power, 217–219
Electromagnets, 4
Ells (cloth measurement), 74
Emerson, Ralph Waldo, 2

Engineering, 2, 43
Engineering architects, 96
Engineering products, 43
Engineering reasoning, 170
Engineers, 26
Enlightened administrators, 71
Enlightenment, The, 13–14, 16–20, 28–29, 44–45, 47, 49, 51–58, 61, 178, 187
Entrepreneurial manufacturers, 24
Entrepreneurialism, 30
Entrepreneurs, 74, 114
Entress, Friedrich, 194
Environmental control, 41–68
Environmental models, 42
Environmental problems, 42
Epcot Center, 136–138
Epistemological ambitions, 164

Factories, 56
Failure, of control, 5–7
Fairthorne, R. A., 231–232
False consciousness, 51
Family harmony concept, 85–86
Family life, 87
Farming, 75
'Felt paths', 60
Feminine disorder theory, 93–94
Films, 27
Financial power, 87
Fire control (arms industry), 210–215
Fire control systems, 212–213
Fit to work theory (Nazi *Arbeitsfähig*), 181
Flânerie (parading), 60
Flexible behavior, 30
Flooding concept, and information flow, 234
Flooring, 91
Flowers, 128–130
Ford, Hannibal, 213
Ford, Henry, 135
Fordism, 26, 42
Ford Rangekeeper, 213
Forrester, Jay, 7–8
Foucault, Michel, 56, 60–61, 173
Fourier, J. B., 20–21
France, 52–53, 69–78, 114–115
    cloth industry, 1780s, 69–80

Frankfurt School (philosophy), 21
Fraud, 77
Free grazing, 124
French, backwardness theory of, 52
French motorists, 115
French Touring Club, 118, 127
Fresh air, 90
Freycinet Plan, 117
Friedlander, Henry, 191, 194
Frontstage, of the home, 94–95
Futuropolis, 135

Gay, Peter, 49
Geddes, Normal Bel, 138
Gender power, 98
Generators, 217
Germany, concentration camps, 177–204
Germany, motoring in, 129
Gerstein, Kurt, 194–195
Giddens, Anthony, 56, 59
Giedon, Siegfried, 27
Gilbreth, Frank, 25
Gilbreth, Lillian, 25
Global temperature, 41
Glücks, Richard, 188, 192
Glushkov, Viktor, 257–259
Goebbels, Josef, 189
Good manners, 125–126
Good taste, 86
Government, as a science, 247–264
Graves, Michael, 138, 142
Greece, 52
Greeks, 49, 51
'Grey card' (motoring), 125
Guilds, 71
Gyroscopes, 4, 20
    see also under Sperry gyroscope

Habermas, Jürgen, 19–20, 61
Hardware, 237
Harmony, domestic, 81–111
    and male power, 86–87
    and order, 85–86
Harvey, William, 59
Hazen, Harold, 218–219
Heimatschutz, 129
Heroic materialism, 55

High Enlightenment, 16, 53
  see also Enlightenment, The
Highway Code, 120–126
Himmler, Heinrich, 180–181, 188
Hiroshima, 5, 260
Histograms, 167
Historians, 129
  of science, 178, 191
Historical buildings, 114
Historical definitions, of control, 21–31
Historical regression, 52
Historical sites, 131–132
Hitler, Adolf, 178
Holloway, David, 254
Holocaust statistics, 181–184
Home Economics Training School, Gothenburg, 85
Home, as a social project, 84–86
Hoover, President Herbert, 26
Horse-drawn carts, 124
Housekeeping books, 100
Human material, 179
Hygiene, 81–105
Hygienic hazard, servants as, 92–94
Hygienism, 90, 96

Ideology, 52–55
  of circulation, 45
  counts, 174–195
Ill-health, 89
Imperialism, 250–251
India, 44
Industrial regulations, 76
Industrial Revolution, 15
Industrial workers, 75
Inertia, 51
Infectious diseases, 90
Information, 5, 7, 8, 15, 28, 58, 73, 140, 145, 155, 226
Information Age, 225
Information anxiety, 225
Information crisis, 227, 229
Information culture, 225
Information economy, 225
Information environment, 225
Information explosion, 229
Information fatigue syndrome, 225, 238
Information overload, 227–228

Information Revolution, 140
Information society, 225
Initiative, 75
Inner housekeeping, 86–87
Innocent III, 3
Inns, 114–115, 126, 128
Insects, 88
Inspector of Concentration Camps, 181–182, 186–188
Inspectors, 74–75, 171
Instrumental reason, 17
Intellectual Revolution, 51
Interchangeability, 157
Internet, 8, 145
Invention, 73
Inventors, 4
Invocative rhetoric, 83
Iron bedsteads, 96, 104
Irving, Washington, 235
Italy, 117, 129

Jameson, Fredric, 61
Johnson, Philip, 143
Junk, 83

Kan, Immanuel, 17
Keller, Evelyn Fox, 6
Kennedy, John F., 235
Keynes, John Maynard, 26–27
Krushchev, Nikita, 258
Kidder, Tracy, 237
Kitchen odours, 92
Kitov, Anatolii, 251–252, 254–257
Knowledge, 58, 155
  see also under Information aspects
Kosygin, Aleksei, 257

Labour Action Office, 183
Lagerstedt, Lotten, 88–89
Laissez-faire economics, 78
Landscape, 44, 61, 129
Landscape preservation, 129
Language, 49, 51
Langedoc, 73
Laplace, M. P. Simon de, 156–157, 164
Laplace-Gauss law, 168
La Turbie, 119
Layered control, 8

Laziness, 75
Le Chatelier, Henry, 156
Liapunov, Aleksei, 251–252, 256, 258
Libraries, 229
Licklider, J. C. R., 229
Ligue de Chauffeurs (Drivers League), 116
Lille, 71
Linguistics, 172
Literary creation, 60
Lithography, 15
Local cuisine, 128
Localism, 61
Local knowledge, 58, 60
Locke, John, 18, 57
Lubar, Steven, 225–226
Lynch, Kevin, 60
Lyon silk workers, 71

Machine Age, 3
Machine civilization, 3
Machines, 43
　in the garden, 2–3
　for production, 3
Maidservants, 100–101
Major housecleaning concept, 99, 102–103
Male power, 86–87
Management theories, 154
Managerial control, 26
Managerial knowledge, 153–154
Manufacturing control, 153–174
Marble shelfs, 96
Markers, 119
Marx, Karl, 14, 21, 46
Marx, Leo, 2, 57
Masculine type of order, 98
Master's room, 98
Master's order, 97–98
Materiality, 173
Mathematics, in France (1780s), 69–80
Maurer, Gerhard, 188–193
Mayr, Otto, 22
McCarthyism, 248
Measurement, 69–78
Mechanism, 2
Media advertising, 30
Mengele, Josef, 191, 195

Merchants, 71–72
Mesopotamia, 13
Metaphor, 173, 233–239
Miasma theory, 90
Michelin, 126–128, 131
Middle classes, 85–86, 89, 104
　European, 15
Mikhailov, A. I., 231
Milan, 117
Military command, 7
Military cybernetics, 253–254
Military engineering, 53
Mill, John Stuart, 99
MIT (Massachusetts Institute of Technology), 217–219
Model order, 87
Moderation, 102–103
Molecular biologists, 7
Montaigne, 22
Moral competence, 101–102
Moral function, 86
Motor car, 113–134
　beginnings of, 114–115
　conflicts, on roadways, 116–118
　manners in operating, 125–126
　road adaptions, 118–120
　written word and, 115–116
　see also Automobile *aspects*

Nagasaki, 5
Napoleon, 71
Napolean III, 129
National grids, 217
National Socialism, 178
　see also Germany, concentration camps
National standards, 71
Nato Air Defense Ground Environment, 7
Nature's gnawing theory, 88
Naval fire control, 210–215
Nazi concentration camps, 177–204
Nazi racial ideology, 177–195
Negative feedback, 6
Nervous control, 25
Netherlands, The, 86, 128
Network, 59
Newton, Isaac, 22
New Urbanism, 135, 138
Nice, 124

Nomadism, 60
North Atlantic Air Defense Command, 7
Nostalgia, and the Disney philosophy, 135–150
Numerical information, 69
   see also under Information *aspects*
Nursing, 93
Nyquist, Harry, 217
Nyquist stability criterion, 217

Objects, 153, 173
   in action, 166–167
   in organization of enterprises, 170–171
   and social process, 171–172
   in support, of rhetoric, 168–170
Oiling, 91
Optimal control, 247–261
Optimal planning, 257–259
Order, and domestic harmony, 81–111
   and male power, 86–87
   objects, 85–86
   significance of, 83–84
   spatial, 94–96
   temporal, 96–97
   threats to, 87–94
Organization, of enterprises, 170–171
Organizational science, 154
Ostentation, 86

Packard, Vance, 30
Paczula, Tadeusz, 191
Panopticon, 56–57
Papermakers, 74
Paris, 70, 124
Paris Bureau of Commerce, 69
Paris-Madrid race, 123
Paris-Rouen-Paris race,114
Paris-Trouville road, 119
Pascal, Blaise, 13–14, 22
Passivity, 51
Pasteur, Louis, 3, 88, 90, 148
Pearsonian theory, 163
Pearson, Karl, 160
Pedestrian rhetoric, 60–61
Pedestrians, 123
Peirce, C. S., 155, 174
Pelli, Cesar, 143–144
Personnel training, 172

Photography, 15
Physiocrats, 48
Picardy, 73
Pigs, 117
Pilots, *see* Aircraft
'Pink card' (motoring), 125
Place
   destruction of, 58–62
   devaluation of, 55–58
Plain board floors, 91
Plater-Zyberk. Elizabeth, 138
Pohl, Oswald, 177–179, 182, 189
Pollution, 5, 41–42
Post-industrial society, 16
Post, Wiley, 205–221
Porches, 140
Povarov, Gelii, 252
Poverty, 75, 85
Power, 57
Precision, 157, 158
Priests, 57
Primary cells, of society, 84
Printing, 49
Prisons, 56–57
   see also under Germany, concentration camps
Proactive administration, 71
Probabilistic Revolution, 156–157
Probability, 22
Product Integraph, 218
'Production/diffusion', of knowledge, 154
Productivism, and concentration camps, 177–181
Progress, 48–52
Property, 86
Proto-industrial Europe, 70
Psychological theory, 15
Public transportation, 173
Public welfare, 72
Pure world concept, 84

Quality, 157–172
Quality control, 155
Quesnay, Francois, 48

Racial hygiene, 180
Racial ideology, 177–181
Radial City, 135

Railroads, 53, 55, 59, 117
Randomness, 158–160
Ras le bois (measurement standard)
Rationalization, 71
Rational cleaning, 89
Rats, 88
Reason, 17, 72
Reception, 166
Regional cuisine, 127
Rhetoric, 168–170
Rickets, 91
Ritual, 97
Road Network Administration, 119
Road repairs, 117
Roads, 46–47, 53, 59, 113
Road signs, 127
Roadways, 44, 59–60, 116–118
Rockwell, Norman, 147
Role objects, 154
Romains, Jules, 59
Romantics, 60
Rome, 46–47, 49, 51–52
Rosenbleuth, Arturo, 6
Rouen Bureau of Encouragement, 76
Rouse, James, 136, 145
Rousseau, Jaques, 17, 29, 47
Routine, 51
Royal Academy, 69
Rozak, Theodore, 226
Rules, 154
Ruling, of the home, 97–98
Rural weavers, 74
  see also Spinning
Rybczynski, Witold, 147

SAGE (Semiautomatic Ground Environment), 7
Saint-Simon, 20–21, 118
Saint-Simonians, 54–55
Salon de l'automobile, Paris, 116
Sanitary engineers, 91
Sanitation, 193
Science, 70–71, 92, 137
  of government, 247
Scientific community, 57
Scientific explanation, 70
Scientific management, 26
Seaside, 138

Second Industrial Revolution, 250
Self control, 25, 125–126
Self mastery, 25
Semiosis, 174
Servants, 92
Servomechanisms, 28–30, 218–219
Sewage, 91
Sexual behavior, and driving, 126
Shannon, Claude, 226
Shewart, W. A., 158–159, 162–166, 169
Shop floor, 170–171
Shop management, 156
Signals, 119, 174
Signs, 155, 174
Silicon Valley, 218
Silk weavers, 71
  see also Spinning
Simplicity, 86
Sims, Admiral, 214
Skinner, B. F., 28–31
Sleeping, 95–96
Smith, Adam, 18, 20–24, 26
Smith, Cecil O., 52–53
Sobolev, Sergei, 251–252
Social construction, of technology, 43
Social context, 42
Social distinction, 85
Social progress, 48–52, 72
Social respectability, 95
Social scientists, 178
Social utility, 72
Social welfare, 73
Society motorists, 120
Sociological models, 42
Socio-technical systems, 1
Sofsky, Wolfgang, 178
Software, 237
Sommer, Karl, 188–189
Soviet All-Union Institution of Scientific and Technical Information, 230
Soviet Cybernetics, 247–261
Space, 55–58
Spatial knowledge, 60
Spatial metaphors, 56
Spatial organizations, 94–95
Speech, 49
Speed limits, 123, 126
Speer, Albert, 178, 182, 189, 195

Sperry, Elmer, 207–210
Sperry gyroscope, 207–210
Spinning, 74
 schools for, 76
 in Soissons, 76
 wheels, 75
Sputnik, 230–231, 238, 247
SS doctors, 188–195
Stalinism, 248, 251–253
Standard cloth measurements, 70
Standardization, 72–73
Statistical distributions, 160–164
Statistical quality control, 164
Statistical thinking, 153–174
'Stepford Wives', 146
Stereotypes, 148
St Maclou workshop, 76
Stockholm, 98
Stocking knitters, Troyes, 75–76
Strategic Air Command Control System, 7
Subservience, 87
Suez Canal, 55
Surveillance, 56
Switzerland, 128–129
Systematic management, 26

Tarascon, 76
Taffetas, 77
Taylor, Frederick Winslow, 25, 156–157
Taylorism, 96
Technological pessimism, 61
Technological systems, 41–62
Technology, 1, 135–149
 and consciousness, 103–105
Techno-nostalgia, 135–149
Telegraph, 5, 15
Telephone, 5, 15, 159
Telephone transmission, 215–217
Teletype, 15
Temporal order, 96–97
Terman, Frederick, 218
Textile guilds, 71
Thermodynamics, 15
Thread, 76
Thrift, 86
Tidiness passion, 102
Tidy Village concept, 128–130

Time, 96–97
Time-motion studies, 26
Tomorrowland, 138, 147
Touring clubs, 116, 119, 126–128, 130
Tourism, 128, 131
Tourist associations, 114
Trachtenberg, Alan, 44
Traditional cooking, 128
Traffic signals, 131
Tuberculosis, 90
Turgot, Anne Marie, 23, 28, 47–50, 53–54, 69–70, 124
Turin, 117
Twentieth Century Machine Age, 3–4
Typewriters, 15

Unanime, 59
Unfit to work (*Arbeitsunfähig*), 181, 184–187
Uniform measures, 78
Universal class, 57
Unregulated production, 73

Vacuum cleaners, 91, 104
Varnishing, 91
Vaucanson's mechanical duck, 24
Vehaeren, Émile, 59
Vélocemen, 117
Venturi, Robert, 142–143
Verne, Jules, 59
Victorian women, 81
Vigilance, 86, 93–94, 124
Vila Viçosa (Portugal), 70
Villages, 'tidy', concept of, 128–130
VINITI, 230–231, 233, 238
Voltaire, 50, 235
Vulnerability, 87–88

Wald, Abraham, 170
Walking, 60–61
Warping board, 75
Wallerstein, Immanuel, 46
Warfare, weapons control, 205–224, 255–257
'Waste not' concept, 86
Watchful eye concept, 89
Water, 92
 as a metaphor, 233–238

Weaponry, 255–257
  see also Warfare, weapons control
Weavers, 74
  see also Spinning
Weber, Max, 19–20
Western Electric, 158, 165, 169
Westinghouse Singing Cascades (New York World Fair), 139
Whirlwind, 7
Wiener, Norbert, 6, 28–31, 229–230, 248–249
Williams, Raymond, 61
Willpower, 15

Wine, 128–129
Winner, Langdon, 173
Women, and domesticity, 93–94
Work skills, 76
World War I, 211, 235
World War II, 5–7, 15–16, 27–28, 148, 208–211, 213–214, 220, 226–230, 234–235, 237
World Wide Military Command and Control System, 7
Writing, 49

Yardstick measure, 72

For Product Safety Concerns and Information please contact our EU representative GPSR@taylorandfrancis.com
Taylor & Francis Verlag GmbH, Kaufingerstraße 24, 80331 München, Germany

www.ingramcontent.com/pod-product-compliance
Ingram Content Group UK Ltd.
Pitfield, Milton Keynes, MK11 3LW, UK
UKHW010813080625
459435UK00006B/61